Security and Privacy in Networks and Multimedia

Security and Privacy in Networks and Multimedia

Editors

Tomasz Rak
Dariusz Rzońca

MDPI

Basel • Beijing • Wuhan • Barcelona • Belgrade • Novi Sad • Cluj • Manchester

Editors

Tomasz Rak
Rzeszow University of Technology
Rzeszów
Poland

Dariusz Rzońca
Rzeszow University of Technology
Rzeszów
Poland

Editorial Office
MDPI AG
Grosspeteranlage 5
4052 Basel, Switzerland

This is a reprint of articles from the Special Issue published online in the open access journal *Electronics* (ISSN 2079-9292) (available at: https://www.mdpi.com/journal/electronics/special_issues/Q5660YL8P2).

For citation purposes, cite each article independently as indicated on the article page online and as indicated below:

Lastname, A.A.; Lastname, B.B. Article Title. *Journal Name* **Year**, *Volume Number*, Page Range.

ISBN 978-3-7258-1861-7 (Hbk)
ISBN 978-3-7258-1862-4 (PDF)
doi.org/10.3390/books978-3-7258-1862-4

Contents

About the Editors

Tomasz Rak

Tomasz Rak is an Assistant Professor at the Rzeszow University of Technology in Poland. He received his Ph.D. in Informatics from the AGH University of Science and Technology in Krakow in 2007. His research interests include Communication/Networking and Information Technology, Software Engineering, and Distributed Component-based Web Systems. He is a member of the R&D staff at SoftSystem (SCC Soft Computer) and was previously a member of the R&D staff at Advanced Technology Systems International (NOVOMATIC Group). He has served as a Guest Editor for various publishers, including Elsevier, Atlantis Press, IET, IGI Global, and MDPI. His research interests also encompass computer engineering, formal modeling and testing, distributed systems, cluster computing, systems modeling, network monitoring, network security, interactive systems, machine learning for the classification of web services, and the architecture of Internet systems.

Dariusz Rzońca

Dariusz Rzońca is an Assistant Professor at the Rzeszow University of Technology, Poland. He received his Ph.D. in Computer Science from the Silesian University of Technology in 2012. His research interests focus on communication, cryptography, cybersecurity, distributed systems, and formal modeling. He is the author or co-author of over 70 journal articles and conference papers.

electronics

Editorial

Security and Privacy in Networks and Multimedia

Tomasz Rak * and Dariusz Rzonca

Department of Computer and Control Engineering, Rzeszow University of Technology, Powstancow Warszawy 12, 35-959 Rzeszow, Poland; drzonca@kia.prz.edu.pl
* Correspondence: trak@kia.prz.edu.pl

1. Introduction

The digital era has significantly transformed the dissemination of information and business operations, creating an intricate web of interconnected systems. As technology continues to advance, so do the complexities of maintaining robust security and privacy across these networks. This Special Issue, "Security and Privacy in Networks and Multimedia", seeks to explore the forefront of research in protecting data networks and multimedia systems against evolving security threats. The articles included in this issue highlight innovative solutions and ongoing research aimed at enhancing security and privacy in various technological environments.

2. Resilient Forecasting and Supply Chain Security

In the realm of smart cities, accurate electricity load forecasting is crucial for grid stability. Mohd Hafizuddin Bin Kamilin and Shingo Yamaguchi present a resilient forecasting network that uses a collective intelligence predictor to mitigate the impact of missing values induced by cyberattacks. This approach decentralizes forecasting processes, achieving remarkable accuracy even under significant data loss scenarios.

Helen C. Leligou and colleagues delve into cybersecurity within supply chain systems, specifically focusing on the farm-to-fork use case. Their FISHY platform integrates machine learning and blockchain technologies to detect security threats and provide evidence for mitigation policies. This innovative approach ensures comprehensive protection for complex supply chain networks.

3. Advanced Detection Methods and Network Privacy

Addressing the threat of jamming in next-generation communication systems, Cem Örnek and Mesut Kartal propose a jamming detection method leveraging the Error Vector Magnitude metric. This method enhances sensitivity and provides critical jammer frequency information, ensuring robust protection for 5G and LTE networks.

Marko Mićović, Uroš Radenković, and Pavle Vuletić explore Format-Preserving Encryption for network layer privacy protection. Their LISPP system, implemented on smart network interface cards, achieves high throughput with minimal delay, proving effective for production networks.

4. Intrusion Detection and AI-Enhanced Security

Hyeon gy Shon and colleagues introduce a semi-supervised alert filtering method for network security. By incorporating semi-supervised clustering, their approach significantly reduces false alerts, conserving resources and improving detection accuracy.

The integration of artificial intelligence in network security is exemplified by Latifah Almuqren and her team's Improved Sine Cosine Algorithm with Deep Learning-Enabled Security Solution (ISCA-DLESS). This method combines feature selection and hyperparameter tuning to enhance anomaly detection, achieving impressive accuracy on benchmark datasets.

Citation: Rak, T.; Rzonca, D. Security and Privacy in Networks and Multimedia. *Electronics* **2024**, *13*, 2887. https://doi.org/10.3390/electronics13152887

Received: 17 July 2024
Accepted: 18 July 2024
Published: 23 July 2024

5. Generative Approaches and Adversary Impact Mitigation

Hao Yang and co-authors tackle the class imbalance problem in Network Intrusion Detection Systems with their SPE-ACGAN method. This resampling approach improves detection performance across various classifiers, addressing the prevalent issue of imbalanced training samples.

Mohd Anjum and his team propose a Permutated Security Framework for IoT security, utilizing end-verifiable keys to manage transactions securely. Their approach adapts to system changes, mitigating adversary impact and service failures while enhancing transaction security.

6. Explainable Security Solutions and Advanced Cryptographic Techniques

Suleiman Y. Yerima and Abul Bashar focus on detecting evasive malicious PDF documents through explainable ensemble learning methods. Their system effectively detects hidden malicious content in PDFs, offering robust security against sophisticated attacks.

Maaz Ali Awan and colleagues discuss the potential of Radio Frequency Fingerprinting in enhancing the cybersecurity of smart grids. Their deployment framework leverages deep learning for effective classification and rogue device detection, bolstering smart grid security.

7. Network Layer Privacy and Anomaly Detection

Raad A. Muhajjar and his team present a hierarchical key management method for wireless sensor networks in medical environments. Their approach ensures data confidentiality and integrity, providing a secure framework for sensitive health data transmission.

Mohammad Jamoos and co-authors introduce a data-balancing approach based on Generative Adversarial Networks for network intrusion detection systems. Their model addresses imbalanced datasets, enhancing the detection rate of minority class attacks.

Saini and Islam focus on the security of the CAN bus, which is widely used in automotive applications. They propose a hardware prototype (FPGA) of an intrusion detection system for the CAN bus, enabling attack detection and response in case of bus-off attacks.

8. Conclusions

The articles in this Special Issue collectively advance the state of the art in network and multimedia security, offering innovative solutions to pressing challenges. From resilient forecasting networks and comprehensive supply chain security to advanced jamming detection and AI-enhanced anomaly detection, these studies contribute significantly to the ongoing efforts in securing our increasingly digital world.

Conflicts of Interest: The authors declare no conflicts of interest.

List of Contributions:

1. Kamilin, M.H.B.; Yamaguchi, S. Resilient Electricity Load Forecasting Network with Collective Intelligence Predictor for Smart Cities. *Electronics* **2024**, *13*, 718. https://doi.org/10.3390/electronics13040718.
2. Leligou, H.C.; Lakka, A.; Karkazis, P.A.; Costa, J.P.; Tordera, E.M.; Santos, H.M.D.; Romero, A.A. Cybersecurity in Supply Chain Systems: The Farm-to-Fork Use Case. *Electronics* **2024**, *13*, 215. https://doi.org/10.3390/electronics13010215.
3. Örnek, C.; Kartal, M. Securing the Future: A Resourceful Jamming Detection Method Utilizing the EVM Metric for Next-Generation Communication Systems. *Electronics* **2023**, *12*, 4948. https://doi.org/10.3390/electronics12244948.
4. Mićović, M.; Radenković, U.; Vuletić, P. Network Layer Privacy Protection Using Format-Preserving Encryption. *Electronics* **2023**, *12*, 4800. https://doi.org/10.3390/electronics12234800.
5. Shon, H.G.; Lee, Y.; Yoon, M. Semi-Supervised Alert Filtering for Network Security. *Electronics* **2023**, *12*, 4755. https://doi.org/10.3390/electronics12234755.
6. Almuqren, L.; Maray, M.; Aljameel, S.S.; Allafi, R.; Alneil, A.A. Modeling of Improved Sine Cosine Algorithm with Optimal Deep Learning-Enabled Security Solution. *Electronics* **2023**, *12*, 4130. https://doi.org/10.3390/electronics12194130.

7. Yang, H.; Xu, J.; Xiao, Y.; Hu, L. SPE-ACGAN: A Resampling Approach for Class Imbalance Problem in Network Intrusion Detection Systems. *Electronics* **2023**, *12*, 3323. https://doi.org/10.3390/electronics12153323.
8. Anjum, M.; Shahab, S.; Yu, Y.; Guye, H.F. Identifying Adversary Impact Using End User Verifiable Key with Permutation Framework. *Electronics* **2023**, *12*, 1136. https://doi.org/10.3390/electronics12051136.
9. Yerima, S.Y.; Bashar, A. Explainable Ensemble Learning Based Detection of Evasive Malicious PDF Documents. *Electronics* **2023**, *12*, 3148. https://doi.org/10.3390/electronics12143148.
10. Awan, M.A.; Dalveren, Y.; Catak, F.O.; Kara, A. Deployment and Implementation Aspects of Radio Frequency Fingerprinting in Cybersecurity of Smart Grids. *Electronics* **2023**, *12*, 4914. https://doi.org/10.3390/electronics12244914.
11. Muhajjar, R.A.; Flayh, N.A.; Al-Zubaidie, M. A Perfect Security Key Management Method for Hierarchical Wireless Sensor Networks in Medical Environments. *Electronics* **2023**, *12*, 1011. https://doi.org/10.3390/electronics12041011.
12. Jamoos, M.; Mora, A.M.; AlKhanafseh, M.; Surakhi, O. A New Data-Balancing Approach Based on Generative Adversarial Network for Network Intrusion Detection System. *Electronics* **2023**, *12*, 2851. https://doi.org/10.3390/electronics12132851.
13. Saini, R.; Islam, R. Reconfigurable CAN Intrusion Detection and Response System. *Electronics* **2024**, *13*, 2672. https://doi.org/10.3390/electronics13132672.

electronics

MDPI

Article

A Perfect Security Key Management Method for Hierarchical Wireless Sensor Networks in Medical Environments

Raad A. Muhajjar [1], Nahla A. Flayh [2] and Mishall Al-Zubaidie [3,*]

1 Department of Computer Science, Faculty of Computer Science and Information Technology, University of Basrah, Basrah 61004, Iraq
2 Department of Computer Information System, Faculty of Computer Science and Information Technology, University of Basrah, Basrah 61004, Iraq
3 Department of Computer Sciences, Education College for Pure Sciences, University of Thi-Qar, Nasiriyah 64001, Iraq
* Correspondence: mishall_zubaidie@utq.edu.iq; Tel.: +964-61-469-869-029

Abstract: Wireless sensor networks (WSNs) have developed during the past twenty years as a result of the accessibility of inexpensive, short-range, and simple-to-deploy sensors. A WSN technology sends the real-time sense information of a specific monitoring environment to a backend for processing and analysis. Security and management concerns have become hot topics with WSN systems due to the popularity of wireless communication channels. A large number of sensors are dispersed in an unmonitored medical environment, making them not safe from different risks, even though the information conveyed is vital, such as health data. Due to the sensor's still limited resources, protecting information in WSN is a significant difficulty. This paper presents a hierarchical key management method for safeguarding heterogeneous WSNs on hybrid energy-efficient distributed (HEED) routing. In the proposed method, the Bloom scheme is used for key management and a pseudo-random number generator (PRNG) to generate keys in an efficient method to keep sensor resources. In addition, using cipher block chaining-Rivest cipher 5 (CBC-RC5) in this method achieved cryptography goals such as confidentiality. A comparison is made between the proposed and existing methods such as dynamic secret key management (DSKM) and smart security implementation (SSI) under the same circumstance to determine the performance of the new method. The data transmission in WSN consumes about 71 percent of a sensor's energy, while encryption computation consumes only 2 percent. As a result, our method reduces the frequency with which data transmissions are made during the key management process. The simulation findings demonstrated that, in comparison to earlier techniques, the proposed method is significantly more secure, flexible, scalable, and energy-efficient. Our proposed method is also able to prevent classifications of node capture attacks.

Keywords: bloom; cipher-block chaining (CBC); HEED protocol; heterogeneous WSN; key management; PRNG; rivest-cipher5 (RC5); WSNs

Citation: Muhajjar, R.A.; Flayh, N.A.; Al-Zubaidie, M. A Perfect Security Key Management Method for Hierarchical Wireless Sensor Networks in Medical Environments. *Electronics* **2023**, *12*, 1011. https://doi.org/10.3390/electronics12041011

Academic Editors: Tomasz Rak and Dariusz Rzońca

Received: 29 January 2023
Revised: 9 February 2023
Accepted: 15 February 2023
Published: 17 February 2023

1. Introduction

Wireless sensor networks have received much interest for using them in a diversity of environments, including health, military, agriculture monitoring, and industrial. However, health systems are making use of WSN technology significantly in recent years. In instances where these sensors are able to communicate with one another wirelessly, the sensed health data from the surrounding area is sent to the sink to be treated either through a single hop if the sink is within the same node's range or through multiple hops if the sink is outside of that range [1,2]. The data routing protocol in the WSN is also important to reduce the load on sensor sources [3]. Because of the medium, any hacker can break through the network and obtain its health data [4]. Passive (eavesdropping hackers on the health data exchanged between the parties without alteration or injecting false data into the network) and active are two kinds of attacks (eavesdropping hackers on the health data and conducting malicious events with it before retransmitting to nodes for the purpose of network destruction or grabbing them) [5,6]. In order to transmit important data in this kind of network, it should keep the information as possible, as well as prevent unwanted parties from providing fake information to sensors. Cryptography technical is used to address some of the health network's security challenges, providing confidence [7] and authentication. The sensors have limited resources from where the processing, energy, and memory [8]. Therefore, considered cryptography traditional techniques are not suitable in WSN because they need additional connection, memory, and processing. As a result, when designing and implementing a key management method, it is important to keep in mind the restricted resources available on these devices [9–12].

Cryptographic mechanisms require efficient key management. Secure communications may be compromised by inadequate key management, resulting in key disclosure to attackers. WSN communication threats and vulnerabilities can be mitigated through key management, which is an essential process. Confidentiality, integrity, and availability are generally considered security requirements [13]. User/patient privacy, customer behavior, message authentication, and control messages are the most important security requirements before sensors are deployed [14]. The security of cryptography keys is a crucial factor in ensuring the confidentiality and integrity of data. To maintain the security of cryptographic keys, health systems must handle secure key management for many devices. A number of studies have been conducted regarding key management systems in recent years. The topics discussed in existing studies on WSNs include architectures, applications, communication, and cyber security. Some studies, on the other hand, dealt with key management systems for WSNs, a very critical area of research that has received surprisingly little attention. In these studies, lightweight encryption was highlighted in addition to a key management scheme that achieves a defensive mission in resistance to WSNs threats [15]. Therefore, an efficient energy-aware secure key management method is significant. The key distribution scheme in a WSN has to satisfy some objectives, such as a low memory requirement, low overhead for computation and communication, and high connectivity and robustness. Managing key information and data delivery in the network is another issue that needs to be addressed. Keys of the same length as the message can be consumed by the encryption process in private key management systems. The key is consequently a limited resource in a WSN. A network with several communication activities will be inefficient and unstable if there is no key management [16]. It should use methods from traditional network research that handle issues such as these to tackle this challenge and improve network activities management.

With the growth of the global population, as remote health monitoring becomes more popular, the demand will increase extremely in the coming years. A major goal of remote health monitoring is to transfer patient information to clinical physicians across the globe [17]. The importance of securing patients' clinical information in this scenario grows, so that unauthorized individuals cannot alter or read it. In terms of encryption, Rivest Cipher (RC5) is a simple and secure cipher. For limited resources environments, such as WSNs, it is considered a suitable block cipher because of its simplicity, fast encryption, low power consumption, easy adaptability and low memory requirements. Key calculation with RC5 is susceptible to attack because of its weak diffusion state. By using RC5 in combination with key management and randomness generators, this can be overcome so that medical data can be ciphered and protected [18] while preventing classifications of node capture attacks.

1.1. Major Contributions

To address all previous issues, we propose a reliable method based on energy-saving routing, lightweight encryption and randomization techniques to achieve efficient key management for WSN. First, the HEED protocol is adopted to support efficient routing and sensor energy conservation, thus supporting energy saving as much as possible to extend the lifetime of the WSN. Second, we use a lightweight RC5 encryption algorithm to maintain medical environment data. Third, we generate unique randomness using PRNG to support RC5 and prevent the attacks from breaching the encryption. Finally, we utilize the Bloom scheme to manage the keys in a secure manner that protects medical environment data. All these techniques are integrated into a single security method to protect patient/provider data and information.

1.2. Research Organization

A description of the research roadmap can be found here: A comprehensive introduction is provided in Section 1. We critique related key management security works in Section 2. The requisite preliminaries are introduced in Section 3. Our proposed method is described in Section 4. Section 5 investigates the proposed method results. Section 6 presents the study's conclusions and future trends.

2. Schemes Related to the Security of WSN Key Management

In this section, we will investigate key management methods and extract their problems and drawbacks.

For a heterogeneous WSN, Li and Wang [19] proposed an effective and hybrid key management strategy. While symmetric methods were used between the cluster's sensors, elliptic curve cryptography was used to generate the key between the cluster heads and the sink. A low-cost, high-level security authentication and key management scheme (AKMS) was intended to be provided as protection from hostile sensors that can appear during networking. Even if the AKMS keys are compromised, attackers cannot utilize the prior keys or the authenticated sensors to cheat. In particular for heterogeneous networks, simulation findings demonstrate that their approach offers effective security with decreased energy usage. However, their scheme is not very safe against cluster head capture attacks. The network model is hierarchical, according to Iwendi et al. [20]. Both the pairwise key between the cluster heads and the base station and the key between sensors and their particular cluster heads have been generated in a symmetric manner employing OR and XOR operations. The approach provides security and makes inefficient use of limited resources, but it also lacks scalability. Zhang and Pengfei [21] purposed approach to secure hierarchical network structures. This method made use of three different sorts of keys. In the first stage, a disposable paired key was formed using the specified function for use in encrypting data exchanged between nodes, and the primary keys were generated using Diffie-Hellman and the specific function. However, the authors do not address issues of secure key storage. Furthermore, their scheme is not suitable for medical environments that require reliable key management and lightweight data encryption. Zhang and Wang [22] suggested a key management method in hierarchical WSNs based on a Bloom scheme with sophisticated advanced encryption standard (AES) and a mesh module for multi-hop packet routing, with high security and scalability. However, their scheme did not provide a mechanism to support random health data encryption. Qin et al. [23] have developed a hybrid key management system (KMS) for multihop WSNs that makes use of secret key-based communication and asymmetric cryptographic approaches to minimize the computational burden on member nodes. KSM's security analysis demonstrates its ability to resist node capture attacks and support node revocation. Data freshness, the number of generated typical keys, throughput, and cost of computation were used to assess KSM's effectiveness. However, their scheme did not provide updating keys for the hierarchical medical WSN.

On off-the-shelf static WSNs, Moara-Nkwe et al. [24] discussed challenges and difficulties experienced during the establishment and application of physical layer secure key generation (PL-SKG) methods. It then suggested a method for generating keys using elliptic curve cryptography (ECC) based on signals from 802.15.4 compliant sensors that could take advantage of the power, simplicity, and diversity of frequency channels available. However, generating keys using asymmetric encryption algorithms will add computation and communication costs to the WSN environment. A key management protocol presented by Chanda et al. [25] is claimed to guarantee the confidentiality, integrity, authenticity, and integrity of wireless sensor networks by handling key generation, distribution, and maintenance. Their proposed method encrypts network information in three levels using three auxiliary keys in addition to the main key. Unfortunately, their protocol fails to provide unique and sufficient keys to protect WSN data and their method is very complicated and resource-consuming for WSN. A network model for the intelligent building energy management system (IBEMS) was developed based on the framework of the WSN [26]. Then, the IBEMS presented a blockchain-based dynamic key approach as well as key management, examining the security of blockchain technology with the Shamir scheme. Experiments were conducted to verify their plan's feasibility. However, the authors did not provide a clear key management technique, they relied on Shamir's secret sharing for key exchange but did not specify the threshold in their method.

Ahlawat and Dave [27] proposed a secure hybrid key pre-distribution scheme (HKP-HD) for WSNs in order to prevent node capture attacks. By combining q-composite and threshold-resistant polynomial schemes, they claimed robustness. Their scheme investigated to make the WSN more solid against the sensor capture threats. There is a presumption that hacker is intelligent and that they frequently develop a matrix of attacks against the network by taking advantage of various weaknesses. It attempts to destroy the whole network with the fewest possible sensors, based on the attack matrix. In order to counteract such vulnerabilities, a comparable threat array was created by the network engineer by investigating sinks as major influencing factors. However, their method was only seeking to decrease the risk of keys being compromised and not to end the problem completely, and this in itself is a security breach. Kumar and Malik [28] examined the keys required to develop resilient and connected WSNs that have a large number of sensors. An improved random key distribution method based on random deployment was presented to increase connectivity and resilience. For the large, medium, and small-scale networks, they investigated the number of keys that are sufficient. However, they did not use a routing protocol to reduce power consumption in WSNs. Recently, Tyagi et al. [29] discovered several security pitfalls in previous methods, such as a man-in-the-middle, an off-line password guessing, and session key attacks. An Internet of Things (IoT) authentication method was created to overcome the pitfalls identified in previous methods. Furthermore, a real-or-random (RoR) model was used to confirm the reliability of their method. Based on computation and communication costs as well as security properties, they evaluated their proposed method against the associated schemes. However, although the authors claimed that their method provides key protection, their method did not provide key security management. Furthermore, although the [30,31] tested their proposed methods against node capture attacks, their approaches are complex and inflexible in handling sensor-transmitted parameters in medical environments.

3. Introductory Details of the Proposed Techniques for the Security of WSN Key Management

In this section, we will outline the fundamental concepts behind the techniques employed in the proposed method.

3.1. *Hybrid Energy-Efficient Distributed Clustering Protocol*

A big crowd of WSN routing approaches addressed the energy conservation issue. Hybrid energy-efficient distributed clustering (HEED) [32,33] and low-energy adaptive clustering hierarchy (LEACH) [34,35] are the most distinguished hierarchical routing-based WSN protocols. However, there is a negative impact on the network's cluster heads' (CHs) in LEACH distribution when carrying out rounds [33]. In addition, the comparison in Table 1 demonstrates that the HEED protocol outperforms the LEACH protocol [36].

Table 1. Comparison of performance properties between HEED and LEACH protocols.

Properties	HEED	LEACH
Balanced clustering	Good	Moderate
Balanced loading	High	Moderate
CH capability	Data aggregation, homogeneous	Data aggregation, homogeneous
Clustering process execution	Iterative	Probabilistic
Cluster overlapping	No	No
Cluster stability	High	Moderate
Delay	Moderate	Very small
Energy efficiency	Moderate	Low
Mobility	Stationary	Stationary
Routing between clusters	Single hop and Multi hop	Single hop
Routing within a cluster	Single hop	Single hop
Scalability	Moderate	Low

HEED protocol maximizes network lifetime by reducing communication costs and utilizing residual energy in sensors. In a set number of iterations, HEED completes the clustering phase, creates well-distributed CHs, reduces control overhead, and optimizes network lifetimes. Sensor distributions or sensor density in a network do not affect HEED [32]. Due to the fact that new CHs are always chosen and clustering starts after each interval of the clustering process time (T_{CP}) + operation time (T_O). Receiving and transmitting messages from neighboring sensors within a defined range is a time-consuming process. HEED defines a fixed percentage of CHs in order to begin clustering. Initial CHs probabilities are set by sensors according to the formula:

$$CH_p = C_p \cdot (E_r / E_m) \quad (1)$$

An initial probability, residual energy, and maximum energy of the sensors are represented by C_p, E_r and E_m, respectively. In order to meet minimum probability (Pmin) = 0.0001, CH_p must not fall below Pmin. Figure 1 shows the clustering approach in HEED protocol.

Sensors periodically communicate with their neighbors about their current status during each round. When sensors identify themselves as CHs or receive an invitation to join from another CH, they are regarded as covered. If a node is running HEED but is still visible, it should declare itself a CH or join the neighboring cluster. As part of the HEED protocol, wireless sensor networks are organized into clusters. An elected sensor from each cluster gathers raw data from its associated sensors and transmits it to the sink. As soon as the sensors for level-1 have been chosen and the HEED protocol has been employed to cluster the network [37]. By reusing the HEED approach, super-elected sensors (level-2) are in this situation elected using a larger cluster radius. A second run of the HEED approach will result in the network being divided into two categories of clusters. An elected sensor of level-1 is part of a cluster comprised of regular sensors within a radius transmission range (Tr1). Super-elected nodes receive data collected from regular nodes within the cluster. After the second HEED protocol execution on the level-1 chosen sensors, the second category of clusters is established. It is made up of a cluster of elected level-1 sensors that are placed close to the Tr2 cluster and an elected level-2 sensor that is in charge of receiving information from the cluster's various members (elected level-1 sensors) and sending it to the sink [37].

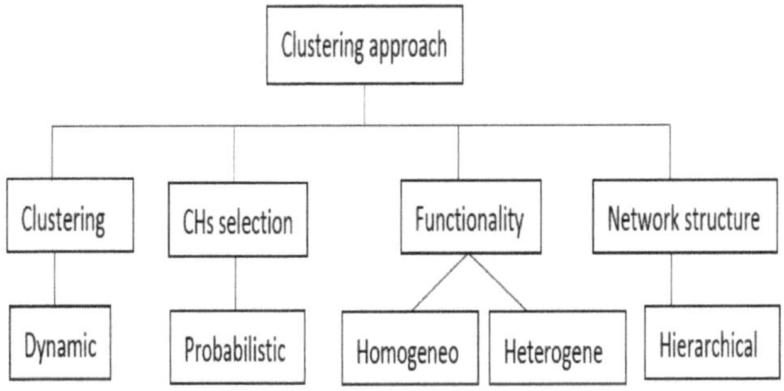

Figure 1. HEED approach of clustering.

3.2. *Rivest Cipher 5*

Ronald Rivest proposed Ron's code or Rivest cipher (RC5) in 1994. This cipher uses block ciphers of symmetric type and is fast. This cipher can be implemented in both hardware and software. Rotation based on data is extensively used in RC5. Both linear and differential cryptanalysis can be prevented by this feature. In this algorithm, the block size and round numbers are parameterized, as well as the key length. As a result, both performance and security are greatly enhanced. A specific RC5 algorithm is the word/round/byte (w/r/b) algorithm. The w bit size is 16, 32 (standard value) and 64. Because RC5 encodes two-word blocks, both plaintext and ciphertext are two words long. Moreover, r values are (0–255), and table (t) = 2 words are included in the expanded keys table. In addition, the number of bytes (b) with values ranging from 0 to 255 specifies the security key. Encryption, decryption and key generation are the three elements of RC5 [38].

A comparison of RC5 with Rivest-Shamir-Adleman (RSA) and Blowfish shows that it is more secure and faster. Sharing secret keys securely remains a challenge with RC5 since it is a symmetric key cryptosystem. This limitation was overcome by combining RC5 encryption with Honey encryption, which had a bigger buffer size and maintained RC5's strengths at the same time [39]. There are three block sizes for encryption: 32 bits, 64 bits, and 128 bits. The best block size is 64-bit. RC5 keys range from 0 to 2040-bit, but 128-bit is most commonly endorsed. Plaintext and ciphertext are stored in two 32-bit registers (A and B). Normally, encryption takes 12 rounds (but it can take as many as 255) [40]. Figure 2 shows RC5 process. In RC5, key operations involves XORing bits, adding words modulo 2w, and shifting left (<<) and right (>>). Due to its flexibility in terms of key size, block size, and rounds, RC5 offers high levels of security and performance.

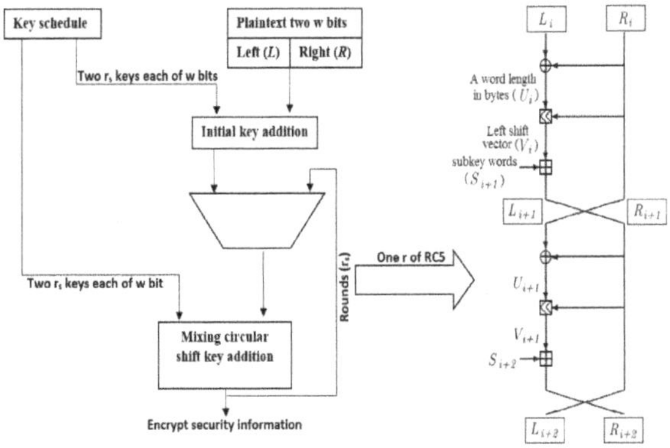

Figure 2. RC5 process.

3.3. Bloom Scheme

To determine whether an element is part of a set, a space-effective probability data structure called a Bloom filter is utilized. Essentially, the matrix starts out with all bits set to 0, it is a bit matrix of length n. A Bloom filter employs k distinct hashes $\{h_1, \ldots h_k\}$ with a range of [0, n 1] to exemplify a set $S = \{x_1, \ldots x_m\}$. The Bloom filter's bits $h_i(x)$ are set to 1 for each element $x \in S$. Multiple instances of setting an index to 1 have no impact; only the initial change does. It is necessary to determine whether all positions of $h_i(x)$ are set to 1 in order to determine whether an element y is in S. Despite the fact that this technique is quick and effective, it is possible to obtain false positives if the bits were accidentally changed from 0 to 1 during the intercalation of another element y where $y \in S$ and $y\ neq\ x$. Figure 3 [41] provides a diagrammatic representation of the general structure of a Bloom filter.

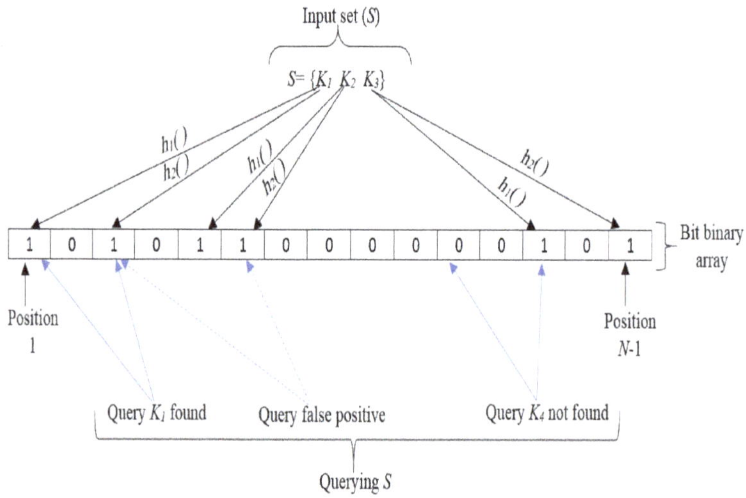

Figure 3. Bloom process.

Bloom filter-based schemes can be made more secure by using a randomly generated key instead of the same key for every filter. Additionally, it is recommended that the key length be at least as long as the Bloom filter, which is the second and most crucial condition. Since absolute secrecy can only be used in specific circumstances, it is crucial to include these qualities. A health biometric protection system based on the Bloom filter that concurrently meets important security needs such as irreversibility and unlinkability with other desired qualities such as recognition effectiveness and data compression. Due to its simplicity and lack of a necessity for pre-aligning the biometric templates, the Bloom scheme quickly gained popularity. Some threats relied mostly on the fact that the encoded templates were somehow tied to the original health biometric information. In order to develop some attacks, it was observed that the original biometric template and the encoded template had the same hamming distance. To reduce extra inputs and outputs induced by checking multiple tables, modern designs use Bloom filters in repositories stores to quickly check the existence of a key pair in an individual table.

The IDs of revoked certificates can be fed into a Bloom filter to condense the revocation list. In probability theory, Bloom determines whether a given element belongs to a set. However, the member's query either returns "possibly in the set" or "definitely not in set", demonstrating the possibility of the Bloom filter finding false positive results. The bit vector for the Bloom filter has a length of m bits and is initially commenced to zero. The certificate serial number $Sensor_i$ is kept in the bit vector for certificate revocation list (CRL) compression after being hashed using the k-hash algorithms. In order to save the element $Sensor_i$, all addresses in the m bit vector that are pointed by the K hashes of the certificate are set to one. The position of the prepared Bloom is compared with the hashed location of the given $Sensor_i$ of the certificate in order to validate it in the vector of the Bloom filter. It is possible that the given $Sensor_i$ of the certificate is on the list if all bit locations are set (matched), otherwise, it is not. In spite of this, it is possible to set the bit to one multiple times since different hashes may point to the same place. The bit vector is made up of the hashes of various $Sensor_i$ for the certificates. As a result, a false match occurs, and the false positive rate is computed as follows. The chance that the location Bi is set to one is given by $(1-(1-1/m)^{KN})^K$, where

$$P(FalsepositiveRate) = (1-(1-1/m)^{KN})^K \tag{2}$$

Consequently, a non-revoked certificate could be interpreted as revoked, which could cause the search to return that is inappropriate for the filter [42].

3.4. Pseudorandom Number Generator

In wireless networks with limited resources, such as WSN, pseudorandom number generators (PRNGs) are a well-liked option for cryptographic methods for key generation. This is partly because of their capacity to produce distinctive sequences from various seeds. Furthermore, these generators can produce long-period sequences devoid of repetitions. For generating key sequences in radio-frequency identification (RFID)/WSN applications, such algorithms have also been considered. In order to assure bit dispersion in the pseudorandom sequence, these PRNGs might include nonlinear filter functions or use different feedback polynomials. However, it should be emphasized that PRNG-based techniques only aid in key generation and management; for authentication, additional methods, such as hash-based or trusted third-party-based procedures, should be used in conjunction with them [43].

Strong foundations are necessary for the existing key management strategies. This might be carried out by improving the basic random number generation procedure that the BS uses to generate initial random numbers. In addition, key randomness techniques based on PRNGs have shown strong initial energy-efficient performance for IoT nodes [44], particularly in health systems. There are several forms of PRNGs that can be applied to clients' health applications. Linear feedback shift registers (LFSRs) are widely utilized as cryptographic primitives, stream ciphers, PRNGs ... etc. because they are highly simple, effective, and reasonably quick circuits. However, because LFSRs are linear, predictable, and dependent on strong seeding, they introduce flaws that have been exploited in previous systems. The Mersenne Twister is another well-liked PRNG since many programs packages use it as a standard PRNG. It has a very long time before repeating since Mersenne Twister relies on the Mersenne prime ($2^{19937}-1$). It is a highly quick and effective PRNG, which leads to its acceptance by many software platforms, including MATLAB, Java, and Python.

With the use of the National Institute of Standards and Technology (NIST) test suite, the unpredictability of these various PRNG stream outputs was evaluated [45]. To show that shorter LFSRs often do not produce better randomness than longer ones, two different-length LFSRs were investigated in the previous method. The 15-tap LFSR failed a number of the tests as was to be expected, demonstrating its lack of security as a PRNG. The Mersenne Twister came extremely near to passing one of the tests, but it did not achieve the minimum acceptable pass rate (98.65% vs. 98.7%). The performance of the PRNG scheme was then assessed using session key streams produced using the same PRNGs. A stream of 80,000 128-bit was created with each PRNG using a random 1024-bit (providing over 10 million bits for statistical testing per PRNG). The NIST test suite was then used to verify the randomness of the produced bits. The weaker LFSR's subpar results were effectively concealed by the PRNG scheme, and all PRNG tests passed. Following the execution of these two test cases, more thorough NIST statistical tests were conducted to look for further trends and defects in the output. All PRNGs worked Absolutely fine, however, the 15-tap LFSR failed. Therefore, using the security scheme, any PRNGs other than the 15-tap LFSR are considered to be adequately safe. An additional layer of security is provided on top of the PRNG depending on the size of the window utilized and the initial starting index (both internal states of the security scheme), which should be preserved safely in the case of a PRNG compromise. A robust PRNG, such as a cryptographically secure PRNG (CSPRNG), however, would still prevent an attacker from generating a correct derived key even if the internal state of the security scheme were exposed [45].

4. The Proposed Method

This paper investigates protecting a heterogeneous WSN in which the sensors have limited capacities and are clustered in diverse ways based on the HEED protocol. Each cluster has a CH who is in charge of the member nodes' communication and collecting the information. The member node performs one task and transmits the information from the surrounding area to the group's leader. Each connected party in the network has to have a secure link in order to protect the transferred information. It should share the secure key to perform cryptographic activities and meet security requirements. Therefore, the proposed method includes key generation, encryption and decryption procedures, key updating and sensor add/delete.

4.1. Key Generation

Prior to placement in the target health area, each sensor is pre-loaded with a key, PRNG, and unique identifier. After the deployment phase, clustering is carried out utilizing the HEED protocol, as previously indicated. The key between each connected node should now be established. The key between neighboring cluster heads and the cluster head and the BS is managed by applying the Bloom scheme, which is utilized in our proposed method in an efficient manner that maintains WSN resources, PRNG is used to generate keys between cluster heads and member nodes.

4.1.1. Key Generation between CH-CH and CH-BS

The Bloom scheme provides high security, and a low amount of overhead achieves a balance in the use of each node's resource and provides scalability. As the standard matrix is a square matrix with zeros and ones to simplify the components similar to the columns, we employed an adjacency matrix to decrease processing and storage. All sensors that are a specific sensor's neighbors are filled with ones in this adjacency matrix, while the other elements are provided with q-1 so that they cannot contain zeros. The adjacency matrix lowers storing the columns in the node's memory. Subsequently, any node can build an adjacency matrix. As Bloom's approach, the prime number is important to produce the keys, which number depends on the required key length. The following array shows the adjacency matrix in its original binary form.

$$\begin{vmatrix} 1 & 1 & 0 & 1 \\ 1 & 0 & 0 & 0 \\ 0 & 0 & 1 & 1 \\ 1 & 0 & 1 & 0 \end{vmatrix}$$

The steps for computing the key are shown below.

- **First step:** We select a prime element from the field GF(q), where q is bigger than the key length and $q > N$. Next, our method constructs a public matrix G based on the sensor neighbors that is $N \times N$ in size relying on the λ, value a number of rows with N columns.
- **Second step:** The BS produces a D symmetric matrix of size $(\lambda + 1) \times (\lambda + 1)$. Then, it computes matrix A by $A = (D \cdot G)^T$.
- **Third step:** All rows of matrix A stored in a memory of the sensors. When sensor i wishes to connect with sensor j, sensor i multiplies the row A_i with the column G_j. Then the result is a secret key. To demonstrate the operation of the modified Bloom's method using an adjacency matrix. For example, the network has 6 sensors,

for instance, $N = 6$, $lambda = 3$ (secure parameter), and $q = 29$ (prime numbers).
Modified adjacency matrix:

$$\begin{vmatrix} 28 & 1 & 1 & 28 & 28 & 28 \\ 1 & 28 & 28 & 1 & 28 & 28 \\ 1 & 28 & 28 & 1 & 1 & 28 \\ 28 & 1 & 28 & 1 & 28 & 28 \\ 28 & 28 & 1 & 28 & 28 & 28 \end{vmatrix}$$

Public matrix (G):

$$\begin{vmatrix} 28 & 1 & 1 & 28 & 28 & 28 \\ 1 & 28 & 28 & 1 & 28 & 28 \\ 1 & 28 & 28 & 1 & 1 & 28 \\ 28 & 1 & 28 & 1 & 28 & 28 \end{vmatrix}$$

Secret semantic matrix (D):

$$\begin{vmatrix} 3 & 5 & 2 & 7 \\ 5 & 6 & 9 & 1 \\ 2 & 9 & 3 & 5 \\ 7 & 1 & 5 & 4 \end{vmatrix}$$

$A = (D.G)^T mod\ 29$:

$$\begin{vmatrix} 26 & 9 & 5 & 24 \\ 3 & 20 & 24 & 5 \\ 18 & 18 & 14 & 26 \\ 22 & 20 & 28 & 14 \\ 16 & 26 & 16 & 22 \\ 12 & 8 & 10 & 12 \end{vmatrix}$$

To suppose two nodes such as sensor 2 and sensor 5, who wish to communicate with one another, we shall multiply sensor 2's private row from matrix A, which is A (2) in sensor 5's public column, by G. (5). In a similar manner, sensor 5 multiplies its private row A (5) in node 2's G public column (2). The previous operation will generate the shared secret key for sensors 2 and 5.

$$K_{5,2} = A_5 \cdot G_2 = |16\ \ 26\ \ 16\ \ 22| = \begin{vmatrix} 1 \\ 28 \\ 28 \\ 1 \end{vmatrix} = 1214\ mod\ 29 = 25$$

$$K_{2,3} = A_2 \cdot G_3 = |3\ \ 20\ \ 24\ \ 5| = \begin{vmatrix} 28 \\ 28 \\ 1 \\ 28 \end{vmatrix} = 808\ mod\ 29 = 25$$

It used the initial key to cluster-head for encrypting any row of the A matrix. The row for a certain cluster head and key ID should transmit. CBC-RC5 is the encryption method utilized. CH will receive a message with its row and key ID and will work to decrypt it before storing it in the sensor's memory. Now each cluster head has its own unique row and key ID. It should be formed as the shared key between CH-CH and BS-CH. Several of CHs directly communicate to the sink over a single hop and others are not directly connected to the sink but through neighboring cluster-head, allowing them to broadcast data across them until they reach the base station. The shared key is calculated as follows:

Shared key N_i = Row of Node i * Public column of Node j;
Shared key N_j = Row of Node j * Public column of Node i.

4.1.2. Key Generation between CH-Sensors

Before deployment, each sensor node was pre-loaded with an initial-key that was utilized to form a key between the CH and the member sensors. The CH and member sensor both through the suggested PRNG generate a shared key, and the authentication key is derived from the shared key using the PRNG. Figure 4 shows the PRNG process to generate the shared key. First, we divide the input initial key value into four parts (K_1, K_2, K_3 and K_4). Second, we use a set of variables X, Y, Z, Q, T, V, F and U and a set of operations such as XOR, not, addition and left shift ($<<$) to obtain high randomness and then store the random result in four registers A, B, C and D to obtain the shared key of 64 bits.

- Step1: The initial key has been split into four parts K_1, K_2, K_3, and K_4
    1. For j from 1 to 32
        - Z [Bit XOR(K_1, K_3)]
        - Y [Bit XOR(K_2, K_4)]
        - For u from 1 to 16
        - V [swapping(Y) $<<$ 5]
        - End
        - X addition(Z,V) module 2^{16}
        - T [Bit XOR(Z,Y)]
        - Q [Bit XOR(V,X)]
        - For i from 1 to 16
        - U [swapping(Q) $<<$ 9]
        - end
        - F addition(T, U) module 2^{16}
        - a_1 [Bit not(X)]
        - end
    2. For j from 1 to 16
        - For s from 1 to 16
        - b_1 [Bit not(V)]
        - end
        - c_1 [Bit XOR(V, F)]
        - For n from 1 to 16
        - d_1 [Bit not(F)]
        - end

    end
    A - [binary Vector to Hex(a_1)]
    B - [binary Vector to Hex(b_1)]
    C - [binary Vector to Hex(c_1)]
    D - [binary Vector to Hex(d_1)]
- Step2: The four registers [A,B,C,D] combine to form the final key (shared key-64 bit).

4.2. *Encryption and Decryption Procedures*

For the sake of providing high security, the security requirements (confidentiality, integrity, and authentication) should be met. A combination of CBC-RC5 is employed in this work to do this task at once as shown in Figure 5. In order to protect the shared key (SH-K), our proposed method inserts this key (generated from the previous random PRNG process) into the CBC-RC5 algorithm for encryption, as this algorithm is known for its ability to block analysis, differential and node capture classifications attacks in addition to its speed in encryption. Hence, the output of this algorithm is the Auth-K key which is used to securely protect the shared key of the health sensors (CH-Sensors).

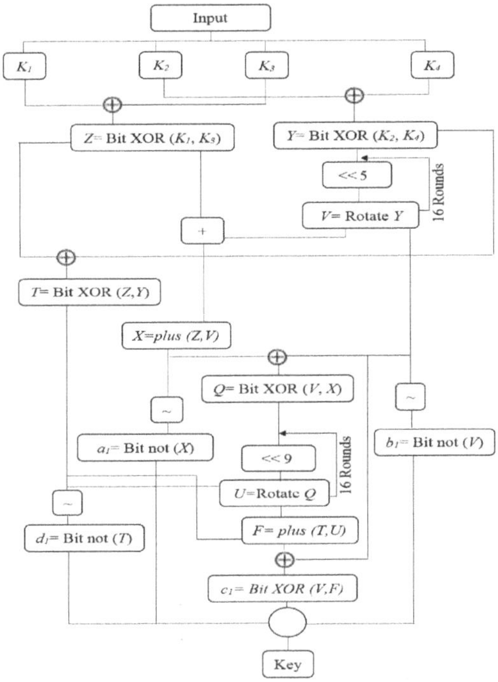

Figure 4. PRNG process with a shared key.

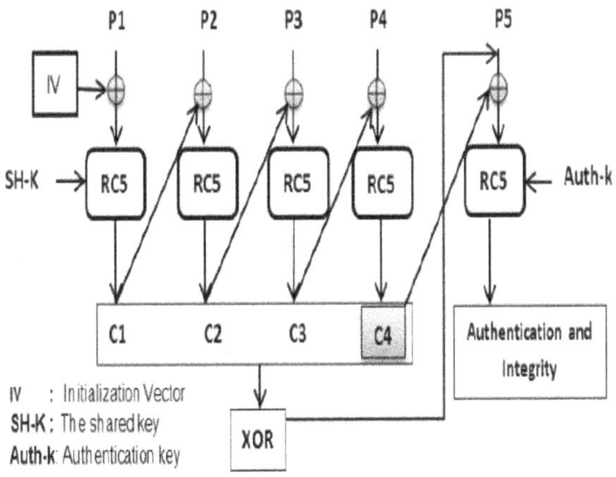

Figure 5. Diagram of CBC-RC5.

4.3. Key-Updating

The key must be updated on a regular basis after time has passed to prevent the hacker from having access to existing key data. The sink transfers a new row and *K*-id encrypted by initial-key to the CH. In this case, our proposed method changes the shared key between CH-CH and CH-BS regularly (see Section 4.1.1). The PRNG is used in conjunction with the previous Auth-*K* to update the Auth-*K*. Additionally, the shared key between CH and sensors needs to be updated. The existing shared key between CH-sensors is used in the PRNG (see Section 4.1.2) to create two new keys: a new SH-*K* and a new Auth-*K* between CH-Sensors.

Sensor Add and Delete

The new sensor can be declared as a cluster head or linked as a member node to another cluster head. If the new sensor becomes a CH, it transfers a need data to the BS, which the sink responds to with a row and key-ID for the new sensor. Then, as mentioned in the key establishment phase, it will generate a shared key. If the additional sensor becomes a member sensor of one CH. To authenticate the new sensor and obtain the new member's initial-key the CH transfers data to the sink, after which the shared key is generated as previously mentioned. If any sensors fail or become compromised, the sink sends out messages to all sensors in the network, instructing them to eliminate the node's ID from the nearby table.

5. Results

This section will explain the performance and security results of our proposed method. Our proposed method focuses on the use of heterogeneous WSN in medical environments with static locations for sensors because firstly these environments are important for people's lives, secondly, the use of this proposal may not be suitable for other environments such as military, natural phenomena such as earthquakes . . . etc. which depend on random distribution of sensors, thirdly, the use of a static distribution of sensors in medical environments makes it easier for us to evaluate performance accurately and without fluctuations.

5.1. Performance Results

The proposed method depends on the distribution of 100 nodes in an area of 100 m × 100 m where the position of the BS is (50.50). The nodes consist of two types: 80 nodes have low resources (0.5 joules of energy, 25 bands, low compute capacity), and 20 nodes have higher resources (2 joules of energy, 40 bands, low compute capacitance). Figure 6 depicts WSN in our proposed method. The distribution of sensors in our method depends on the static distribution because it is applied in a healthy environment. Furthermore, MATLAB 2020b was used to perform the simulation under Windows7 64 bits operating system with CPU i5-2540M @ 2.60 GHz and RAM 4.0 GB. Examining the performance of our method is very important but it is very difficult to find similar methods to our method under the same conditions and parameters. Therefore, we tried to find the closest existing method and compare it with our method to prove its superiority and acceptability. Our proposed method is compared with the two existing methods: A highly dynamic secret key management (DSKM) [21] method and a smart security implementation (SSI) [22] method for WSN nodes under the same circumstance. Figure 7 displays the clustering stage utilizing the HEED approach. This figure shows the correlation of the sensors to the closest CH based on the HEED method.

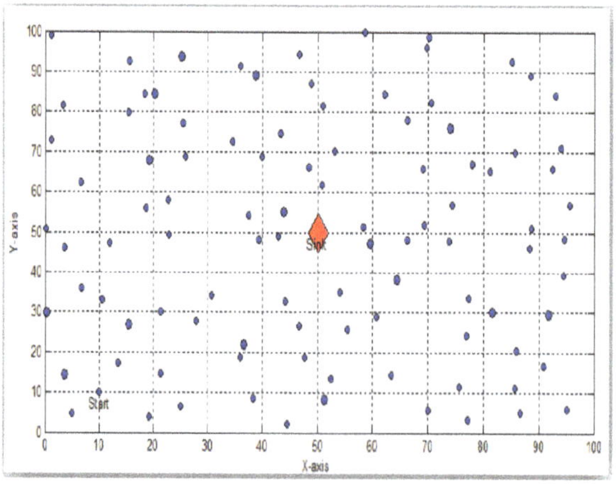

Figure 6. Random-distribution of the sensors.

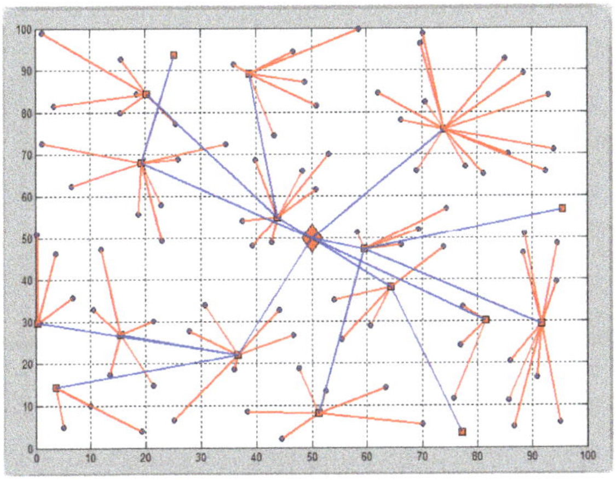

Figure 7. The clustering stage.

Figure 8 displays the energy expendable for generation shared keys comparison with the DSKM and SSI methods. The result is that our method expends less energy than the others. Where we note that our method performs significantly better than SSI in energy conservation and slightly better than DSKM, this means that the sensors in our method will collect health data for a longer period.

Figure 9 displays the size of memory used in cluster heads. In terms of memory expenditure in CHs, we notice in this figure that SSI is less memory expenditure compared with our method and DSKM but DSKM suffers from the different and unstable fluctuation of memory expenditure, generally, our method is relatively stable in memory usage compared with SSI and DSKM.

Figure 10 displays the size of memory used in the member sensor. In terms of memory expenditure in sensors, our method has a size of memory used less than precedent methods, which require twice the amount of memory. Figure 10 shows that SSI and DSKM are very memory-consuming compared to our method. Where we notice that the sensors in SSI are very memory-consuming compared to DSKM and our method. However, this figure shows that the health sensors in our method do not require large memory expenditures because we use lightweight techniques to generate shared keys and authentication keys. Figure 11 displays the processing time for generation shared keys. In comparison to the two existing methods, our proposed method to generate keys is lightweight. The use of HEED, RC5, Bloom and PRNG achieves fast and lightweight operations, which makes the processing time of our method very fast compared to SSI and DSKM. As we notice from Figure 11 that DSKM requires a very large processing time compared to SSI and our method. Finally, we note that our method is superior to SSI and DSKM in terms of energy consumption, memory expenditure in CHs, memory expenditure in sensors and processing time. However, there are some limitations to the proposed method. First, if the sensors are randomly distributed (in environments other than healthy ones), the results may differ. In a healthy environment, we can put the sensors in static locations, which provides the ability to control the stability of the results, but if they are used in a different environment, for example, the military, which requires random distribution, which may lead to different results. Data size and key sizes also can affect the results (it is left for future work). Duplicate data/information or decryption without detection of the encryption breach could consume WSN resources which are not addressed in this proposal.

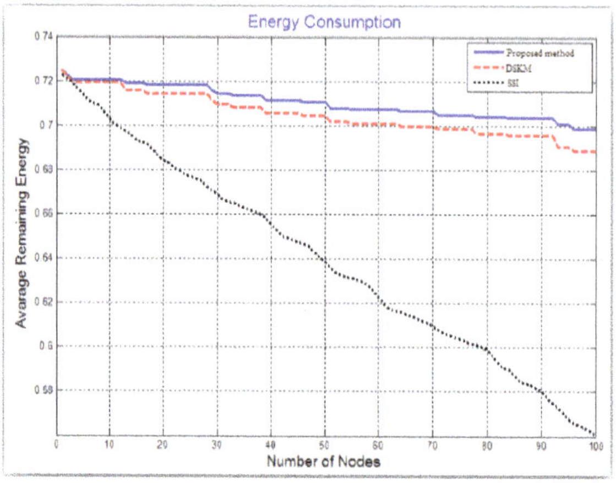

Figure 8. Energy consumption test.

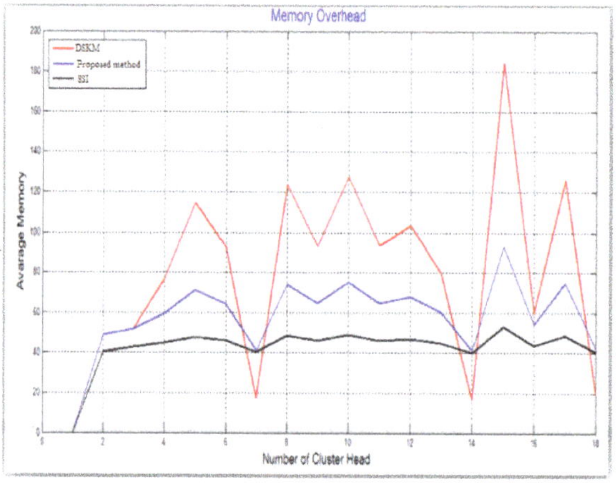

Figure 9. Size of memory used in cluster-heads.

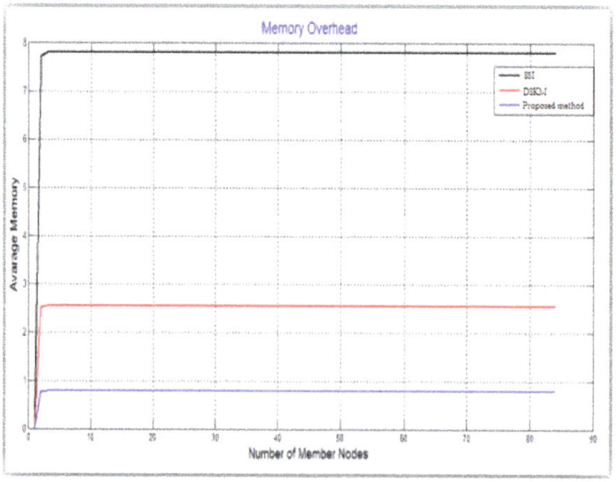

Figure 10. Size of memory used in the member sensors.

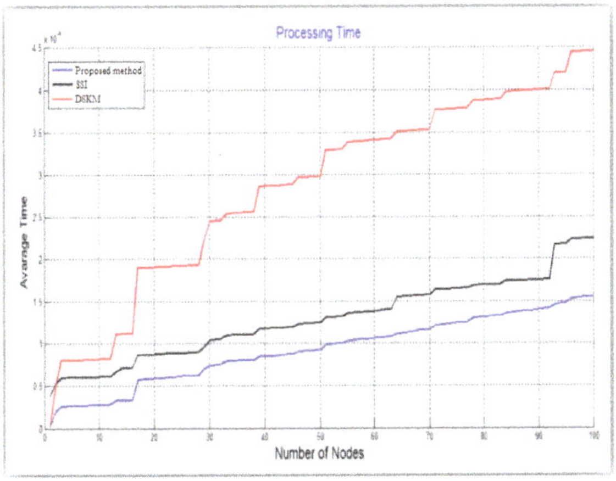

Figure 11. Processing time for generating shared keys.

Theoretically, a comparison of some recent references with our proposed method shows the superiority of our proposal in terms of performance. Refs. [24,25] used asymmetric cryptographic algorithms such as ECC for the key generation which will add significant costs to WSN resources while our method is based on PRNG which is a lightweight key generation method. In addition, in [26,27], the authors did not specify a suitable method for routing information, parameters, and sharing keys within the sensor network while our method relies on HEED which provides a suitable routing for sensor energy conservation. Moreover, the ref. [28] depends on randomly distributing the sensors, this leads to different distances between the sensors which will lead to a significant increase in the sensor communication expenses, this problem is avoided in our proposal due to the static distribution of the sensors in the medical environment. Finally, the ref. [29] relies on hash-intensive operations as well as encryption operations in key management which will negatively affect the computational costs of the sensors while our method relies on Bloom-lightweight key management.

5.2. Security Results

The attacker attempts to use node capture attacks to compromise information and WSN parameters. Five classifications of capture attacks are possible: sensor node, sensor CH, BS, and more than one sensor node or CH. These attacks try to penetrate shared keys, authentication keys and availability. The flaws that were exploited by capture attacks include problems with dictionary and forward secrecy, improper parameter distribution, irrational design purpose, ineffective verification, and unsecured parameter communication.

- Sensor node capture attack: When the attacker compromises a sensor and obtains some previous parameters such as SK-*K*, they try to use that key in future sessions of the WSN. In our proposed method, all sensors use new SH-*K* for each session. Therefore, when a hacker performs a sensor node capture attack on our WSN it will not affect the confidential information of other sensors.
- Sensor CH capture attack: When the hacker succeeds in executing this attack on CH. It tries to use the previous Auth-*K* to make all its sensors trust it and send all data and information to that hacker in that session. In our proposed method, the sensors within the cluster do not handle the old Auth-*K*. Thus, this attack cannot compromise WSN information by relying on a single CH, namely, our proposed method resists the CH capture attack.
- BS capture attack: We assume that BS is safe against capture attacks. However, assuming that the hacker was able to penetrate the BS either remotely or by stealing the BS device. The hacker will not benefit from the previous information of sensors or CHs because all security parameters (such as SH-*K* and Auth-*K*) in our proposed

method are generated instantly/unique by PRNG and Bloom and are hidden by RC5. However, the hacker may find some data collected by sensors. First, we assume that the data is transferred periodically to a central server so that even if the hacker tampered with this data, the original copy will be safe. Second, our research focuses on security key management and not the data collected. Therefore, our method is able to block BS capture attacks.

- A capture attack of more than one sensor node: If the hacker was able to compromise two or three sensors. Then he tried to analyze the obtained security parameters (such as Auth-*Ks*) for these sensors. The hacker cannot use these current parameters to hack network information in the current session. Because our method uses the RC5 algorithm, which has the advantage of preventing analysis and differential risks. Therefore, our method prevents this attack from extracting security parameters from Auth-*Ks*.
- A capture attack of more than one CH: When a hacker can compromise two CHs or three CHs. It tries to use the security parameters available from the compromised CHs. However, our proposed method uses a Bloom filter between BS and CHs to manage and verify the exchanged keys. The hacker cannot use the old parameters to communicate with the BS because these parameters will be rejected by the BS. Therefore, our proposed method is able to prevent this attack.

Table 2 shows the comparison of security features between the proposed method and the security key management methods in WSN. Where Sym is symmetric encryption and Asym is asymmetric encryption. The [23,27,30] methods are not discussed for classifications of node capture attacks. This indicates that their methods can be an easy target for various classes of node capture attacks. While our method and ref. [31] investigated different classes of these attacks. However, ref. [31] did not discuss compromising multiple sensors and CHs, nor did they specify countermeasures. Moreover, our method provides high randomness (by using PRNG) to the shared keys which is superior to existing methods that use low or medium randomness. The high randomness gives Auth-*Ks* and SH-*Ks* keys resistance to analysis and deferential threats. Finally, our method uses a flexible manner such as the Bloom scheme to manage security keys where Bloom is not used in the existing methods.

Table 2. Comparison of key management methods with our proposed method.

Security Feature	Qin et al. [23]	Ahlawat and Dave [27]	Liu et al. [30]	Wang et al. [31]	Proposed Method
Anti node capture attacks	One	One	One	Many	Many
Encryption type	Sym		Sym/Asym	Sym	Sym
Flexibility					Yes
Forward secrecy		Yes	Yes	Yes	Yes
Info. hiding	Yes		Yes		Yes
Keys randomness	Low	Medium	Low	Medium	High
Scalability	Low	Low	Medium	Low	High

6. Conclusions and Future Trends

For continuous data collection and monitoring, a wireless sensor network generally comprises sensor nodes dispersed in areas sensitive to data, such as the health sector. All sensor nodes gather data, which is then transmitted either directly or indirectly to the base station. Due to the nature and variety of applications of WSNs, Security has constantly been a serious problem. In a heterogeneous/hierarchical WSN, for securing connections in all hops a security method has been proposed. This approach provides strong security by attaining confidentiality (RC5), management (Bloom) and randomness (PRNG), in which the information is encrypted/decrypted and authenticated in each stage until it reaches the target node, resulting in increased secrecy of the transmitted message. In addition, this approach has great scalability and flexibility. Furthermore, our proposed method provides high node capture resistance, as the attacker must capture (λ + 1) of cluster heads to compromise the cluster heads' keys. Whereas the capture of a member node has no impact on the other nodes because each member node possesses key information that is unique. However, the sensor's resource is employed in an inequality manner to ensure network balance, resulting in a WSN method that is both efficient and secure. For future directions, we intend to investigate more Bloom filters to support key management. In addition, the accountability requirement will support the robustness of the key management scheme if added to all network devices, which will enhance the security of network device key ownership. Furthermore, we plan to extend the use of WSN within IoT applications to quickly transmit sensor-collected health data anywhere but this will require more attack testing and performance evaluation with more modern methods.

Author Contributions: Contributions to the paper were made by all authors. Conceptualization, R.A.M., N.A.F. and M.A.-Z.; methodology, R.A.M., N.A.F. and M.A.-Z.; software, R.A.M., N.A.F. and M.A.-Z.; validation, M.A.-Z.; formal analysis, M.A.-Z.; investigation, R.A.M. and M.A.-Z.; writing—original draft preparation, R.A.M., N.A.F. and M.A.-Z.; writing—review and editing, M.A.-Z.; project administration, R.A.M. and M.A.-Z. All authors have read and agreed to the published version of the manuscript.

Funding: This research received no external funding.

Conflicts of Interest: The authors declare no conflict of interest.

Abbreviations

This paper uses the following abbreviations:

Auth-*K*	Authentication key
BS	Base station
CBC-RC5	Cipher block chaining-Rivest cipher 5
CH	Cluster head
HEED	Hybrid energy efficient distributed
ID	Identifier
PRNG	Pseudo-random number generator
SH-*K*	Shared key
Sym/Asym	Symmetric/Asymmetric

References

1. Patil, H.K.; Szygenda, S.A.; Szygenda, S.A. *Security for Wireless Sensor Networks Using Identity-Based Cryptography*; CRC Press: Boca Raton, FL, USA, 2013.
2. Huanan, Z.; Suping, X.; Jiannan, W. Security and application of wireless sensor network. *Procedia Comput. Sci.* **2021**, *183*, 486–492. [CrossRef]
3. Awaad, M.H.; Jebbar, W.A. Study to analyze and compare the LEACH protocol with three methods to improve it and determine the best choice. *J. Comput. Sci. Control. Syst.* **2014**, *7*, 5.
4. Al-Zubaidie, M.; Zhang, Z.; Zhang, J. REISCH: Incorporating lightweight and reliable algorithms into healthcare applications of WSNs. *Appl. Sci.* **2020**, *10*, 2007. [CrossRef]
5. Banerjee, A.; De, S.K.; Majumder, K.; Das, V.; Giri, D.; Shaw, R.N.; Ghosh, A. Construction of effective wireless sensor network for smart communication using modified ant colony optimization technique. In *Advanced Computing and Intelligent Technologies*; Springer: Singapore, 2022; pp. 269–278.
6. Khalaf, O.I.; Romero, C.A.T.; Hassan, S.; Iqbal, M.T. Mitigating hotspot issues in heterogeneous wireless sensor networks. *J. Sens.* **2022**, *2022*, 7909472. [CrossRef]
7. Al-Zubaidie, M.; Zhang, Z.; Zhang, J. RAMHU: A new robust lightweight scheme for mutual users authentication in healthcare applications. *Secur. Commun. Netw.* **2019**, *2019*, 3263902. [CrossRef]
8. Majid, M.; Habib, S.; Javed, A.R.; Rizwan, M.; Srivastava, G.; Gadekallu, T.R.; Lin, J.C.W. Applications of wireless sensor networks and internet of things frameworks in the industry revolution 4.0: A systematic literature review. *Sensors* **2022**, *22*, 2087. [CrossRef]
9. Al-Zubaidie, M. Implication of lightweight and robust hash function to support key exchange in health sensor networks. *Symmetry* **2023**, *15*, 152. [CrossRef]
10. Sastry, A.S.; Sulthana, S.; Vagdevi, S. Security threats in wireless sensor networks in each layer. *Int. J. Adv. Netw. Appl.* **2013**, *4*, 1657.
11. Lee, C.-C. Security and privacy in wireless sensor networks: Advances and challenges. *Sensors* **2020**, *20*, 744. [CrossRef]
12. Barati, H. A hierarchical key management method for wireless sensor networks. *Microprocess. Microsyst.* **2022**, *90*, 04489.
13. Al-Zubaidie, M.; Zhang, Z.; Zhang, J. PAX: Using pseudonymization and anonymization to protect patients' identities and data in the healthcare system. *Int. J. Environ. Res. Public Health* **2019**, *16*, 1490. [CrossRef] [PubMed]

14. Al-Zubaidie, M.H.A. Incorporating Security into Electronic Health Records Based Healthcare Systems with Wireless Sensor Networks. Ph.D. Dissertation, University of Southern Queensland, Darling Heights, QLD, Australia, 2020.
15. Ghosal, A.; Conti, M. Key management systems for smart grid advanced metering infrastructure: A survey. *IEEE Commun. Surv. Tutor.* **2019**, *21*, 2831–2848. [CrossRef]
16. Zhou, H.; Lv, K.; Huang, L.; Ma, X. Security assessment and key management in a quantum network. *arXiv* **2019**, arXiv:1907.08963.
17. Al-Zubaidie, M.; Zhang, Z.; Zhang, J. User authentication into electronic health record based on reliable lightweight algorithms. In *Handbook of Research on Cyber Crime and Information Privacy*; IGI Global: Hershey, PA, USA, 2021; pp. 700–738.
18. Shahzadi, R.; Anwar, S.M.; Qamar, F.; Ali, M.; Rodrigues, J.J. Chaos based enhanced RC5 algorithm for security and integrity of clinical images in remote health monitoring. *IEEE Access* **2019**, *7*, 52858–52870. [CrossRef]
19. Li, L.; Wang, X. A high security dynamic secret key management scheme for wireless sensor networks. In Proceedings of the Third International Symposium on Intelligent Information Technology and Security Informatics, Jinan, China, 2–4 April 2010; pp. 507–510.
20. Iwendi, C.; Allen, A.; Offor, K. Smart security implementation for wireless sensor network nodes. *J. Wirel. Sens. Netw.* **2015**, *1*, 1.
21. Zhang, Y.; Pengfei, J. An efficient and hybrid key management for heterogeneous wireless sensor networks. In Proceedings of the 26th Chinese Control and Decision Conference (2014 CCDC), Changsha, China, 31 May 2014–2 June 2014; pp. 1881–1885.
22. Zhang, X.; Wang, J. An efficient key management scheme in hierarchical wireless sensor networks. In Proceedings of the 2015 International Conference on Computing, Communication and Security (ICCCS), Pointe aux Piments, Mauritius, 4–5 December 2015; pp. 1–7.
23. Qin, D.; Jia, S.; Yang, S.; Wang, E.; Ding, Q. A lightweight authentication and key management scheme for wireless sensor networks. *J. Sens.* **2016**, *2016*, 1547963. [CrossRef]
24. Moara-Nkwe, K.; Shi, Q.; Lee, G.M.; Eiza, M.H. A novel physical layer secure key generation and refreshment scheme for wireless sensor networks. *IEEE Access* **2018**, *6*, 11374–11387. [CrossRef]
25. Chanda, A.; Sadhukhan, P.; Mukherjee, N. Key management for hierarchical wireless sensor networks: A robust scheme. *EAI Endorsed Trans. Internet Things* **2020**, *6*, 23. [CrossRef]
26. Jia, C.; Ding, H.; Zhang, C.; Zhang, X. Design of a dynamic key management plan for intelligent building energy management system based on wireless sensor network and blockchain technology. *Alex. Eng. J.* **2021**, *60*, 337–346. [CrossRef]
27. Ahlawat, P.; Dave, M. An attack resistant key predistribution scheme for wireless sensor networks. *J. King Saud Univ.-Comput. Inf. Sci.* **2021**, *33*, 268–280. [CrossRef]
28. Kumar, V.; Malik, N. Enhancing the connectivity and resiliency of random key pre-distribution schemes for wireless sensor network. *Int. J. Syst. Assur. Eng. Manag.* **2022**, *13*, 92–99. [CrossRef]
29. Tyagi, P.; Kumari, S.; Alzahrani, B.A.; Gupta, A.; Yang, M.H. An enhanced user authentication and key agreement scheme for wireless sensor networks tailored for IoT. *Sensors* **2022**, *22*, 8793. [CrossRef]
30. Liu, J.; Liu, L.; Liu, Z.; Lai, Y.; Qin, H.; Luo, S. WSN node access authentication protocol based on trusted computing. *Simul. Model. Pract. Theory* **2022**, *117*, 102522. [CrossRef]
31. Wang, C.; Wang, D.; Tu, Y.; Xu, G.; Wang, H. Understanding node capture attacks in user authentication schemes for wireless sensor networks. *IEEE Trans. Dependable Secur. Comput.* **2020**, *19*, 507–523. [CrossRef]
32. Ullah, Z. A survey on hybrid, energy efficient and distributed (HEED) based energy efficient clustering protocols for wireless sensor networks. *Wirel. Pers. Commun.* **2020**, *112*, 2685–2713. [CrossRef]
33. Gupta, P.; Sharma, A.K. Clustering-based optimized HEED protocols for WSNs using bacterial foraging optimization and fuzzy logic system. *Soft Comput.* **2019**, *23*, 507–526. [CrossRef]
34. Awaad, M.H.; Jebbar, W.A. Prolong the lifetime of WSN by determining a correlation nodes in the same zone and searching for the best not the closest CH. *Int. J. Mod. Educ. Comput. Sci.* **2014**, *6*, 31. [CrossRef]
35. Mishall Hammed, A. Improve the effectiveness of sensor networks and extend the network lifetime using 2BSs and determination of area of CHs choice. *J. Comput. Sci. Control. Syst.* **2014**, *7*, 15.
36. Anitha, G.; Vijayakumari, V.; Thangavelu, S. A comprehensive study and analysis of LEACH and HEED routing protocols for wireless sensor networks—With suggestion for improvements. *Indones. J. Electr. Eng. Comput. Sci.* **2018**, *9*, 778–783. [CrossRef]
37. Boudhiafi, W.; Ezzedine, T. Optimization of multi-level HEED protocol in wireless sensor networks. In *Communications in Computer and Information Science: International Conference on Applied Informatics*; Springer: Cham, Swidzerland, 2021; pp. 407–418.
38. Jamil, A.S.; Rahma, A.M.S. Image encryption based on multi-level keys on RC5 algorithm. *iJIM* **2022**, *16*, 101.
39. Raj, K.V.; Ankitha, H.; Ankitha, N.G.; Hegde, L.K. Honey encryption based hybrid cryptographic algorithm: A fusion ensuring enhanced security. In Proceedings of the 2020 5th International Conference on Communication and Electronics Systems (ICCES), Coimbatore, India, 10–12 June 2020; pp. 490–494.
40. Alenezi, M.N.; Alabdulrazzaq, H.; Mohammad, N.Q. Symmetric encryption algorithms: Review and evaluation study. *Int. J. Commun. Netw. Inf. Secur.* **2020**, *12*, 256–272.
41. Sadhya, D.; Sing, S.K. Providing robust security measures to bloom filter based biometric template protection schemes. *Comput. Secur.* **2017**, *67*, 59–72. [CrossRef]
42. Lim, K.; Liu, W.; Wang, X.; Joung, J. SSKM: Scalable and secure key management scheme for group signature based authentication and CRL in VANET. *Electronics* **2019**, *8*, 1330. [CrossRef]
43. Sampangi, R.V.; Sampalli, S. Metamorphic framework for key management and authentication in resource-constrained wireless networks. *Int. J. Netw. Secur.* **2017**, *19*, 430–442.

44. Ghorpade, S.; Zennaro, M.; Chaudhari, B.S. Towards green computing: Intelligent bio-inspired agent for IoT-enabled wireless sensor networks. *Int. J. Sens. Netw.* **2021**, *35*, 121–131. [CrossRef]
45. Mcginthy, J.M.; Michaels, A.J. Further analysis of prng-based key derivation functions. *IEEE Access* **2019**, *7*, 978–995. [CrossRef]

electronics

MDPI

Article

Identifying Adversary Impact Using End User Verifiable Key with Permutation Framework

Mohd Anjum [1], Sana Shahab [2], Yang Yu [3,*] and Habib Figa Guye [4]

1 Department of Computer Engineering, Aligarh Muslim University, Aligarh 202002, India
2 Department of Business Administration, College of Business Administration, Princess Nourah Bint Abdulrahman University, P.O. Box 84428, Riyadh 11671, Saudi Arabia
3 Centre for Infrastructure Engineering and Safety (CIES), University of New South Wales, Sydney, NSW 2052, Australia
4 Department of Information Science, College of Informatics, Bule Hora University, Hagere Maryam 144, Ethiopia
* Correspondence: yang.yu@uts.edu.au

Abstract: In the Internet of Things (IoT), security is a crucial aspect that ensures secure communication, transactions, and authentication for different applications. In IoT security, maintaining the user interface and platform security is a critical issue that needs to be addressed due to leaky security distribution. During communication, synchronisation and security are important problems. The security problems are caused by the adversary impact and vulnerable attacks, leading to service failure. Therefore, the Permutated Security Framework (PSF) is designed to manage security in the IoT by providing secure communication, transactions, and authentication for different applications. The PSF uses time intervals to manage transaction security. These intervals are secured using end-verifiable keys generated using the conventional Rivest–Shamir–Adleman (RSA) technique in IoT-based communication-related applications. In this approach, the key validity is first provided for the interval, and in the latter, the access permitted time modifies its validity. The security of transactions is managed by dividing time into smaller intervals and providing different levels of security for each interval. By using time intervals, the framework is adaptable and adjustable to changes in the system, such as user density and service allocation rate, adapting parallel transactions per support vector classifications' recommendations. The proposed framework aims to synchronise interval security, service allocation, and user flexibility to mitigate adversary impact, service failures, and service delays while improving the access rate and transactions. This allows for more flexibility and better management of transaction security. The proposed framework reduces adversary impact (10.98%), service failure (11.82%), and service delay (10.19%) and improves the access rate by 7.73% for different transactions.

Keywords: Internet of Things; RSA; security; support vector machine; wireless sensor networks

Citation: Anjum, M.; Shahab, S.; Yu, Y.; Guye, H.F. Identifying Adversary Impact Using End User Verifiable Key with Permutation Framework. *Electronics* **2023**, *12*, 1136. https://doi.org/10.3390/electronics12051136

Academic Editors: Tomasz Rak and Dariusz Rzońca

Received: 28 January 2023
Revised: 23 February 2023
Accepted: 23 February 2023
Published: 26 February 2023

1. Introduction

The Internet of Things (IoT) is a rapidly growing technology that connects everyday devices to the internet, allowing them to collect and share data. It encompasses a wide range of devices, from smartphones and laptops to home appliances, industrial equipment, and even automobiles. It helps to increase the communication process among users and organisations. This technology has the potential to revolutionise many industries by enabling more efficient and automated processes, improved decision-making, and new business models. IoT is widely used in smart applications to enhance the system's overall performance and provide a better user experience [1]. As the number of connected devices grows, so do security and privacy concerns. Additionally, IoT systems are distributed and open; therefore, they are vulnerable to various security threats such as hacking, data breaches, and unauthorised access. IoT nodes transfer lightweight data among the users and provide a better authentication process. Security is a major concern in IoT due to the large amount of data that needs to be managed. IoT is used in smart devices and wireless sensor networks (WSN) to enhance user services [2]. Proper authentication processes are used to address security issues, such as Authentication and Key Agreement (AKA) schemes. AKA schemes are applied in IoT to identify unauthorised persons from accessing the personal information of users [3]. A secret session key is shared with the users for the authentication process, and authentication will be declined without the key. AKA helps protect users from attackers by providing a better authentication process and maximising the system's performance by ensuring users' security and privacy. WSN is also used in security issues to find the users' exact location and identify intruders. While authenticating, a device's current location is traced, which helps to finalise the authentication process [4,5].

The IoT is widely utilised in various applications to improve communication among organisations and users and provide better services. Data processing is one of the main tasks in IoT, which helps to improve user performance [6]. IoT enables users to transfer data or information from one person to another using smart devices. Data transaction or transfer allows users to send information from their current location without travelling [7]. However, data transfer may also cause some security threats, and a proper authentication process is needed to ensure a secure data processing system [8]. Privacy and security are major concerns in IoT while transferring data. To address these concerns, technologies such as radiofrequency identification (RFID) are used in IoT to enhance security and privacy. RFID interacts with tags of the information and provides a better solution to security issues [9]. RFID tags have electronic product codes for each transaction, which helps to track the exact whereabouts of the data being transferred. WSN is also used in IoT and has nodes that identify the information's frequency and bandwidth. Using WSN in IoT applications makes the users' communication process safe and secure [4,10].

Synchronised security measures play a crucial role in every IoT application. WSN is used in IoT to ensure the users' security and prevents data processing errors. Independent nodes identify security errors and eliminate unwanted threats [11]. WSN captures the users' location by analysing the network's frequency and bandwidth, which plays a vital role in the authentication process. WSN also synchronises the security process by reducing the latency rate in services and providing better services to the users at the needed time [12,13]. RFID is also used in IoT for communication, where interacting tags are identified based on the device's frequency and ensure the users' security [14]. Electronic Product Code is used in every transaction process to track the data's whereabouts and secure the users' information from attackers. RFID is analysed by a classification, which is performed based on certain features of the users. AKA ensures the security of the users in a synchronised form by providing a secret session key to users from the device for an authentication process. The session key helps users to prevent cyber-attacks [11,15].

The proposed PSF aims to manage security in the IoT by providing secure communication, transactions, and authentication for different applications. One of the key features of the PSF is the use of time intervals, which are secured by end-verifiable keys generated using the conventional RSA technique. In this approach, the PSF creates unique security keys for different IoT devices and applications by permuting the elements of a set. These keys are then used to encrypt communications and authenticate transactions between devices. However, unlike the conventional approach, the proposed approach uses time intervals to generate the permuted keys. The keys are generated at specific time intervals and are valid for a limited period. This approach allows for more frequent updates to the security keys, which helps keep communications and transactions more secure. Even if a key is compromised, it will only be valid for a short time and will be replaced by a new key shortly. The RSA technique is used to generate the keys so that they are end-verifiable, which means that the authenticity of the key is verified at the end of the communication, which ensures that the communication is secure. It is a widely used and widely accepted encryption method. It creates a pair of public and private keys used to encrypt and decrypt data, respectively. In the PSF, the same pair of keys is used to encrypt and decrypt the data. This ensures that only authorised devices can communicate with each other and that transactions are secure. Additionally, the PSF includes a mechanism for updating and revoking the keys at regular intervals, which allows for the secure management of IoT devices over time. The time intervals at which the keys are generated and updated can be adjusted based on the specific requirements of the application. This allows for a more dynamic and adaptive approach to security, which can help keep communications and transactions more secure overall.

The paper is structured into the following sections: Section 1 introduces an overview of the problems of security in the IoT and the need for a framework to manage security in this context, introducing the PSF and its key features, such as the use of time intervals and end-verifiable keys generated using the RSA technique. Section 2 illustrates the review of existing research on security in the IoT and identifies the key contributions. Section 3 provides a detailed description of the proposed PSF, including the initial system setup and the RSA algorithm used to generate the keys, and explains how the PSF synchronises interval security, service allocation, and user flexibility to improve the access rate and transactions. Section 4 presents the results of the research, including the performance parameters of the PSF, such as adversary impact, service failure, service delay, access rate, and service transactions, and compares the PSF with the existing system based on all the performance parameters and analyses the results. Lastly, Section 5 summarises the key findings of the research and describes the contributions of the PSF to managing security in the IoT.

2. Related Works

This literature survey explores the various studies and research conducted on IoT security and privacy issues. With the increasing popularity of IoT and the integration of interconnected devices and systems, it has become imperative to address the concerns surrounding the security and privacy of data transmitted over these networks. Various authentication solutions have been proposed to address these concerns, but they often fall short in terms of efficiency and practicality as compared to the proposed model. In this related work, we will delve into the various studies conducted in this field and examine the proposed solutions and their effectiveness. We will also explore the potential of new technologies, such as blockchain and elliptic curve cryptography, in addressing these issues and the challenges that still need to be addressed.

Biswas et al. [16] proposed a scalable blockchain framework for secure IoT transaction processes using a peer network. One of the biggest challenges of combining IoT and blockchain technology is the scalability of the ledger and the speed at which transactions can be executed within a blockchain system. The network's scalability is improved by balancing the ledger and execution time during the transaction process. A peer network assists the system in understanding every detail of the transaction and identifying the gap between ledger bridges. The proposed solution addresses the scalability issues associated with integrating IoT and blockchain by implementing a scalable local ledger that limits the number of transactions entering the global blockchain while maintaining peer validation at both the local and global levels. Experiment results show that the proposed framework increases transaction security while decreasing network storage size and blockchain weight. Currently, smart home environments are vulnerable to security breaches; therefore, Yu et al. [17] created a secure and efficient three-factor authentication protocol for IoT-enabled smart homes to address the security weaknesses found in Kaur and Kumar's protocol. Elliptic curve cryptosystems are used in the proposed protocol to ensure the users' security and privacy. The formal and informal security analysis process is done in the proposed framework for improving users' privacy. Compared with other existing privacy-preserving protocols, the proposed framework increases the users' overall security and improves the system's efficiency. Asheralieva et al. [18] designed a mobile edge computing network mechanism for IoT-based applications to provide system security and scalability. The proposed method uses the peer technique to identify the blocks of the shared nodes and provide better communication to the users during the transaction process. The proposed system uses a new consensus mechanism in which each peer votes on the outputs of each block task in its shard, using a reputation-based coalitional game model (RBCGM). RBCGM is also used here to improve the overall services of the system. Huang et al. [19] introduced a new efficient revocable large universe multi-authority attribute-based encryption to address the security issues related to controlling access to data in constantly changing IoT environments. This method supports user-attribute, which is used in a security process. Integrating a cloud computing system also increases the network's overall security. The proposed scheme supports user-attribute revocation, prevents collusion attacks, and protects against the collusion attack of revoked and non-revoked users. It satisfies both forward and backward security requirements, making it suitable for large-scale collaborations across multiple domains in the dynamic and cloud-assisted IoT. It increases the overall performance of the network by ensuring the security of the users from attackers.

Sadri et al. [20] proposed an anonymous two-factor authentication protocol for preserving the integrity and confidentiality of the transmitted messages in WSNs for the IoT that addresses the security vulnerabilities of the existing state-of-the-art protocol proposed by Wu et al. [21]. A WSN is used in the proposed protocol to extend the system's lifetime. The proposed method analyses formal and informal problems to secure the authenticating user process and provide better communication services and are secure against various known attacks such as sensor and user trace, sensor capture, offline password guessing, and replay attacks. Dorri et al. [22] established a lightweight, scalable blockchain method for IoT applications that address traditional blockchain technology's computational and scalability limitations. The proposed blockchain method uses a distributed time-based consensus algorithm, which helps reduce latency and system delay rates. It helps to manage blockchain delays and provides better services to users. Compared with other methods, the proposed lightweight, scalable blockchain method strongly protects from various security attacks. Simulation studies indicate that it reduces packet overhead and delay and increases the overall performance and blockchain scalability compared to relevant baselines. Vishwakarma et al. [23] developed a novel communication and authentication method for providing identification, authentication, secure communication, and data integrity in the IoT network. Blockchain and a hybrid cryptosystem technique are used in the proposed scheme to enhance the security system of the applications. Angular distance based on the cluster approach is used here to analyse the system's securities. Analytical results show that the proposed secure communication and authentication method reduced the computation time and protected systems from various cyberattacks such as impersonation, message replay, man-in-the-middle, and botnet attacks.

Peneti et al. [24] introduced a method for managing security, privacy, and confidentiality in next-generation networks such as IoT and 6G by combining blockchain and a grey wolf-optimised modular neural network approach. The proposed method creates user-authenticated blocks to manage security and privacy properties, and the neural network is used to optimise latency and computational resource utilisation in IoT-enabled smart applications. A simulation study is performed to display the over-efficiency of the system with respect to the multi-layer perceptron and deep learning networks, and it is shown to have low latency and high security (99.12%). Majumder et al. [25] introduced a constraint application protocol based on elliptic curve cryptography. It establishes a secure session key between IoT devices and a remote server using lightweight elliptic curve cryptography to overcome the limitations of key management and multicast security in constraint application protocol, which is used for communication between lightweight resource constraint devices in an IoT network. The proposed approach provides a constraint application protocol implementation for authentication in IoT networks, and it is found to be lightweight and secure after analysing various cryptographic attacks. Lin et al. [26] introduced a new settlement model for IoT data exchange services that use blockchain technology to overcome the limitations of traditional centralised models. The proposed model includes a Bitcoin-based time commitment scheme and an optimised practical Byzantine fault-tolerant consensus protocol named ReBFT to ensure fairness and accountability in the decentralised network. It also ensures users a safe and secure transaction process and prevents unauthorised authentication. Several experiments are conducted to verify the feasibility of the proposal. Compared with existing protocols, the proposed scheme raises the feasibility and service efficiency.

Attarian et al. [27] proposed a communication protocol for secure and anonymous mHealth transactions using a combination of onion routing, blockchain smart contracts, and the user datagram protocol to protect the security and privacy of clients' identities. The blockchain approach is used in the proposed protocol to ensure the structure and architecture of the application. The proposed protocol aims to address challenges of anonymity, untraceability, unlinkability, and unforgeability in healthcare transactions and can detect malicious clients who send false data and helps to eliminate those details from the database. The proposed protocol ensures the security and privacy of the users while transacting data. Experimental outcomes and privacy proofs show that the proposed protocol has a reasonable computational cost and provides sufficient protection for IoT-based mHealth transactions. Yazdinejad et al. [28] discussed the challenges of IoT, such as security and energy consumption. They proposed a solution to mitigate these challenges by combining blockchain and software-defined networks in IoT networks. The proposed architecture uses a cluster structure with a new routing protocol. It utilises both public and private blockchains for peer-to-peer communication between IoT devices and software-defined network controllers, which eliminates proof-of-work and uses an efficient authentication method, making it suitable for resource-constrained IoT devices. Software-defined network controller plays a vital role in this protocol, which helps ensure the users' security while processing data. The experimental results show that this proposed architecture performs better throughput, delay, and energy consumption than other routing protocols. Compared with other security methods, the proposed protocol increases users' scalability, security, and privacy and reduces the computation cost with the help of the blockchain technique.

Srinivas et al. [29] proposed a new lightweight chaotic map-based authenticated key agreement protocol (CMAKAP) for the industrial environment that aims to increase security using a fuzzy extractor technique for biometric verification. The authentication process is done based on the user's biometrics, personal information, and smart cards, which help to prevent the users from being unauthorised. The real-or-random method is used here to analyse the security issues in the applications. The scheme also supports adding new devices, changing passwords/biometrics, and revoking smart cards. Formal security analysis and simulation studies were conducted, and it was found that the proposed scheme provides superior security compared to other existing methods. Pham et al. [30] introduced a mutual privacy-preserving authentication protocol (MPPAP) by using an elliptic curve cryptography approach to improve security and protect the privacy of IoT devices while also being efficient in resource consumption. It helps to provide better communication services to the users. A secret session key is shared with the users for the authentication process, ensuring the users' security and privacy. The proposed model extends previous works and includes a distributed network architecture and secure communications. The protocol has been formally proven correct, is resilient to attacks, and has low energy consumption. Then, the overall summary of the existing works is summarised in Table 1.

Table 1. Summary of the related works.

Reference	Method(s)	Purpose	Efficiency
Biswas et al. [16]	Scalable blockchain framework	To address the scalability issues associated with integrating IoT and blockchain.	Increases transaction security while decreasing network storage size and blockchain weight.
Yu et al. [17]	Three-factor authentication protocol	To address the security weaknesses found in Kaur and Kumar's protocol.	Increases the users' overall security and improves the system's efficiency.
Asheralieva et al. [18]	Reputation-based coalitional game model (RBCGM)	To identify the blocks of the shared nodes and provide better communication.	Improves the overall services of the system.
Huang et al. [19]	Revocable large universe multi-authority attribute-based encryption	To address the security issues related to controlling access to data in constantly changing IoT environments.	Ensures the security of the users from attackers.
Sadri et al. [20]	Anonymous two-factor authentication protocol	To address the security vulnerabilities.	Preserves the integrity and confidentiality of the transmitted messages.
Wu et al. [21].	Three-factor authentication protocol	To analyse both formal and informal problems to secure the authenticating user process.	Manages data security and confidentiality.
Dorri et al. [22]	A lightweight, scalable blockchain method	To address the computational and scalability limitations of traditional blockchain technology.	Reduces latency and system delay rates.
Vishwakarma et al. [23]	Blockchain and a hybrid cryptosystem technique	To resolve integrity and security-related issues.	Reduces the computation time and protect systems from various cyberattacks.
Peneti et al. [24]	Blockchain and grey wolf-optimised modular neural network approach	To optimise latency and computational resource utilisation.	Low latency and high security
Majumder et al. [25]	Constraint application protocol	To overcome the limitations of key management and multicast security in a constraint application protocol.	Secures the information from different cryptographic attacks.
Lin et al. [26]	Byzantine fault-tolerant consensus protocol	To overcome the limitations of traditional centralised models.	Ensures users a safe and secure transaction process and prevents unauthorised authentication
Attarian et al. [27]	Combination of onion routing, blockchain smart contracts	To protect the security and privacy of clients' identities	Addresses challenges of anonymity, untraceability, unlinkability, and unforgeability in healthcare transactions and can detect malicious clients
Yazdinejad et al. [28]	Blockchain and software-defined networks	To propose a solution to mitigate these challenges by combining blockchain and software-defined networks.	Better performance in throughput, delay, and energy consumption than other routing protocols.
Srinivas et al. [29]	Lightweight chaotic map-based authenticated key agreement protocol (CMAKAP)	To increase security by using a fuzzy extractor technique for biometric verification.	The proposed scheme provides superior security compared to other existing methods.
Pham et al. [30]	mutual privacy-preserving authentication protocol (MPPAP)	To improve security and protect the privacy of IoT devices	Proven correct and resilient to different attacks while having low energy consumption.

3. Proposed Permuted Security Framework

The design goal of PSF is to improve the user flexibility rate of the IoT applications by reducing adversary fewer services in IoT combined end-user applications. This platform provides secure transactions, authentication, and communication for various end-user industrial applications. Its experience in controlling security is synchronising the IoT platform and user interface. It provides different security threats to be distributed for secure and dependable transactions through the IoT network. The proposed PSF is illustrated in the IoT environment as in Figure 1. The cloud and security have the connections that are used to manage data security. Here, security techniques are utilised to manage data security.

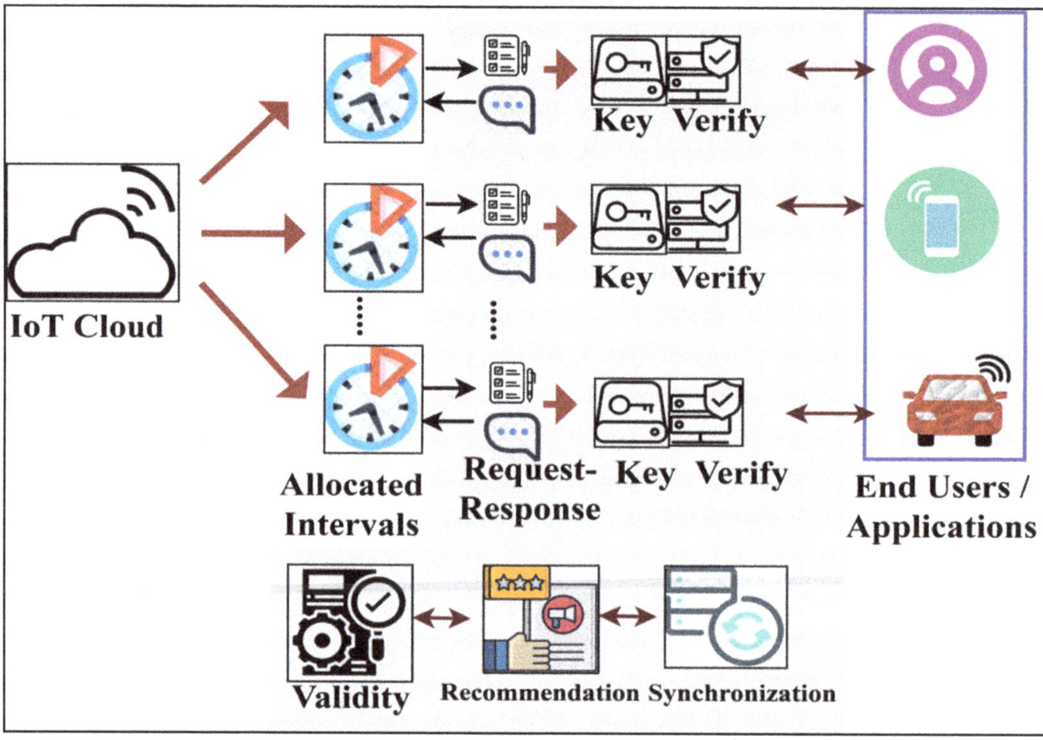

Figure 1. Proposed permuted security framework in IoT environment.

The proposed framework can provide secure data collection and security distribution for synchronisation between end-user applications and the platform using transaction time intervals. In this manner, the data transactions, authentication, and communication through the IoT platform are secured from permitting adversary fewer services to improve user flexibility harmoniously and the service allocation rate of smart end-user applications, as shown in Figure 1. The function of PSF assisted in providing a secure data collection and distribution security. Data collection from the IoT cloud and user side is performed, and security is the distribution to both sender and receiver. The applications and processing centres are linked through IoT. Permutated security in the IoT platform and the user interface is administered to prevent leaky security distribution, adversary fewer services, and service failures. The IoT environment ensures data transactions between the applications and processing centres. The operations of the IoT cloud and user interface in the platform are used for synchronisation, transactions, and authentication. Synchronising fewer services for the applications and processing centres is processed and analysed using learning.

Initial System Setup

The IoT network is determined using two terminals: the IoT cloud and the user interface. The IoT cloud terminals collect data, and user interface terminals administer security and another mitigating adversary impact. The IoT cloud terminals communicate with $I_{oT} = \{1, 2, \ldots, z\}$ set of services that can access data from all the end-user applications from the smart technology. The above I_{oT} transmits various quantities of data in the different time interval $D_T = \{1, 2, \ldots, T\}$. Let n represent the number of adversaries and fewer services in the end-user applications. Based on the above definition, the number of data transfers per unit of time is i such that the collection of secure data transaction $\exists_i$ is estimated as:

$$\left.\begin{array}{c} \exists_i = \begin{cases} I_{oT} \times i \times T \,\forall\, I_{oT} \rightarrow D_T, \text{ if } n = 0 \\ A_{fs} \times \frac{z-n}{I_{oT}} \times T \,\forall\, (I_{oT}, n) \rightarrow D_T, \text{ else } n \neq 0 \end{cases} \\ \text{such that} \\ I_{oT} \rightarrow D_T = \prod_{i=1}^{I_{oT}} i_n \\ \text{and} \\ (I_{oT}, n) \rightarrow D_T = \sum_{i=1}^{s} i_n - A_{fs} \sum_{i=1}^{n} i_n \\ \text{and} \\ A_{fs} = \frac{A_{ft}}{A_{ft}+i} \end{array}\right\} \quad (1)$$

In Equation (1), the variables A_{fs} and A_{ft} denote the adversary's fewer service rate and data transmission in D_T. The expressions $I_{oT} \rightarrow D_T$ and $(I_{oT}, m) \rightarrow D_T$ show the mapping of the IoT cloud and the user interface terminals at the different time interval D_T. The data synchronisation or information from the IoT architecture is concealed into two levels: IoT cloud network for security. The IoT cloud terminal, the transmission of data, and $\exists_i$ are the sum-up metrics for securing the collection for the mapped D_T, where it satisfies. For data collection, the user interface terminal provides synchronisation and secure authentication. The synchronisation of data between $I_{oT} \in$ i and n are operated with the help of their mapping and transaction time. According to Equation (1), the given condition $n > I_{oT}$ specifies less and insufficient data from the IoT network. The different time mapping for the IoT cloud and the sequential process $\exists_i$ rely upon $(z \times i)$, which is the evaluating condition for synchronisation.

$$T_n = \prod_{i=1}^{s} \frac{\mu_n}{T_i}; \text{ where } \urcorner\exists_i = \frac{\exists_i}{(i-n)} - (\mu - A_{ft}) \quad (2)$$

Based on the above equation, variables T_n and $\urcorner\exists_i$ represent the different mapping time instances and sequential collection of data. The above-derived equations are the reliable synchronisation of the security distribution (S_r), where it is evaluated for each access level of D_T. This estimation is observed for identifying the function $n \neq 0$ and $n = 0$ for all D_T using the conventional RSA technique. This RSA cryptography analysis is an approach to public-key cryptography, and it is based on random contours over each access level in that network. The collection of the secured data sequence βT_n and $\urcorner\exists_i$ such that the S_r is defined for all the output for the centre level O_u. The linear output of security distribution of $\urcorner\exists_i$ in T_n is the synchronising observation for augmenting $(z \times i)$. The O_u and result (Z) are important in defining S_r. The different instances of IoT cloud inputs for the determination of $\urcorner\exists_i$ for both $I_{oT} \rightarrow D_T$ and $(I_{oT}, n) \rightarrow D_T$ include different mappings sequences. If the IoT cloud is accessed in the mapping time, it is one; otherwise, it is zero. Figure 2 presents the synchronisation mapping for linear access.

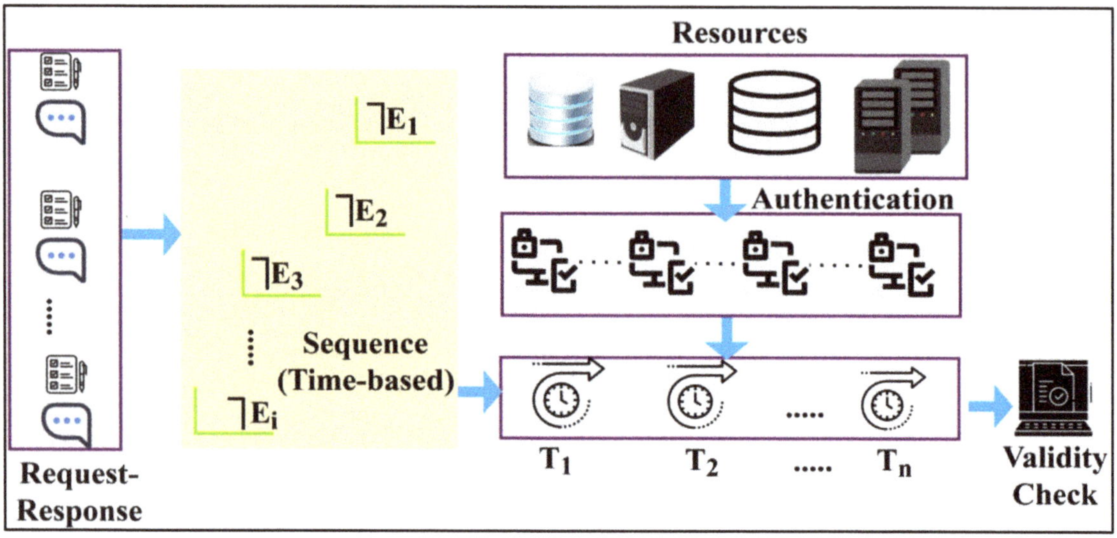

Figure 2. Synchronisation mapping process.

The proposed framework performs a mapping based on $\daleth E_i$ indifferent transactions T_n. The available resources are authenticated using sequence-based validations to improve the transactions. The proposed framework performs a validity check if the transaction is authenticated. Therefore, the mapping process is performed for $I_{oT} \rightarrow D_T$, whereas synchronisation is achieved as $(I_{oT}, n) \rightarrow D_T$, as shown in Figure 2. This is performed to achieve a solution until $n \neq 0$. The solution of the centre-level access output in the first mapping $I_{oT} \rightarrow D_T$ produces a linear limitable result whereas $(I_{oT}, n) \rightarrow D_T$ extracts solution of z with $n \neq 0$. The following equation shows the centre-level access output, and the final result of Z for $I_{oT} \rightarrow D_T$ is estimated. These estimations have functioned for both the conditions of A_{ft} and the conditional estimation of $\mu = 1$ or $\mu = 0$ in D_T. Hence, the output is accessed for the entire distributed time instance D_T. From the above mapping condition, n serves as an IoT cloud input, and the synchronisation of A_{fs} in $I_{oT} \rightarrow D_T$ mapping is given as:

$$\left.\begin{array}{c} O_u{}^1 = \daleth\exists_{i1} T_1 + n_1 \mu_1 \\ O_u{}^2 = \daleth\exists_{i2} T_2 - A_{ft1} + \daleth\exists_{i1} \mu \\ O_u{}^3 = \daleth\exists_{i3} T_3 - A_{ft2} + \daleth\exists_{i2} \mu \\ \vdots \\ O_u{}^t = \daleth\exists_{it} T_{D_T} - A_{ftT-1} + \daleth\exists_{itT-1} \mu \end{array}\right\} \tag{3}$$

Instead,

$$\left.\begin{array}{cc} Z_1 = O_u{}^1 & Z_1 = \daleth\exists_{i1} T_1 + n_1 \mu_1 \\ Z_2 = O_u{}^2 - A_{fs1} i_2 & Z_2 = \daleth\exists_{i2} T_2 - A_{ft1} + \daleth\exists_{i2} \mu - A_{fs1} i_2 \\ Z_3 = O_u{}^3 - A_{fs2} i_3 & Z_3 = \daleth\exists_{i3} T_3 - A_{ft2} + \daleth\exists_{i3} \mu - A_{fs2} i_3 \\ \vdots & \vdots \\ Z_{D_T} = O_u{}^t - A_{fsT-1} i_{T-1} & Z_{D_T} = \daleth\exists_{it} T_{D_T} - A_{ftT-1} + \daleth\exists_{itT-1} \mu - A_{fsT-1} i_{T-1} \end{array}\right\} \tag{4}$$

From the above equation, the linear access solution for each level of data transactions is determined as $Z = \daleth\exists_i T - A_{ftt} + \daleth\exists_i \mu - A_{fs} n$ and $n = 0$, then $\mu = 1$ and $\daleth\exists_{it} T_{D_T} = n\exists_i$ and therefore, $Z = n\exists_i T + n\exists_i = n\exists_i (T + 1)$ is the reliable solution and $S_r = 1$. Here, the synchronisation of such IoT cloud systems is retained at once. The secure transaction requires $\{S_r, \beta, I_{oT}\}$ for each level of access D_T and this data provides security for the IoT information. Therefore, $(I_{oT}, n) \rightarrow D_T$ mediate solution and results are estimated as in the following equations, respectively.

$$\left.\begin{array}{c} O_u{}^1 = \exists_{i1} \\ O_u{}^2 = \exists_{i2} - A_{fs1} - \mu_{i_1} i_1 \\ O_u{}^3 = \exists_{i3} - A_{fs2} + \mu_{i_2} i_2 \\ \vdots \\ O_u{}^t = \exists_{iT} - A_{fsT-1} - \mu_{i_{T-1}} i_{T-1} \end{array}\right\} \tag{5}$$

where in Equation (5), Equation (6) is derived.

$$\left.\begin{array}{c} Z_1 = O_u{}^1 = \exists_{i1} \\ Z_2 = O_u{}^2 + T_{n_1} - \urcorner\exists_{i1} = \exists_{i2} - A_{fs1} - i_1 + T_{n_1} - \urcorner\exists_{i1} \\ Z_3 = O_u{}^3 + T_{n_2} - \urcorner\exists_{i2} = \exists_{i3} - A_{fs2} i_2 + T_{n_2} - \urcorner\exists_{i2} \\ \vdots \\ Z_T = O_u{}^t + T_{n_t} - \urcorner\exists_{iT} = \exists_{iT-1} - A_{fsT-1} - \mu_{i_{T-1}} + T_{n_{t-1}} - \urcorner\exists_{iT-1} \end{array}\right\} \tag{6}$$

The solution, as in the above-derived equations, is obtained by verifying the functions $\urcorner\exists_i = (z-n)\exists_i$ and $\mu = 1$ or $\mu = 0$ in each level-by-level manner. If $\mu = 0$, then $Z_T = \exists_i - \mu_{i_{T-1}} i_t - \urcorner\exists_i$ is the final output, and if $\mu = 1$, then $A_{ft} = 0$, and therefore, the output is $Z = \exists_i + T_n - \urcorner\exists_i$. Hence, if $I_{oT} \rightarrow D_T$, then $Z = n\,\exists_i(T+1)$ is the output and $Z = \exists_i + T_n - \urcorner\exists_i$ is the segregated result. From this output, $S_r = \left[\frac{\mu - A_{fs} \times A_{ft}}{n}\right]$ is the synchronisation value, and this can be updated with all the outputs of $O_u{}^t$ and Z_T in Equations (5) and (6). This condition is not relevant for the first estimation as in Equations (4) and (5) because it depends upon all mapped I_{oT} to the D_T. Therefore, the S_r together with β and I_{oT} is accessed by the IoT platform, and hence it remains consistent. The following instance of collecting data S_r on its existing D_T defines the leaky security distribution of acquiring data. In this condition, the consequence of transactions is observed in $n > i$, and then the collection from $z \in I_{oT}$ is halted to prevent each data access level from sender and receiver in the synchronisation, recommendation, and validation process. The security distributions in the synchronisation of information from the IoT network pass it on to the end-verifiable key to their participation in the D_T. This overcomes permitting adversaries fewer services and PSF by collecting unwanted or incorrect data. At the same time, user flexibility is high. The controlled PSF makes certain service delays data synchronisation within the IoT architecture. In the data synchronisation process, the transaction follows the synchronisation of user interface terminals. The user interface depends on (β, S_r, I_{oT}) for synchronising data through end-user applications and the IoT platform. This data security distribution is administered based on the synchronisation recommendation and S_r Simultaneously. In this distribution of security process, the end-to-end verifiable authentication, the keys are distributed between the terminals. Using the RSA algorithm, the following steps are to generate an end-verifiable key:

1. Select two large prime numbers X and Y such that $X \neq Y$, randomly and autonomous of each other.
2. Compute

$$z = XY \tag{7}$$

3. Compute the quotient function

$$\varnothing(z) = (X-1)(Y-1) \tag{8}$$

4. Select an integer ε such that $I_{oT} < \varepsilon < \varnothing(z)$, which is relatively prime to $\varnothing(z)$.
5. Compute $\mathfrak{C}_d$ such that

$$\mathfrak{C}_d \varepsilon \equiv 1; (\text{mod } (\varnothing(z))) \tag{9}$$

The key generation process for T_n is illustrated in Figure 3.

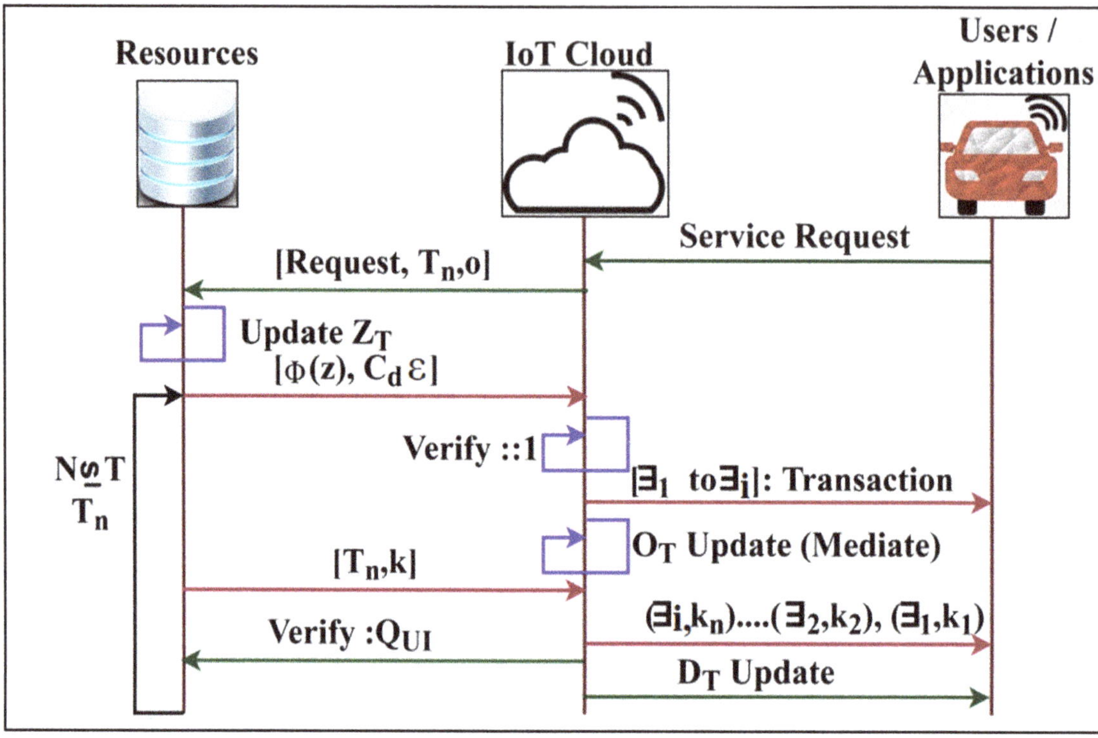

Figure 3. Key generation process in T_n.

The key generation process pursues in Equations (7)–(9) for the requests through the IoT cloud. The secure transactions for $\exists_i$ is verified for $N = T_n$ such that O_T is a mediate update. Based on this update, the D_T is performed by verifying Q_{UI} such that $[T_n, k]$ is true, and hence the key assigning is sequential. This ensures maximum authentication for the T_n for which D_T is updated using the β factor, depicted in Figure 3. The public key consists of the z, the modulus, and ε is the variable representing the public exponent for sometimes performing encryption, whereas the private key consists of z, the modulus, and ε for the private exponent and sometimes performs decryption, which can be hidden. The transmission of data from sender to receiver keeps the private key secret. X and Y are exposed since the factors of z and allow computation of $\mathbb{C}_r$ have given ε.

$$\left.\begin{array}{c} Q_{ICT} = \mathbb{C}_r \times \mu_i \times I_{oT} \text{ and } Q_{UI} = \mathbb{C}_r \times \beta \\ \text{such that,} \\ Q_{ICT} :\rightarrow D_T \text{ and } D_T :\rightarrow \beta\ \forall\, I_{oT} \\ Q_{ICT} :\rightarrow D_T \text{ and } D_T : \rightarrow (\beta - \mu_i \times z) \forall\, (z - n) \end{array}\right\} \quad (10)$$

Based on the above equation, $\mathbb{C}_r$ is the random number computation from which the two large prime numbers $\mathbb{C}_f$ are fetched for synchronisation. Equation (10) differentiates the rationality of D_T for either I_{oT} Or $(z - n)$ as classified by the support vector classifications. Now, each level of session access keys k is distributed as:

$$k = Q_{ICT} * P_{UI} * |\mathbb{C}_f| = Q_{UI} * P_{ICT} |\mathbb{C}_f| \quad (11)$$

Each level of accessing this session key is valid until the condition $T \in D_T$ after which K is synchronised based on $\mathfrak{C}_r$. Here, the key validity is generated as:

$$\left.\begin{array}{c} K(\exists_i) = G(S_r|\beta|\exists_i|\mathfrak{C}_f|K) \\ \text{and} \\ \text{Security distribution} = \{(Q_{ICT} \oplus K(\exists_i) \oplus \mathfrak{C}_f \oplus D_T),\ z\} \end{array}\right\} \quad (12)$$

Equation (12) specifies the security distribution relies on the condition of $z \in I_{oT}$ and β in the D_T. These metrics turn into verifying sequences in the end-user applications. Here, D_T is linked with the k; hence, the changes of D_T is existing in $\mathfrak{C}_r$. The user side verifies entire security features to improve overall efficiency. The analysed synchronising data is valid if the $T \in D_T$ is access level. This access level is computed in different points, such as permitting overlapping and pursued instances of the following sessions. The classification process is presented in Figure 4.

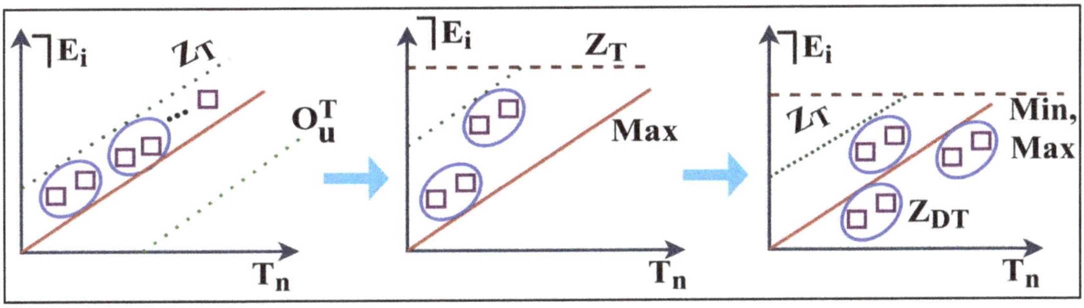

Figure 4. Classification process.

In the classification process, the access level is defined based on the previous $\exists_i$ such that Z_T defines the updates and maximum deviation. This process is differentiated based on Z_{DT} and Z_T for which the classifier performs min-max alignment. The process is restricted for T_n that is stuck under Z_T updates wherein D_T is true. This is required in the other processes to reflect multiple instances and improve access levels, as shown in Figure 4. In this IoT framework, user access level authentication is prohibited from decreasing the complexity of communication and extra service delay. The user interface terminal performs a synchronisation verification check as in the following equation. This security verification check makes certain appropriate k, D_T, and $\exists_i \in z \in I_{oT}$ is synchronised.

$$\left.\begin{array}{c} [(I_{oT} \rightarrow D_T) \oplus T_n \oplus Y \oplus \mathfrak{C}_f] = \left[\exists_i \oplus T \in D_T \oplus \frac{\mathfrak{C}_r}{I_{oT}} \oplus \beta\right],\ \forall\, \exists_i \text{ in } D_T \\ [Q_{ICT} \oplus \mu \oplus I_{oT}] = [Q_{UI} \oplus \beta \oplus \mathfrak{C}_r],\ \forall\, z \in I_{oT} \rightarrow D_T \\ G(S_r|\beta|\exists_i) = H(\neg\exists_i \oplus T_n \oplus S_r),\ \forall\, \neg\exists_i = z\exists_i \end{array}\right\} \quad (13)$$

The authentication and key verification process, Equations (12) and (13), adapts for $I_{oT} \rightarrow D_T$ where the grouping changes as in Equation (1) do not match for the above condition. Therefore, the mediate output of $O_u{}^t$ decides the different data transmission intervals and, therefore, the mapping. Based on the integrity of the end-user applications is verified and IoT cloud service instances and autonomous authentication are not lined up properly; therefore, the delay does not happen. The concurrent sequence and instances-related data integrities are verified by PSF without requiring extra computations. In addition, concurrency and integrity-related synchronisation minimise the number of computations during the verification. The classification procedure maximises the IoT cloud and user-side integrity and check. On the processing side, sequences are denoted by the user interface terminal, and security check S_r is utilised to improve the process. In the IoT cloud process, it is performed as the getting terminal by synchronising X and Y as per β and K. This synchronisation minimises the adversary impact, service failures, and service delays in the end-user application of the IoT terminal. In Table 2, the required sessions for different transactions are tabulated.

Table 2. Required sessions for transactions.

Transactions	Mapping Instances	Access Level	Required Sessions
40	53	0.27	29
80	93	0.36	69
120	174	0.41	121
160	316	0.68	158
200	210	0.52	136
240	355	0.93	162

Table 2 presents the required sessions for different transactions. As the transactions increase, T_n is augmented based on $\urcorner\exists_i$ and O_u^t. This improves the synchronisation in mapping based on Z_1 to Z_T updates. The RSA-based authentication provides high Q_{ICT} in determining the session validity. As the mapping instances increase, the access is open for high users, varying the required sessions, permitting diverse T_n. Table 3 presents the session validity (%) under different access level rates.

Table 3. Session validity (%) for different access levels.

Access Level	Generated Keys	Actual Session Time (s)	Validity (%)
0.2	40	62.3	80.7
0.4	117	324.15	85.16
0.6	165	547.37	89.62
0.8	249	625.69	91.15
1	328	710.4	94.2

Table 3 presents the session validity for the proper access level from the observed data. The active sessions require keys in O_u^1 to O_u^t updates for which $\varnothing(Z)$ are required. This increases the key validity until the session is closed. Hence, $\forall\urcorner\exists_i$, the Z generation and $k(\exists_i)$ is retained at a maximum level using $I_{oT} \rightarrow D_T$ validation. Therefore, a maximum validity (%) for the allocated access level is generated for different keys. Figure 5 presents the self-analysis for mapping and updating instances and verification checks observed under different transactions.

An analysis of instances (mapping and update) and verification checks for different transactions are presented in Figure 5. The O_u^1 to O_u^t is assigned for different $\urcorner\exists_i$ and is mapped with the available resources for which Z_1 to Z_T is provided. However, Z_1 to Z_T is interrupted based on mediate O_u^t solution and hence Z_1 to Z_{DT} is updated in different instances. This is enhanced if the mapping is pursued at a high rate in $k(\exists_i)$ maximised instances. The $I_{oT} \rightarrow D_T$ is performed for Z_T to Z_{DT} modified update for improving precise response. Therefore, the verification checks are extended for the session validity and $k(\exists_i)$ instances. This is performed under different Q_{ICT} in Z_T to Z_{DT} chances requiring high verification checks. In Figure 6, the session validity for different access levels is presented.

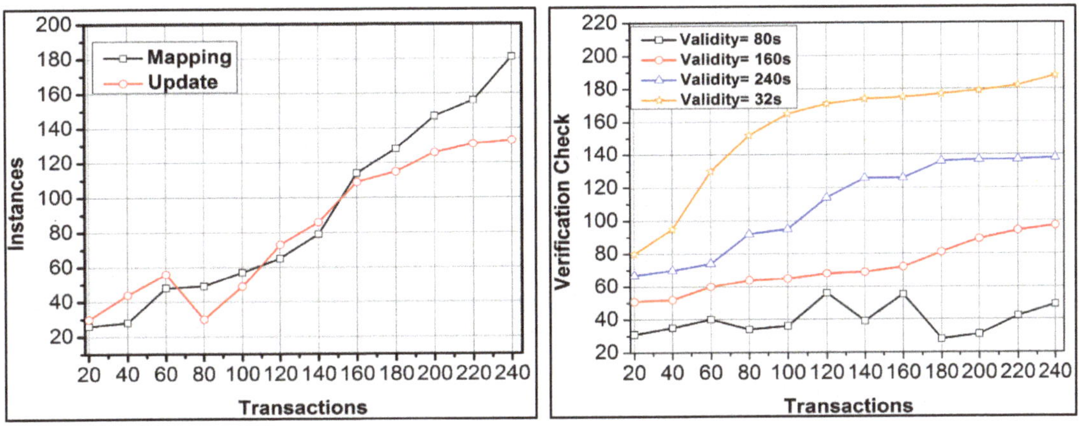

Figure 5. Mapping and updating, and verification checks under different transactions.

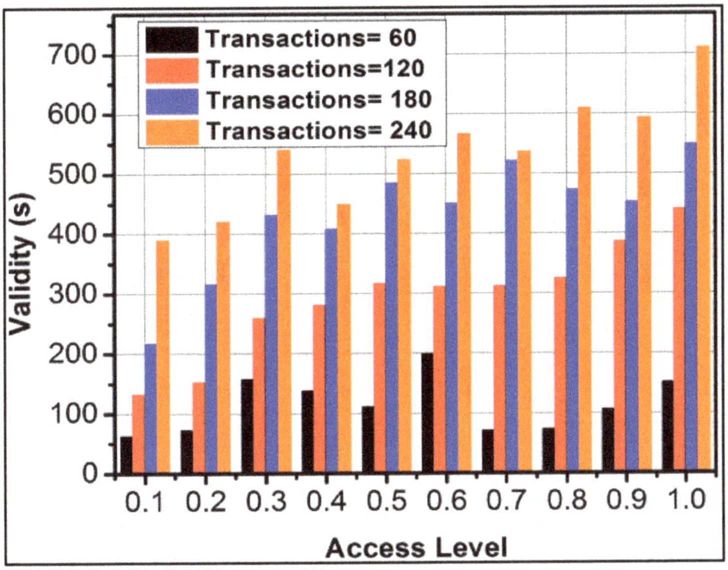

Figure 6. Session validity for different access levels.

The varying access levels require high validity as the transaction increases. In the proposed framework, the $k(\exists_i)$ is performed in different Q_{ICT}. This increases the O_u^1 to O_u^t for $I_{oT} \rightarrow D_T$ instances, increasing the validity. The notable feature is the synchronisation of Z_{DT} and Z_T in multiple instances (access) increases the validity requirement. Hence, the consecutive sequence is required to improve Z_{DT} and service distribution. Moreover, the adverse impact is reduced for extended validity-based verification checks (refer to Figure 6).

4. Results and Discussion

This section elucidates the proposed framework's performance verified using OPNET simulations. In this simulation, 80 IoT users performed 20–240 transactions through six resource servers. The request-to-response rate is varied between 0.7 and 1 with a mean transaction delay of 120 ms. This experimental scenario considers a man-in-the-middle attack for deceiving the transactions. With this setup, the metrics of adversary impact, service failure, service delay, access rate, and service transactions are compared for analysis. In the comparative analysis, the following methods are considered: CMAKAP [29], RBCGM [18], and MPPAP [30]. The NETMASTER CXC-150 modem is utilised for internet access, Linux IPTables Firewall, Microsoft DNS server, Linux open VPN server, web server, Windows 2008-IIS 7.0.

4.1. Adversary Impact

The comparative analysis for adversary impact is presented in Figure 7 with the existing methods. The $T_n \forall \neg\exists_i$ is assessed for $n \neq 0$ and $n = 0$ conditions under different transactions for reducing the adversary impact. In the proposed framework, the synchronisation is performed for S_r and βT_n. The synchronisation is performed to prevent $(Z \times i)$ augmentation that injects the adversaries. However, the different instances for the above augmentation are classified using support vectors based on k and P_{UI}. Therefore, the adversary injecting instances in Z_{DT} are updated from which O_u^t is split, and new allocations are made. The classifications performed for $\exists_i$ and $(A_{fs} - \mu)$ such that the consecutive occurrence is reduced. Therefore, the classification is instigated until Z_1 to Z_T is performed for O_u^1 to O_u^t such that Z_{DT} is true. The authentication using RSA performs secured transactions without breaching $\neg\exists_i$ and hence the impact is less. Moreover, for k, $K(\exists_i)$ is induced by balancing $I_{oT} \rightarrow D_T$ in retaining T_n. Therefore, for T_n and O_u^1 to O_u^t, validity is improved in defining less adversary impact for transactions and access levels.

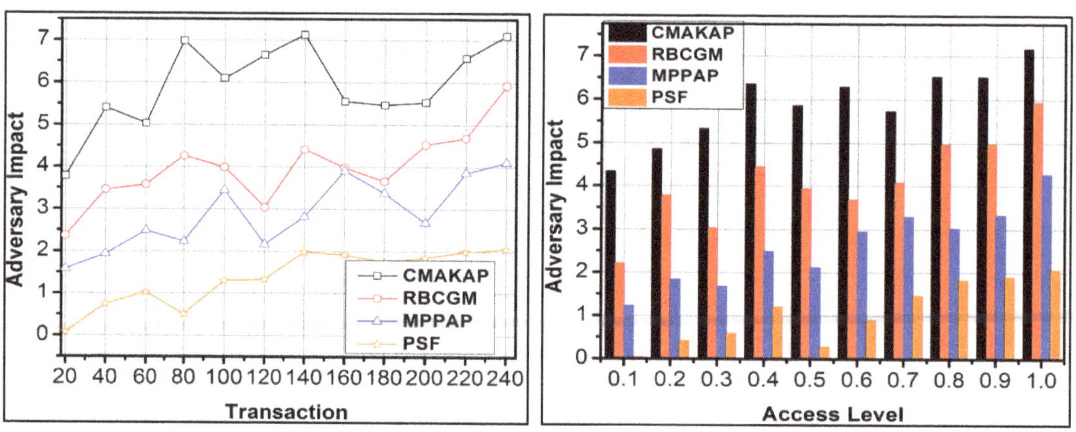

Figure 7. Adversary impact analysis.

4.2. Service Failure

In Figure 8, the efficiency analysis of service failure under various transactions and access levels is presented. The proposed framework reduces service failure based on Z_1 to Z_T and k verification. First, the (i_{T-1}) in O_u^t is identified as improving T_n and $\exists_i$. If the Z_T is outraged by Z_{DT}, then the classification process is instigated, for which Q_{ICT} is performed. The classification for $-\neg\exists_i$ and $\mu = 1$ condition distinguishes multiple adversaries impacted $\neg\exists_i$. Hence, $K(\exists_i)$ is extended $\forall\ (T+1)$ in $Z = n$, and hence the sessions are secured. In this process, Z_{DT} is performed, requiring new $z \in I_{oT}$ such that T_n is retained. As the T_n is retained, the available instances improve the Q_{ICT} for the consecutive $n > i$ interval. Hence, (β, S_r, I_{oT}) are consecutively shared in retaining the session. Therefore, the change in $\neg\exists_i$ or $\varnothing(Z)$ requires a high k, to prevent the failure of the session. This is recursive for S_r in different transactions, preventing additional failures. The security is administered by validating $C_{d\varepsilon} \equiv 1$ such that $Q_{ICT} :\rightarrow D_T$ is verified under different users as well. Therefore, the service failures are reduced in the proposed framework, achieving fair results.

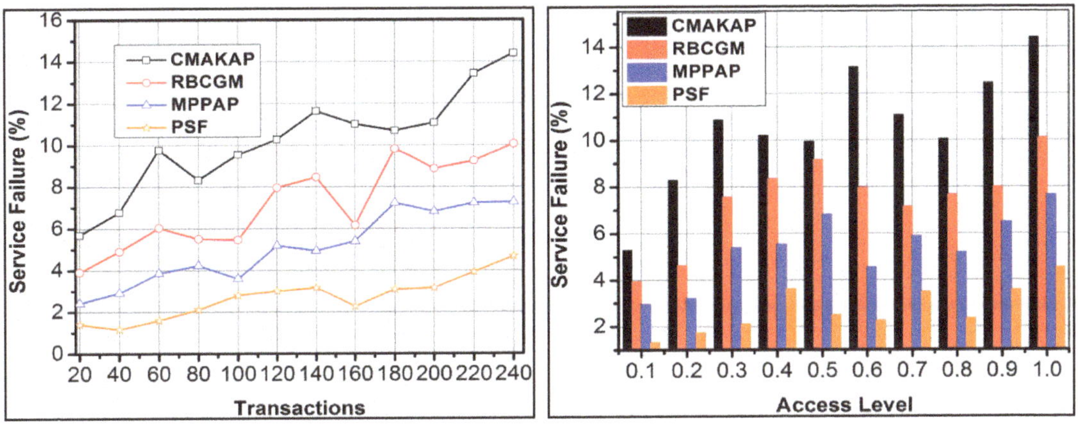

Figure 8. Service failure analysis.

4.3. Service Delay

The proposed framework achieves less service delay compared to the other methods. In the proposed framework, $\exists_i$ maximised by reducing failures, and hence reassignment (resource) is less required. The Z_1 to Z_T based on O_u^1 to O_u^t as in Equation (3) shows up as delay without increasing failures. In the Q_{ICT} definition, $Q_{UI} = C_r \times \beta$ and $(\beta - \mu_i)$ are first validated for conventional service allocations. Contrarily, if a failure occurs, then $(\mu_i \times z)\ \forall(z - n)$ is validated for detecting the time requirement. The classifier learning devices Z_1 to Z_T as in Equation (6) for Z_{DT} for identifying S_r. Based on S_r, the allocations are performed. In this allocation, two conditions are verified, namely $\neg\exists_i = n\exists_i$ and $n = 0$, and hence the allocations are validated. These validations improve the swiftness in $\exists_i$, in a concurrent manner, under T_n, reducing additional time. The classifier instance now relies on Z_1 to Z_T as in Equation (6) for improving the response. Therefore, the delay is confined $\forall\ \mu = 1$ verified for the above conditions. This is common for different transactions and access levels, achieving less delay, as presented in Figure 9.

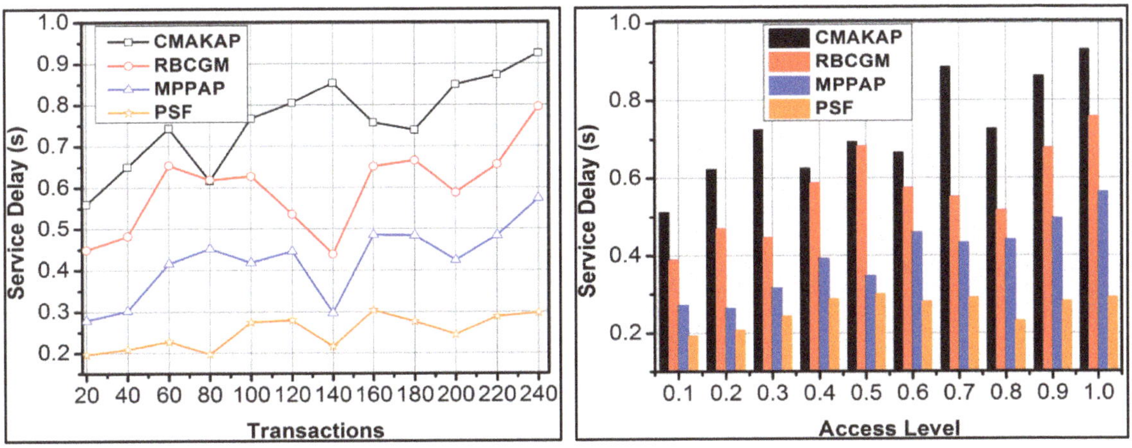

Figure 9. Service delay analysis.

4.4. Access Rate

The proposed framework achieves a high access rate for different transactions and access levels, which is shown in Figure 10. The adversary impacts are mitigated based on

Z and $\varnothing(Z)$ processes for securing access and service distributions. The O_u^1 to O_u^t based classifications using support vectors are performed to identify Z_{DT} in Z_1 to Z_T iterations. Further, the $K(\exists_i)$ is analysed for improving the access rate beyond the extended $\daleth\exists_i - n\exists_i$ and hence the $I_{oT} \rightarrow D_T$ is improved. In different T_n, the $\daleth\exists_i$ is analysed for detecting mediates in O_u^T as in Equation (5). Therefore, Z_1 to Z_T is modified depending on Q_{CT}, this modification has to satisfy two distinct conditions for retaining the access rate. First, $n \neq 0$ in either $\mu = 1$ or $\mu = 0$ such that D_T is retained. For the retained D_T, S_r is performed based on $z \in I_{oT}$, and hence the $n > i$ is achieved. If this condition is satisfied, then classification is improved to reduce the adversary impact. In the second condition, $I_{oT} < \varepsilon < \varnothing(z)$ and the authentication modes and their access levels are defined. In the proposed framework, the defined $\varnothing(z)$ is used for C_r and ε validation for maximising the access level. This leads to further access delegation regardless of the users and T_n.

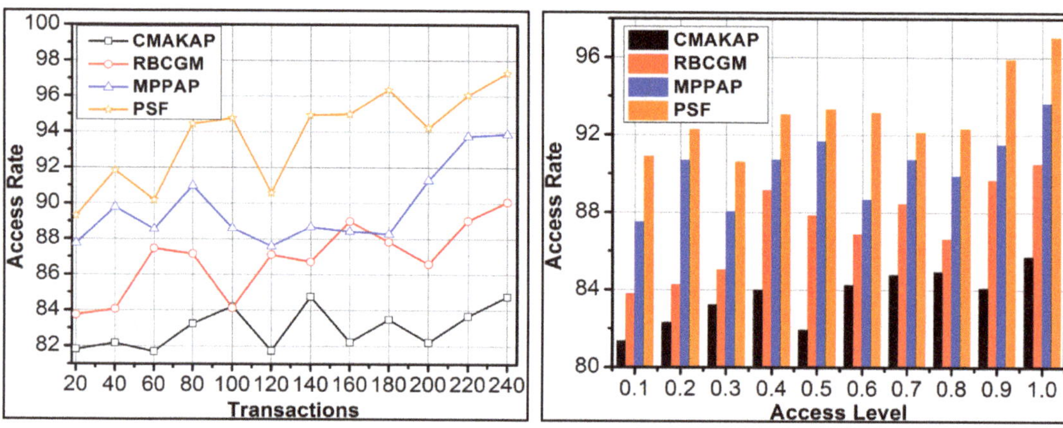

Figure 10. Access Rate Analysis.

4.5. Service Transactions

The proposed framework achieves high service transactions for different access levels, which is depicted in Figure 11. The initial T_n is required for improving service distributions without reducing the change in service allocation. In the proposed framework, $D_T = \{1 \text{ to } T\}$ is augmented to improving $\exists_i$ and hence the $n \neq 0$ is achieved. In this case, the change in T_n is achieved for multiple iterations as classified by the learning process. The Z_{DT} update in different instances is required for $(T+1)$ for $S_r = 1$, and hence the $\exists_i$ are improved. The classifier performs $(C_r \times \beta)$ and $(z - n)$ differentiation for improving service transactions. In the proposed framework, the validation is performed under different instances for $|C_f|$. The $D_T :\rightarrow \beta \; \forall \; I_{oT}$ mapping increases T_n for leveraging the distribution. Therefore, for varying access levels, the transactions are improved without increasing the overhead. The procedure is general for various Z_{DT} overwhelming service failures. Then, various transactions and access level-related comparative analyses are shown in Tables 4 and 5.

The proposed framework reduces adversary impact, service failure, and service delay by 10.98%, 11.82%, and 10.19%, respectively. Contrarily, it improves the access rate by 7.73%.

The proposed framework achieves 11.16% less adversary impact, 12.34% less service failure, 10.19% less service delay, 7.1% high access rate, and 10.12% high service transaction.

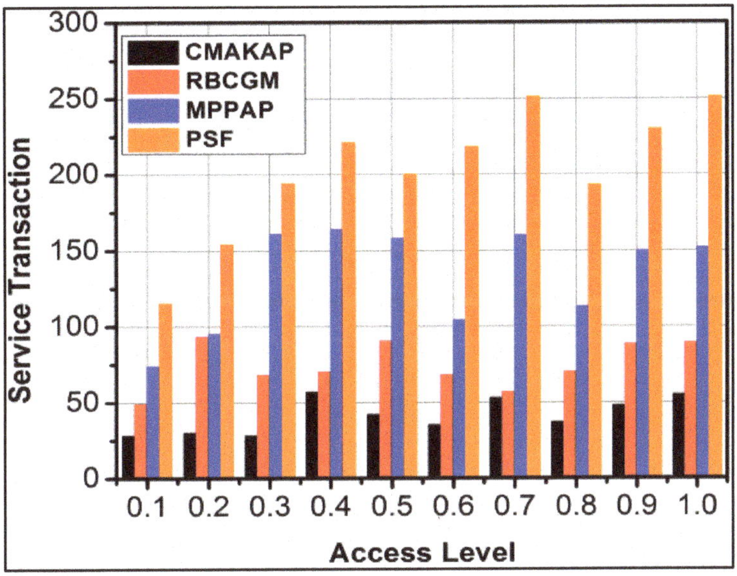

Figure 11. Service transaction analysis.

Table 4. Comparative analysis summary for transactions.

Metrics	CMAKAP	RBCGM	MPPAP	PSF
Adversary Impact	7.11	5.93	4.11	2.0575
Service Failure (%)	14.4	10.07	7.28	4.673
Service Delay (s)	0.925	0.795	0.574	0.2967
Access Rate	84.77	90.08	93.89	97.311

Table 5. Comparative analysis summary for access level.

Metrics	CMAKAP	RBCGM	MPPAP	PSF
Adversary Impact	7.17	5.94	4.27	2.0718
Service Failure (%)	14.39	10.11	7.65	4.547
Service Delay (s)	0.931	0.758	0.564	0.2918
Access Rate	85.68	90.48	93.62	97.037
Service Transaction	55	89	152	251

5. Conclusions

This article presents an access and transaction adaptable PSF for mitigating the adversary impact over dense IoT services. The secure transaction sequence between the users/applications and the resources through the cloud is linearly mapped and synchronised for providing high-level access. The sessions are distinguished based on access time intervals and authenticated using RSA. In the classification process, support vectors are employed for handling linear and synchronised access between the users. The proposed framework fits the user and transaction flexibility without deviating from data collection and update. For ease of service allocation, the classifications are performed based on failing and mapping updates. This is considered by the classifier for improving the end-to-end verification checks. Based on the verification validity, the session intervals are modified, and hence the synchronisation is retained. The proposed framework reduces adversary impact, service failure, and service delay by 10.98%, 11.82%, and 10.19%, respectively. Contrarily, it improves the access rate by 7.73% for different transactions.

Author Contributions: Conceptualisation, M.A. and S.S.; methodology, M.A. and S.S.; software, M.A. and S.S.; validation, M.A., S.S. and Y.Y.; formal analysis, Y.Y.; resources, S.S.; data curation, H.F.G. and S.S.; writing—original draft preparation, M.A. and S.S.; writing—review and editing, H.F.G., S.S. and Y.Y.; visualisation, S.S. and Y.Y. All authors have read and agreed to the published version of the manuscript.

Funding: This research was supported by the Princess Nourah bint Abdulrahman University Researchers Supporting Project number (PNURSP2023R259), Princess Nourah bint Abdulrahman University, Riyadh, Saudi Arabia.

Data Availability Statement: Experiments are performed on simulator for real scenarios.

Conflicts of Interest: The authors declare no conflict of interest.

References

1. Mishra, S.; Tyagi, A.K. The role of machine learning techniques in internet of things-based cloud applications. In *Artificial Intelligence-Based Internet of Things Systems*; Springer: Berlin, Germany, 2022; pp. 105–135.
2. Li, Y.; Cao, B.; Peng, M.; Zhang, L.; Zhang, L.; Feng, D.; Yu, J. Direct Acyclic Graph-Based Ledger for Internet of Things: Performance and Security Analysis. *IEEE/Acm Trans. Netw.* **2020**, *28*, 1643–1656. [CrossRef]
3. Tournier, J.; Lesueur, F.; Le Mouël, F.; Guyon, L.; Ben-Hassine, H. A survey of IoT protocols and their security issues through the lens of a generic IoT stack. *Internet Things* **2021**, *16*, 100264. [CrossRef]
4. Javanmardi, S.; Shojafar, M.; Mohammadi, R.; Nazari, A.; Persico, V.; Pescapè, A. FUPE: A security driven task scheduling approach for SDN-based IoT–Fog networks. *J. Inf. Secur. Appl.* **2021**, *60*, 102853. [CrossRef]
5. Li, D.; Cai, Z.; Deng, L.; Yao, X.; Wang, H.H. Information security model of block chain based on intrusion sensing in the IoT environment. *Clust. Comput.* **2019**, *22*, 451–468. [CrossRef]
6. Xu, X.; Wang, X.; Li, Z.; Yu, H.; Sun, G.; Maharjan, S.; Zhang, Y. Mitigating Conflicting Transactions in Hyperledger Fabric-Permissioned Blockchain for Delay-Sensitive IoT Applications. *IEEE Internet Things J.* **2021**, *8*, 10596–10607. [CrossRef]
7. Shamieh, F.; Wang, X.; Hussein, A.R. Transaction Throughput Provisioning Technique for Blockchain-Based In-dustrial IoT Networks. *IEEE Trans. Netw. Sci. Eng.* **2020**, *7*, 3122–3134. [CrossRef]
8. Wang, J.; Wei, B.; Zhang, J.; Yu, X.; Sharma, P.K. An optimised transaction verification method for trustworthy blockchain-enabled IIoT. *Ad Hoc Netw.* **2021**, *119*, 102526. [CrossRef]
9. Lee, K.; Yim, K. Study on the transaction linkage technique combined with the designated terminal for 5G-enabled IoT. *Digit. Commun. Netw.* **2021**, *8*, 124–131. [CrossRef]
10. Li, H.; Pei, L.; Liao, D.; Wang, X.; Xu, D.; Sun, J. BDDT: Use blockchain to facilitate IoT data transactions. *Clust. Comput.* **2021**, *24*, 459–473. [CrossRef]
11. Rachit; Bhatt, S.; Ragiri, P.R. Security trends in Internet of Things: A survey. *Sn Appl. Sci.* **2021**, *3*, 121. [CrossRef]
12. Al-Otaibi, Y.D. Distributed multi-party security computation framework for heterogeneous internet of things (IoT) devices. *Soft Comput.* **2021**, *25*, 12131–12144. [CrossRef]
13. Djedjig, N.; Tandjaoui, D.; Medjek, F.; Romdhani, I. Trust-aware and cooperative routing protocol for IoT security. *J. Inf. Secur. Appl.* **2020**, *52*, 102467. [CrossRef]
14. Hodgson, R. Solving the security challenges of IoT with public key cryptography. *Netw. Secur.* **2019**, *2019*, 17–19. [CrossRef]
15. Oh, M.-K.; Lee, S.; Kang, Y.; Choi, D. Wireless Transceiver Aided Run-Time Secret Key Extraction for IoT Device Security. *IEEE Trans. Consum. Electron.* **2019**, *66*, 11–21. [CrossRef]
16. Biswas, S.; Sharif, K.; Li, F.; Nour, B.; Wang, Y. A Scalable Blockchain Framework for Secure Transactions in IoT. *IEEE Internet Things J.* **2018**, *6*, 4650–4659. [CrossRef]
17. Yu, S.; Jho, N.; Park, Y. Lightweight Three-Factor-Based Privacy-Preserving Authentication Scheme for IoT-Enabled Smart Homes. *IEEE Access* **2021**, *9*, 126186–126197. [CrossRef]
18. Asheralieva, A.; Niyato, D. Reputation-Based Coalition Formation for Secure Self-Organized and Scalable Sharding in IoT Blockchains with Mobile-Edge Computing. *IEEE Internet Things J.* **2020**, *7*, 11830–11850. [CrossRef]
19. Huang, K. Secure Efficient Revocable Large Universe Multi-Authority Attribute-Based Encryption for Cloud-Aided IoT. *IEEE Access* **2021**, *9*, 53576–53588. [CrossRef]
20. Sadri, M.J.; Asaar, M.R. An anonymous two-factor authentication protocol for IoT-based applications. *Comput. Netw.* **2021**, *199*, 108460. [CrossRef]
21. Wu, F.; Li, X.; Xu, L.; Vijayakumar, P.; Kumar, N. A Novel Three-Factor Authentication Protocol for Wireless Sensor Networks with IoT Notion. *Ieee Syst. J.* **2021**, *15*, 1120–1129. [CrossRef]

22. Dorri, A.; Kanhere, S.S.; Jurdak, R.; Gauravaram, P. LSB: A Lightweight Scalable Blockchain for IoT security and anonymity. *J. Parallel Distrib. Comput.* **2019**, *134*, 180–197. [CrossRef]
23. Vishwakarma, L.; Das, D. SCAB-IoTA: Secure communication and authentication for IoT applications using block-chain. *J. Parallel Distrib. Comput.* **2021**, *154*, 94–105. [CrossRef]
24. Peneti, S.; Kumar, M.S.; Kallam, S.; Patan, R.; Bhaskar, V.; Ramachandran, M. BDN-GWMNN: Internet of Things (IoT) Enabled Secure Smart City Applications. *Wirel. Pers. Commun.* **2021**, *119*, 2469–2485. [CrossRef]
25. Majumder, S.; Ray, S.; Sadhukhan, D.; Khan, M.K.; Dasgupta, M. ECC-CoAP: Elliptic curve cryptography based constraint application protocol for internet of things. *Wirel. Pers. Commun.* **2021**, *116*, 1867–1896. [CrossRef]
26. Lin, W.; Yin, X.; Wang, S.; Khosravi, M.R. A Blockchain-enabled decentralised settlement model for IoT data exchange services. In *Wireless Networks*; Springer: Berlin, Germany, 2020; pp. 1–15.
27. Attarian, R.; Hashemi, S. An anonymity communication protocol for security and privacy of clients in IoT-based mobile health transactions. *Comput. Netw.* **2021**, *190*, 107976. [CrossRef]
28. Yazdinejad, A.; Parizi, R.M.; Dehghantanha, A.; Zhang, Q.; Choo, K.K.R. An energy-efficient SDN controller ar-chitecture for IoT networks with blockchain-based security. *IEEE Trans. Serv. Comput.* **2020**, *13*, 625–638. [CrossRef]
29. Srinivas, J.; Das, A.K.; Wazid, M.; Kumar, N. Anonymous Lightweight Chaotic Map-Based Authenticated Key Agreement Protocol for Industrial Internet of Things. *IEEE Trans. Dependable Secur. Comput.* **2018**, *17*, 1133–1146. [CrossRef]
30. Pham, C.D.; Dang, T.K. A lightweight authentication protocol for D2D-enabled IoT systems with privacy. *Pervasive Mob. Comput.* **2021**, *74*, 101399. [CrossRef]

electronics

Article

A New Data-Balancing Approach Based on Generative Adversarial Network for Network Intrusion Detection System

Mohammad Jamoos [1,2], Antonio M. Mora [1], Mohammad AlKhanafseh [3] and Ola Surakhi [4,*]

1 Department of Signal Theory, Telematics and Communications, University of Granada, 18012 Granada, Spain; jamoos@staff.alquds.edu (M.J.); amorag@ugr.es (A.M.M.)
2 Department of Computer Science, School of Science and Technology, Al-Quds University, Jerusalem P.O. Box 51000, Palestine
3 Department of Computer Science, Birzeit University, West Bank, Birzeit P.O. Box 14, Palestine; malkhanafseh@birzeit.edu
4 Department of Computer Science, American University of Madaba, Madaba 11821, Jordan
* Correspondence: o.surakhi@aum.edu.jo

Abstract: An intrusion detection system (IDS) plays a critical role in maintaining network security by continuously monitoring network traffic and host systems to detect any potential security breaches or suspicious activities. With the recent surge in cyberattacks, there is a growing need for automated and intelligent IDSs. Many of these systems are designed to learn the normal patterns of network traffic, enabling them to identify any deviations from the norm, which can be indicative of anomalous or malicious behavior. Machine learning methods have proven to be effective in detecting malicious payloads in network traffic. However, the increasing volume of data generated by IDSs poses significant security risks and emphasizes the need for stronger network security measures. The performance of traditional machine learning methods heavily relies on the dataset and its balanced distribution. Unfortunately, many IDS datasets suffer from imbalanced class distributions, which hampers the effectiveness of machine learning techniques and leads to missed detection and false alarms in conventional IDSs. To address this challenge, this paper proposes a novel model-based generative adversarial network (GAN) called TDCGAN, which aims to improve the detection rate of the minority class in imbalanced datasets while maintaining efficiency. The TDCGAN model comprises a generator and three discriminators, with an election layer incorporated at the end of the architecture. This allows for the selection of the optimal outcome from the discriminators' outputs. The UGR'16 dataset is employed for evaluation and benchmarking purposes. Various machine learning algorithms are used for comparison to demonstrate the efficacy of the proposed TDCGAN model. Experimental results reveal that TDCGAN offers an effective solution for addressing imbalanced intrusion detection and outperforms other traditionally used oversampling techniques. By leveraging the power of GANs and incorporating an election layer, TDCGAN demonstrates superior performance in detecting security threats in imbalanced IDS datasets.

Keywords: Generative Adversarial Network; Intrusion Detection System; imbalanced dataset; machine learning; unsupervised learning

Citation: Jamoos, M.; Mora, A.M.; AlKhanafseh, M.; Surakhi, O. A New Data-Balancing Approach Based on Generative Adversarial Network for Network Intrusion Detection System. *Electronics* **2023**, *12*, 2851. https://doi.org/10.3390/electronics12132851

Academic Editors: Dariusz Rzońca and Tomasz Rak

Received: 29 May 2023
Revised: 16 June 2023
Accepted: 17 June 2023
Published: 28 June 2023

1. Introduction

The process of data science comprises multiple stages, starting with the collection of a dataset, followed by its preparation and exploration, and eventually modeling the data to yield solutions. However, since different problem domains have varying datasets, the data-gathering process may uncover various issues within the dataset that must be addressed and rectified before proceeding with data modeling. Successfully handling these problems can significantly impact the model's accuracy.

One application where machine learning methods are widely used is intrusion detection systems (IDSs) [1,2]. IDS is employed to monitor the network traffic and identify any unauthorized efforts to access that network through the analysis of incoming and outgoing actions, with the aim of detecting indications of potentially harmful actions [3].

Machine learning (ML) methods, such as supervised network intrusion detection, have demonstrated satisfactory effectiveness in identifying malicious payloads within network traffic datasets that are annotated with accurate labeling. Nevertheless, the substantial growth in network scale and the proliferation of applications processed by network nodes have led to an overwhelming volume of data being shared and transmitted across the network. Consequently, this has given rise to significant security threats and underscored the urgency to enhance network security. As a result, numerous researchers have focused their efforts on enhancing intrusion detection systems (IDSs) by improving the detection rate for both novel and known attacks, while concurrently reducing the occurrence of false alarms (false alarm rate or FAR) [1]. Unsupervised intrusion detection techniques have emerged as a solution that eliminates the need for labeled data [4]. These methods can effectively train using samples from a single class, typically normal samples, aiming to identify patterns that deviate from the training observations. However, the accuracy of these unsupervised learning approaches tends to decline when faced with imbalanced datasets, where the number of samples in one class significantly exceeds or falls short of the number of samples in other classes.

To tackle the issue of imbalanced datasets, oversampling techniques are frequently employed. Traditional approaches utilize interpolation to generate samples among the nearest neighbors, such as the synthetic minority oversampling technique (SMOTE) [5] and the adaptive synthetic sampling technique (ADASYN) [6]. However, a novel generative model called the generative adversarial network (GAN) has emerged, providing a fresh framework for sample generation [7]. GAN allows the generator to effectively learn data features by engaging in a game-like interaction with the discriminator to simulate data distributions. GAN has demonstrated remarkable advancements in generating images, sounds, and texts [8–10]. As a result, researchers from various domains are increasingly incorporating this method into their research endeavors.

This paper proposes a new oversampling technique based on GAN applied for IDS considering the viewpoint of imbalanced data. The new model is called the triple discriminator conditional generative adversarial network (TDCGAN). This model consists of one generator and three discriminators with an added layer at the end for election. The TDCGAN employs a structure comprising a single generator and three discriminators. The generator utilizes random noise from a latent space as input and produces synthetic data that closely resemble real data, with the intention of evading detection by the discriminators. Each discriminator is a deep neural network with distinct architecture and parameter settings. Their primary task is to extract features from the generator's output and classify the data with varying levels of accuracy, differing for each discriminator. A new layer called the election layer is incorporated at the end of the TDCGAN architecture. This layer receives the outputs from the three discriminators and conducts an election procedure to determine the optimal outcome, selecting the result that achieves the highest classification accuracy. This process resembles an ensemble method, where multiple inputs are combined to produce a superior result. The generator model is designed as a deep multi-layer perceptron (MLP) comprising an input layer, an output layer, and four hidden layers. The initial hidden layer consists of 256 neurons, while an embedded layer is employed between the hidden layers to effectively map input data from a high-dimensional space to a lower-dimensional space. The second hidden layer comprises 128 neurons, followed by a third hidden layer with 64 neurons, and a final hidden layer with 32 neurons. The ReLU activation function is applied to all of these layers, and a regularization dropout of 20% is included to prevent overfitting. The output layer is activated using the Softmax activation function, with 14 neurons corresponding to the number of features in the dataset. Each discriminator within the TDCGAN architecture is implemented as an MLP model, featuring distinct configurations in terms of hidden layers, number of neurons, and dropout percentages. The first discriminator consists of three hidden layers, with each layer containing 100 neurons and a dropout regularization of 10%. The second discriminator includes five hidden layers with varying neuron counts—64, 128, 256, 512, and 1024—for each respective layer. A dropout percentage of 40% is applied in this case. The last discriminator is composed of four hidden layers with 512, 256, 128, and 64 neurons per layer, accompanied by a 20% dropout percentage. The LeakyReLU activation function with an alpha value of 0.2 is employed for the hidden layers in all discriminators. Two output layers are utilized for each discriminator, with the Softmax activation function applied to one output layer and the sigmoid activation function to the second output layer. The model is trained using two loss functions: binary cross entropy for the first output layer and categorical cross-entropy loss for the second output layer. The output from each discriminator is extracted and fed into the final layer of the model, where the election process takes place to determine the best result. The dataset [11] used in this paper to evaluate and test our model is the UGR'16 dataset. There are many datasets for IDSs, such as KDD CUP 99-1998, CICIDS2017, DARPA-1998 and more [12]; we chose UGR'16, because it is built with real traffic and up-to-date attacks.

This paper makes two main contributions. Firstly, it addresses the issue of high-class imbalance by analyzing the UGR'16 dataset. Secondly, it conducts evaluations on this dataset using several commonly used machine learning algorithms for data balancing.

The rest of this paper is organized as follows: Section 2 presents some of the relevant studies in this topic. Section 3 gives an overview about IDS and UGR'16 dataset. Section 4 proposes the TDCGAN model. The design, execution and results are given in Section 5. Finally, Section 6 gives the conclusion and future works.

2. Related Works

The impact of data resampling on machine learning model performance has been analyzed in multiple studies, since this issue can result in diminished predictive capabilities of the model.

The concept of employing GAN models to address the class imbalance problem is introduced by Lee and Park in reference [13]. In general, GAN is an unsupervised learning technique rooted in deep learning and generates synthetic data that closely resembles the existing data. The authors in this work used GAN to effectively tackle fitting issues, class overlaps, and noise through the process of resampling by explicitly defining the desired rare class. To evaluate the classifier's performance, the re-sampled data are trained using the widely adopted machine learning technique called random forest (RF). The proposed solution demonstrates superior performance compared to the methods currently utilized. Hajisalem and Babaie in the study referenced in [14] apply swarm intelligence optimization heuristics, specifically artificial fish swarm (AFS) and bee colony optimization (BCO), for the anomaly detection process. The detection approach proposed in that research focuses on reducing the subset of characteristics.

The study referenced in [15] presents a novel solution that applies an optimum allocation technique to efficiently manage large datasets by selecting the most representative samples. This approach aims to develop a new network intrusion detection system (NIDS) based on the least support vector machine (LSVM). The samples are arranged based on the desired confidence interval and the number of observations. The authors in [16] aimed to tackle the problem arising from the increasing quantity and diversity of network attacks, which leads to insufficient data during the training phase of machine learning-based intrusion detection systems (IDSs). The authors addressed this issue by examining a considerable number of network datasets from recent years. Each dataset's limitations, such as a shortage of attack instances and other issues, are identified. As a result, Kumar, et al. proposed a new dataset that aims to resolve, or at least alleviate, the encountered problem [17].

The authors introduced a new IDS system designed to address five common conventional attacks. In this solution, the author constructs a new dataset that surpasses the UNSW-NB15 dataset. A misuse-based strategy is employed to create a fresh dataset, and a gain information technique is applied to collect features from the original UNSW-NB15 dataset.

Another IDS solution based on GAN was proposed in [18]. Due to the limited number of known attack signatures for vehicle networks, the author employ the concept of generating unknown attacks during the training process to enable the IDS to effectively handle various types of attacks. In the context of vehicle IDS, accuracy is of the utmost importance to ensure driver safety, as any false positive error could have serious consequences. Traditional IDS approaches are inadequate for dealing with numerous new and undiscovered attacks that may arise. The proposed GAN-based IDS solution successfully detects four previously unknown attacks. The authors in [19] propose a novel method by combining ADASYN and RENN techniques. This approach aims to tackle the imbalances between negative and positive instances in the initial dataset, as well as addressing the issue of feature redundancy. The RF algorithm and Pearson correlation analysis are employed to select the most relevant features. In conclusion, the studies presented in this section cover various approaches and techniques for addressing challenges in machine learning-based intrusion detection systems (IDSs) and class imbalance problems. The introduction of GAN models in reference [13] offers a promising solution by generating synthetic data to tackle class imbalances, resulting in improved model performance. The utilization of swarm intelligence optimization heuristics, such as artificial fish swarm (AFS) and bee colony optimization (BCO), for anomaly detection as described in reference [14] focuses on reducing the subset of characteristics to enhance detection accuracy. Another study referenced in [15] introduced an innovative approach that efficiently manages large datasets for network intrusion detection systems (NIDS). Addressing the problem of insufficient data during the training phase of IDS, the study mentioned in [16] examined multiple network datasets and proposed a new dataset to alleviate the limitations caused by increasing network attacks. Overall, these studies contribute valuable insights and propose effective solutions to enhance the performance and capabilities of intrusion detection systems in the face of various challenges, such as class imbalance, limited data, and emerging attack types.

3. UGR'16 Dataset

In this paper, the UGR'16 dataset [11] is used to test the performance of the proposed model and achieve data balancing. The data are sourced from multiple netflow v9 collectors that are strategically positioned within the network of a Spanish ISP. An ISP is an Internet Service Provider. It provides access to Internet for many different hosts (most of them inside private networks, like homes or companies). The main aspects of the ISP network infrastructure are as follows:

- Netflow probes are set up on the outgoing network interfaces of two redundant border routers, BR1 and BR2, which enable access to the Internet. This configuration allows for the collection of all incoming and outgoing connections.
- The ISP has two different subnetworks. One is termed the core network, where the services that are not protected by a firewall are located. The second is the inner network, where firewall services are provided to the clients.
- At the highest level, there is a network of attacker machines consisting of five units, designated as A1–A5.
- Within the core network, five victim machines specifically for dataset collection purposes are set up. These machines, named V11–V15, are located alongside genuine clients in an existing network referred to as victim network V1.
- In relation to the inner network, a collective of 15 additional victim machines is positioned across three separate existing networks, with each network consisting of 5 machines. These networks are designated as victim network V2 (machines V21–V25), victim network V3 (machines V31–V35), and victim network V4 (machines V41–V45).

The entire dataset comprises two distinct sets: a calibration and a testing set. The calibration set is used in constructing and adjusting the machine learning models. This set contains attacks, but they are not controlled, nor labeled, and data that were recorded between March and June 2016. it includes inbound and outward ISP network traffic. The testing set, acquired in July and August of 2016, is used to evaluate the models in the detection process. For both the calibration and test sets, the collected files are consolidated into a single file per week for each of the two capture periods. These files are typically compressed tar files with an average size of approximately 14 GB. The calibration set consists of 17 files, while the test set comprises 6 files. To anonymize the IP addresses of the machines in the dataset, the CryptoPan prefix-preserving anonymization technique [20] was applied. This anonymization process is carried out using the nfanon tool [21]. Table 1 contains the list of different attacks with their corresponding labels in the UGR'16 dataset.

Table 1. List of attacks in UGR'16 dataset.

Attack	Label Description	
DoS11	DoS	One-to-one DoS (denial of service) attack, where the attacker A1 attacks the victim V21
DoS53s	DoS	The five attackers A1–A5 attack three of the victims, each one at a different network
DoS53a	DoS	The attacks are executed as in DoS53s, but now every victim is sequentially selected
Scan11	Scan11	One-to-one scan attack, where the attacker A1 scans the victim V41
Scan44	Scan14	Four-to-four scan attack, where the attackers A1, A2, A3 and A4 initiate a scan at the same time to the victims V21, V11, V31 and V41
Botnet	Nerisbotnet	Mixing botnet captures recorded elsewhere in a controlled environment with our background traffic
IP in blacklist	Blacklist	It is an attack of class signature
UDP Scan	Anomaly-udpscan	Depending on the source port of the connection, each victim host is scanned through a specific range of 60 ports
SSH Scan	Anomaly-sshscan	An anomaly attack
SPAM	Anomaly-spam	An anomaly attack

The artificial attack traffic was generated in 2h batches, during which all attack variants were executed. There are two possible scheduling patterns for the execution of the attack variants within each batch:

1. Planned scheduling: every attack within the batch is executed at a predetermined and known time, which is determined by an offset from the initial batch time, denoted as t0.
2. Random scheduling: the initial time for the execution of each of the attacks is randomly selected between t0 + 00h00m and t0 + 01h50m, thus restricting the total duration of the batch to a maximum of 2h.

The UGR'16 is created based on packet and flow data. It contains 16,900,000,000 anonymous network traffic flows. The network flow features are derived from actual network traffic, and these features are detailed in Table 2.

The UGR'16 dataset is divided into 23 compressed files, each of which is assigned to a particular week. Based on this, 16 of the files are assigned to the calibration class of datasets, and the remaining 6 to the test class. The size of each file is around 14 GB in the compressed format, and they can be downloaded in the csv format.

Table 2. UGR'16 dataset network flow features.

Number	Feature Name	Type
1	Timestamp	date-time
2	Flow duration	continuous numeric
3	Source IP address	categorical
4	Source IP address	categorical
5	Source port number	discrete numeric
6	Destination port number	discrete numeric
7	Protocol	categorical
8	Flag	categorical
9	Forwarding status	numeric
10	Source type of service	discrete numeric
11	Total number of packets	Continuous numeric
12	Total number of bytes	Continuous numeric
13	Class (Label)	categorical

4. Proposed Model

4.1. Data Preparation

The UGR-16 dataset used in this paper contains 16.9 billion records. While the deep learning algorithms require high hardware resources, such as CPU, memory and GPU for data processing and training, a subset of data points that cover all types of normal and anomalous traffic from UGR'16 dataset was selected. The subset selection, which included all types of attacks, was conducted using specific measures to prevent imbalanced distributions and bias. The following measures were put in place:

- Stratified sampling: The subset selection process employed stratified sampling techniques to ensure proportional representation of each type of attack. This approach helped maintain a balanced distribution of attacks in the subset.
- Class balancing: Additional steps were taken to balance the representation of different attack types in the subset. This might include oversampling the minority classes or undersampling the majority classes to mitigate the imbalanced distribution.
- Randomization: To minimize any potential bias, randomization techniques were applied during the subset selection process. This ensured that the selection was not influenced by any specific order or predetermined biases.

By implementing these measures, the subset selection aimed to create a representative subset of attacks that avoided imbalanced distributions and potential biases, enabling a more reliable analysis of the dataset.

This subset was then pre-processed, including cleaning it from the missing values and removing the duplicate instances. The details of the selected subset are shown in Table 3.

Table 3. UGR'16 subset details.

From	To	Class Label	Counts	Percentage
27 July 2016	31 July 2016	background	197,185	98.5%
27 July 2016	31 July 2016	dos	1169	0.6%
27 July 2016	31 July 2016	scan44	578	0.3%
27 July 2016	31 July 2016	blacklist	545	0.3%
27 July 2016	31 July 2016	nerisbotnet	227	0.1%
27 July 2016	31 July 2016	anomaly-spam	170	0.1%
27 July 2016	31 July 2016	scan11	126	0.1%

Within the context of network security, normal traffic tends to occur more often than malicious traffic, leading to imbalanced class proportions and an imbalanced dataset [22]. This poses a challenge for machine learning, as learning from imbalanced data is a common issue. In order to address this problem, one potential solution is to either undersample the majority class or oversample the minority classes.

In this paper, dataset records with class labels equal to the background are major. The other class labels are oversampled to obtain a balanced subset of the UGR'16 dataset. The original number of records and classes of the selected subset is given in Table 3.

Since machine learning algorithms work with numerical data, some features in the dataset need to be encoded: protocol, source IP, destination IP and class label. One-hot encoded is used to convert these features. The dataset is then scaled using MinMaxScaler from the Scikit-learn library to scale the values to the interval [0,1].

Random forest classifier is used to explore the features importance based on mean decrease in impurity (MDI). The calculation for a given feature's importance involves summing the number of splits that incorporate the feature across all trees, proportional to the number of samples that it splits. Figure 1 shows the highest numerical features of the UGR'16 dataset based on MDI value. In the proposed model, all the features are included in the process, being the most important feature is the Source_IP.

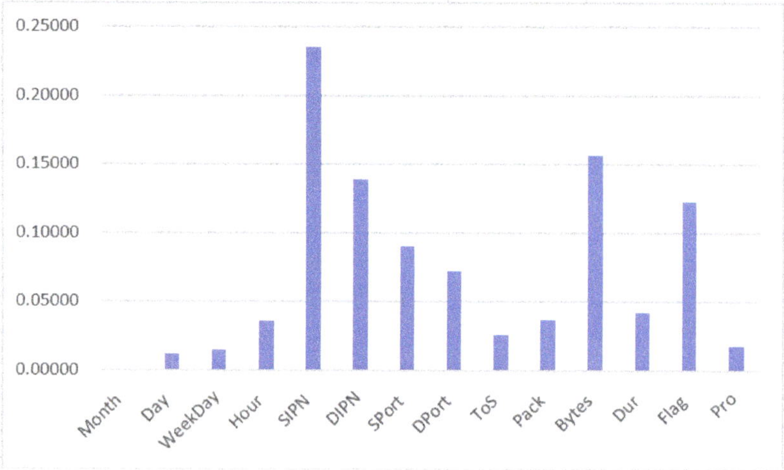

Figure 1. The highest numerical features of the UGR'16 dataset based on the mean decrease in impurity (MDI).

4.2. Setup of Proposed Model

The generative adversarial network (GAN) is a machine learning-based deep learning method used to generate new data. It is an unsupervised learning task that involves learning from input data to produce new samples from the original dataset. GAN is used in the literature in many applications, such as computer vision [23], time-series applications [24], health [25] and more, making significant advancement and outperformance in data generation. As many improvements and versions for the GAN are proposed, in order to fit it with the application domain and increase the performance and model accuracy [26,27], this paper proposes a new version of GAN called triple discriminator conditional generative adversarial networks (TDCGANs) as an augmentation tool to generate new data for the UGR'16 dataset with the aim to restore balance in the dataset by increasing minor attack classes.

In the TDCGAN, the architecture consists of one generator and three discriminators. The generator takes random noise from a latent space as input and generates raw data that closely resemble the real data, aiming to avoid detection by discriminators. Each discriminator is a deep neural network with different architecture and different parameter settings. Each discriminator's role is to extract features from the output of the generator and classify the data with varying levels of accuracy for each them. An election layer is added to the end of TDCGAN architecture that obtains the output from the three discriminators and performs an election procedure to achieve the best result with the highest classification accuracy in a form of the ensemble method. The model aims to classify data into two groups: normal flows for the background traffic with 0 representation, and anomaly flows for the attack data with 1 representation. Additionally, in the case of anomaly flow, the model classifies it as its specific class type. Figure 2 shows the workflow of the proposed TDCGAN model. The setting details of generator and each discriminator are given below.

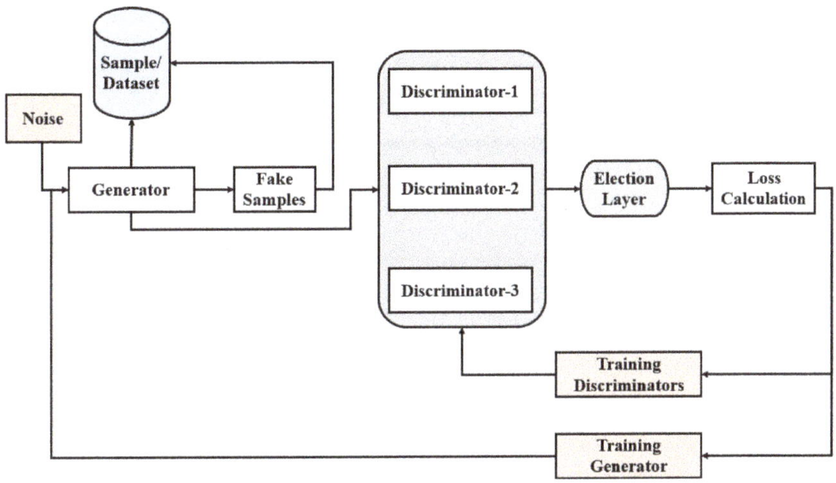

Figure 2. Workflow of TDCGAN model.

The model of the generator is a deep multi-layer perceptron (MLP) composed of an input layer, output layer and four hidden layers. Initially, the generator takes a point from the latent space to generate new data. The latent space is a multi-dimensional hypersphere normal distributed points, where each variable is drawn from the distribution of the data in the dataset. An embedded layer in the generator creates a vector representation for the generated point. Through training, the generator learns to map points from the latent space into specific output data, which are different each time the model is trained. Taken a step further, new data are then generated using random points in the latent space. So, these points are used to generate specific data. The discriminator distinguishes the new data generated by the generator from the true data distribution.

GAN is an unsupervised learning method. Both the generator and discriminator models are trained simultaneously [28]. The generator produces a batch of samples, which, along with real examples from the domain, are fed to the discriminator. The discriminator then classifies them as either real or fake. Subsequently, the discriminator undergoes updates to improve its ability to distinguish between real and fake samples in the subsequent round. Additionally, the generator receives updates based on its success or failure in deceiving the discriminator with its generated samples.

In this manner, the two models engage in a competitive relationship, exhibiting adversarial behavior in the context of game theory. In this scenario, the concept of zero-sum implies that when the discriminator effectively distinguishes between real and fake samples, it receives a reward, or no adjustments are made to its model parameters. Simultaneously, the generator is penalized with significant updates to its model parameters.

Alternatively, when the generator successfully deceives the discriminator, it receives a reward, or no modifications are made to its model parameters. Whereas, the discriminator is penalized. This is the generic GAN approach.

In the proposed TDCGAN model, the generator takes as input points from the latent space and produces data for the data distribution of the real data in the dataset. This is done through fully connected layers with four hidden layers, one input layer and one output layer. The discriminators try to classify the data into their corresponding class, which is done through a fully connected MLP network.

MLP has gained widespread popularity as a preferred choice among neural networks [29,30]. This is primarily attributed to its fast computational speed, straightforward implementation, and ability to achieve satisfactory performance with relatively smaller training datasets.

In this paper, the generator model learns how to generate new data similar to the minor class in the URG'16 dataset, while discriminators try to distinguish between real data from the dataset and the new one generated by generator. During the training process, both the generator and discriminator models are conditioned on the class label. This conditioning enables the generator model, when utilized independently, to generate minor class data within the domain that corresponds to a specific class label. The TGCGAN model can be formulated by integrating both the generator and three discriminators' models into a single, larger model.

The discriminators undergo separate training, where each of the model weights are designated as non-trainable within the TDCGAN model. This ensures that solely the weights of the generator model are updated during the training process. This trainability modification specifically applies when training the TDCGAN model, not when training the discriminator independently. So, the TDCGAN model is employed to train the generator's model weights by utilizing the output and error computed by the discriminator models.

Thus, a point in the latent space is provided as input to the TDCGAN model. The generator model creates the data based on this input, which is subsequently fed into the discriminator model. The discriminator then performs a classification, determining whether the data are real or fake, and in the case of fake data, the model classifies them to their corresponding type of attck.

The generator takes a batch of vectors (z), which are randomly drawn from the Gaussian distribution, and maps them to G(z), which has the same dimension as the dataset. The discriminators take the output from the generator and try to classify it. The loss is then evaluated between the observed data and the predicted data and is used to update the weights of the generator only to ensure that only generator weights are updated. The difference between the observed data and the predicted data is estimated using the cross-entropy loss function, which is expressed in the following equation:

$$[LOSS]_C E = -1/N \sum_{n=1}^{N} y_i.log(p(y_i)) + (1 - y_i).log[(1 - p(y_i)] \quad (1)$$

where y_i is the true label (1 for malicious traffic and 0 for normal traffic) and $p(y_i)$ is the predicted probability of the observation (i) calculated by the sigmoid activation function. N is the number of observations in the batch.

The generator model has four hidden layers. The first one is composed of 256 neurons with a rectified linear unit (ReLU) activation function. An embedded layer is used between hidden layers to efficiently map input data from a high-dimension to lower-dimension space. This allows the neural network to learn the data relationship and process it efficiently. The second hidden layer consists of 128 neurons, the third has 64 neurons and the last one has 32 neurons, with the ReLU activation function used with them all, and a regularization dropout of 20% is added to avoid overfitting. The output layer is activated using the Softmax activation function with 14 neurons as the number of features in the dataset.

After defining the generator, we define the architecture of each discriminator in the proposed model. Each discriminator is a MLP model with a different number of hidden layers, different number of neurons and different dropout percentage. The first discriminator is composed of 3 hidden layers with 100 neurons for each and 10% dropout regularization. The second has five hidden layers with 64, 128, 256, 512, and 1024 neurons for each layer, respectively. The dropout percentage is 40%. The last discriminator has 4 hidden layers with 512, 256, 128, and 64 neurons for each layer and 20% dropout percentage.

The LeakyReLU(alpha = 0.2) is used as an activation function for the hidden layers in the discriminators. Two output layers are used for each discriminator with the Softmax function as an activation function for one output layer and the Sigmoid activation function for the second output layer. The model is trained with two loss functions, binary cross entropy for the first output layer, and categorical cross-entropy loss for the second output layer. The output is extracted from each discriminator and is then fed to the last layer in the model, where the election is performed, to obtain the best result.

The TDCGAN model can be defined by combining both the generator model and the three discriminator models into one larger model. This large model is used to train the weights in the generator model, using the output and error calculated of the discriminators. The discriminators are trained separately by taking real input from the dataset.

The model is then trained for 1000 epochs with a batch size of 128. The optimizer is Adam with a learning rate equal to 0.0001. The proposed model allows the generator to train until it produces a new set of data samples that resembles the real distribution of the original dataset.

Nevertheless, this training strategy frequently fails to function effectively in various application scenarios. This is due to the necessity of preserving the relationships within the feature sets of the generated dataset by the generator, while the dataset used by the discriminator may differ from it. This disparity often leads to instability during the training of the generator.

In numerous instances, the discriminator quickly converges during the initial stages of training, thereby preventing the generator from reaching its optimal state. To tackle this challenge in network intrusion detection tasks, we adopt a modified training strategy, where three discriminators with different architectures are used. This approach helps preventing the early emergence of an optimal discriminator, ensuring a more balanced training process between the generator and discriminators.

4.3. Training Phase

The primary objective of the training methodology employed in the GAN framework is for the generator to generate fake data that closely resemble real data, and for the discriminator to acquire sufficient knowledge to differentiate between real and fake samples. Both the generator and discriminator are trained until the discriminator can no longer distinguish real data from fake data. This mean that the generated network can estimate the data sample's distribution and achieve Nash equilibrium.

In order to assess the performance of our model with precision, it is customary to divide the data into training and test sets to produce accurate predictions on unseen data. The training set is utilized for model fitting, while the test set is employed to measure the predictive precision of the trained model. The dataset is split into 70% for training and validation and 30% for testing. The training set is divided into minor class data and other class data. The TDCGAN model uses the minor class to generate data. The generator is trained to model the distribution of the anomaly data (minor class), while fixing the discriminator. The output from the generator is fed as input to the discriminator to predict it. The error is estimated, and the generator's weight is then updated. The training continues until the discriminator cannot distinguish if the input data come from the generator's output or from the real anomaly dataset. In the training process, we make sure that all architectures undergo an equal number of epochs and that the weights from the final epoch are selected to generate artificial attack samples.

We begin by adhering to this iterative training procedure and ultimately utilize the generator to produce attack samples. Eventually, we incorporate the generated attack samples into the training set.

By this, we oversample minor classes in the dataset during the training phase. The test dataset is then used to test the model performance.

5. Experimental Results

Within this section, we methodically plan and execute a sequence of experiments, and subsequently analyze the obtained results.

5.1. Experimental Setup

Our experiments were carried out on the Python Colab Jupyter notebook that runs in the browser with the integrated free GPUs and freely installed Python libraries. The system setup is shown in Table 4.

Table 4. System environment specifications.

Unit	Description
Processor	Intel® Xeon®
CPU	2.30 GHz with No.CPUs 2
RAM	12 GB
OS	
Packages	TensorFlow 2.6.0

5.2. Performance Metrics

To assess the effectiveness of our proposed model, we employ performance metrics, such as classification accuracy, precision, recall, and F1 score.

We utilize the metric of accuracy (Acc) to quantify the correct classification of data samples, considering all predictions made by the model as measured by the following equation:

$$Acc = (TP + TN)/(TP + TN + FP + FN) \tag{2}$$

where TP is the true positive, which represents the number of truly predicted anomalies; TN is the true negative, which indicates the number of truly predicted normal instances; FP is the false positive indicator that denotes the number of normal instances that are incorrectly classified as anomalies; and FN is the false negative indicator that indicates the number of the number of anomalies that are misclassified as normal.

Precision is employed to assess the accuracy of the correct predictions, calculated as the ratio of accurately predicted samples to the total number of predicted samples for a specific class as given in the following equation:

$$Precision = TP/(TP + FP) \tag{3}$$

Recall, which is known as the true positive rate (TPR), is used to determine the ratio of correctly predicted samples of a particular class to the total number of instances within the same class as given by the following equation:

$$TPR(Recall) = TP/(TP + FN) \tag{4}$$

Finally, the F1 score computes the balance between precision and recall, evaluating the trade-off between the two metrics as given in the following equation:

$$F1 = 2 * ((Precision * Recall)/(Precision + Recall)) \tag{5}$$

5.3. Experimental Results and Analysis

The performance of the TDCGAN model is evaluated on the testing dataset. The previous metrics are used to evaluate and compare the results. The results after training the TDCGAN model for URG'16 dataset balancing are given in Table 5.

Table 5. Performance evaluation metrics score for TDCGAN model.

Accuracy	Precision	F1 Score	Recall
0.95	0.94	0.94	0.96

Figure 3 shows the loss function while training the model for different numbers of epochs: 200, 400, 600, 800 and 1000.

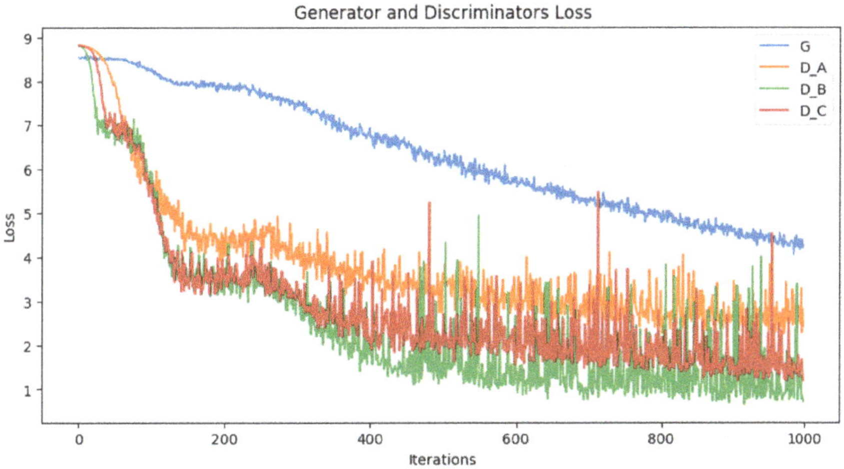

Figure 3. The loss function of G: generator, D_A: first discriminator, D_B: second discriminator and D_C: third discriminator in the TDCGAN model.

We compare the performance of the TDCGAN model for data balancing on the testing dataset with some resampling methods. The methods are as follows: (1) The synthetic minority oversampling technique (SMOTE) is a method for oversampling that produces artificial instances from the minor class. Its purpose is to create a training set that is either synthetically balanced or close to balance in terms of class distribution, which is subsequently utilized for classifier training. We use the implementations provided in the imbalanced-learn Python library, which provides a range of resample techniques that can be combined for evaluation comparison. (2) Random oversampling is used, which randomly duplicates the instances from the minor class. (3) Then, we combine SMOTE with edited nearest neighbor (ENN) SMOTEENN. (4) With Borderline-SMOTE (oversample technique using Borderline-SMOTE), the minority instances which are near the borderline are oversampled. (5) SVMSMOTE combines the support vector machine (SVM) with SMOTE. (6) We oversample using SMOTE-Tomek Links. Tomek Links denotes a technique used to detect pairs of closest neighbors within a dataset that exhibit dissimilar classes. Eliminating either one or both instances from these pairs, particularly those from the majority class, results in a reduction in noise or ambiguity within the decision boundary of the training dataset. (7) SMOTE_NC (synthetic minority over-sampling technique for nominal and continuous) is used to oversample data with categorical features. (8) CGAN (conditional generative adversarial network) is a conditional GAN that generates data under a conditional generation. Lastly, (9) CTGAN (conditional tabular generative adversarial networks) models tabular data using CGAN. The results are listed in Table 6 and shown in Figure 4.

Table 6. Performance evaluation metrics score for TDCGAN model and other resampling methods.

Model	Accuracy	Precision	F1 Score	Recall
SMOTE	0.88	0.86	0.87	0.91
Random Oversampling	0.85	0.89	0.90	0.88
SMOTEENN	0.86	0.89	0.90	0.89
The Borderline SMOTE	0.84	0.87	0.87	0.88
SVMSMOTE	0.89	0.90	0.91	0.89
SMOTE-Tomek Links	0.90	0.87	0.89	0.87
SMOTE_NC	0.85	0.88	0.86	0.85
CGAN	0.83	0.83	0.83	0.83
CTGAN	0.76	0.76	0.76	0.76
TDCGAN	0.95	0.94	0.94	0.96

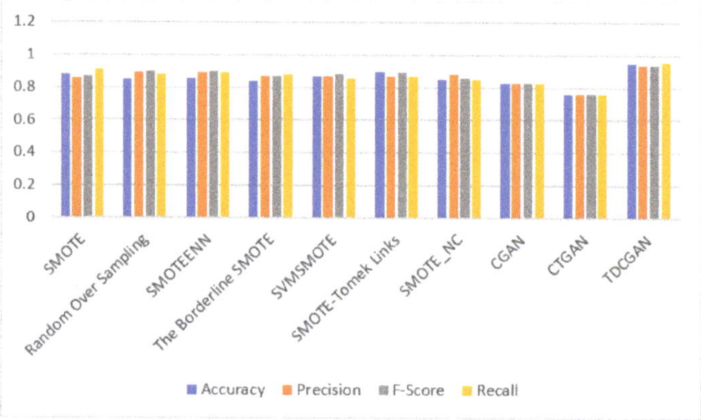

Figure 4. Performance evaluation metrics score for TDCGAN model and other resampling methods.

After conducting extensive experiments on UGR'16 dataset, our proposed model showcases its remarkable effectiveness in generating synthetic network traffic datasets, which in turn aids in the identification of anomalous network traffic. Through benchmarking, our model surpassed other similar generative models, achieving an impressive accuracy of over 0.95%.

6. Conclusions and Future Works

The imbalanced distribution of attacks in historical network traffic presents a significant challenge for intrusion detection systems (IDSs) based on traditional machine learning methods. These methods often struggle to effectively address the issue of imbalanced learning. In response, this paper introduces a novel technique called TDCGAN, a technology based on generative adversarial networks (GANs), specifically designed to tackle the problem of imbalanced datasets in IDS. The proposed TDCGAN model consists of a generator and three discriminators, all implemented using multi-layer perceptron (MLP) networks. This architecture allows the generator to generate synthetic data closely resembling real network traffic, while the discriminators aim to differentiate between genuine and attack traffic. To further enhance the TDCGAN framework, an additional layer is added at the end of the network to select the optimal outcome from the outputs produced by the three discriminators, enhancing the overall performance of the model. To evaluate the effectiveness of the proposed approach, the UGR'16 dataset, widely used in IDS research, is utilized for testing and evaluation purposes. A subset of the dataset is extracted and divided into training and testing sets. The experimental results showcase the outstanding performance of the proposed TDCGAN model across various evaluation metrics, including accuracy, precision, F1 score, and recall. Additionally, a comparison is made with other oversampling machine learning techniques, highlighting the superiority of the proposed method. While balancing datasets can be beneficial, it is important to note that it might not always be necessary or feasible, especially in cases where the class imbalance reflects the real-world distribution. In many real-world applications, the distribution of classes is often imbalanced. For instance, fraud detection, disease diagnosis, or rare event prediction typically involve imbalanced datasets. By balancing the dataset during training, the model learns to handle these imbalances and becomes more effective in addressing real-world scenarios. Additionally, balancing the dataset should be performed carefully to avoid introducing artificial patterns or losing valuable information from the original data.

As for future work, the proposed TDCGAN model shows promise for application in IDS within a vehicle ad hoc network (VANET) environment to detect unknown attacks. This opens up avenues for further research and development in leveraging the capabilities of TDCGAN for enhanced intrusion detection in dynamic vehicular networks.

Author Contributions: Conceptualization, O.S., M.A. and M.J.; methodology, O.S., M.A. and M.J.; software, M.J.; validation, O.S., M.A., M.J. and A.M.M.; formal analysis, O.S., M.A., M.J. and A.M.M.; investigation, O.S.; resources, M.J.; data curation, M.J.; writing—original draft preparation, O.S. and M.A.; writing—review and editing, O.S. and M.A.; visualization, O.S., M.A. and M.J.; supervision, A.M.M.; project administration, O.S., M.A. and M.J.; funding acquisition, M.J. and A.M.M. All authors have read and agreed to the published version of the manuscript.

Funding: This work was partially funded by projects PID2020-113462RB-I00, PID2020-115570GB-C22 and PID2020-115570GB-C21 granted by Ministerio Español de Economía y Competitividad; as well as project TED2021-129938B-I0, granted by Ministerio Español de Ciencia e Innovación.

Data Availability Statement: Not applicable.

Conflicts of Interest: The authors declare no conflict of interest.

References

1. Surakhi, O.M.; García, A.M.; Jamoos, M.; Alkhanafseh, M.Y. A Comprehensive Survey for Machine Learning and Deep Learning Applications for Detecting Intrusion Detection. In Proceedings of the 2021 22nd International Arab Conference on Information Technology (ACIT), Muscat, Oman, 21–23 December 2021; pp. 1–13.
2. AlKhanafseh, M.Y.; Surakhi, O.M. VANET Intrusion Investigation Based Forensics Technology: A New Framework. In Proceedings of the 2022 International Conference on Emerging Trends in Computing and Engineering Applications (ETCEA), Karak, Jordan, 23–24 November 2022; pp. 1–7.
3. Susilo, B.; Sari, R.F. Intrusion detection in IoT networks using deep learning algorithm. *Information* **2020**, *11*, 279. [CrossRef]
4. Schlegl, T.; Seeböck, P.; Waldstein, S.M.; Schmidt-Erfurth, U.; Langs, G. Unsupervised anomaly detection with generative adversarial networks to guide marker discovery. In Proceedings of the Information Processing in Medical Imaging: 25th International Conference, IPMI 2017, Boone, NC, USA, 25–30 June 2017; Springer: Cham, Switzerland, 2017; pp. 146–157.
5. Chawla, N.V.; Bowyer, K.W.; Hall, L.O.; Kegelmeyer, W.P. SMOTE: synthetic minority over-sampling technique. *J. Artif. Intell. Res.* **2002**, *16*, 321–357. [CrossRef]
6. He, H.; Bai, Y.; Garcia, E.A.; Li, S. ADASYN: Adaptive synthetic sampling approach for imbalanced learning. In Proceedings of the 2008 IEEE International Joint Conference on Neural Networks (IEEE World Congress on Computational Intelligence), Hong Kong, China, 1–8 June 2008; pp. 1322–1328.
7. Goodfellow, I.; Pouget-Abadie, J.; Mirza, M.; Xu, B.; Warde-Farley, D.; Ozair, S.; Courville, A.; Bengio, Y. Generative adversarial nets Advances in neural information processing systems. *arXiv* **2014**, arXiv:1406.2661.
8. Ledig, C.; Theis, L.; Huszár, F.; Caballero, J.; Cunningham, A.; Acosta, A.; Aitken, A.; Tejani, A.; Totz, J.; Wang, Z.; et al. Photo-realistic single image super-resolution using a generative adversarial network. In Proceedings of the IEEE Conference on Computer Vision and Pattern Recognition, Honolulu, HI, USA, 21–26 July 2017; pp. 4681–4690.
9. Su, H.; Shen, X.; Hu, P.; Li, W.; Chen, Y. Dialogue generation with gan. In Proceedings of the AAAI Conference on Artificial Intelligence, New Orleans, LA, USA, 2–7 February 2018; Volume 32.
10. Zhu, J.Y.; Park, T.; Isola, P.; Efros, A.A. Unpaired image-to-image translation using cycle-consistent adversarial networks. In Proceedings of the IEEE International Conference on Computer Vision, Venice, Italy, 22–29 October 2017; pp. 2223–2232.
11. Maciá-Fernández, G.; Camacho, J.; Magán-Carrión, R.; García-Teodoro, P.; Therón, R. UGR'16: A new dataset for the evaluation of cyclostationarity-based network IDSs. *Comput. Secur.* **2018**, *73*, 411–424. [CrossRef]
12. Abdulrahman, A.A.; Ibrahem, M.K. Toward constructing a balanced intrusion detection dataset based on CICIDS2017. *Samarra J. Pure Appl. Sci.* **2020**, *2*, 132–142.
13. Lee, J.; Park, K. GAN-based imbalanced data intrusion detection system. *Pers. Ubiquitous Comput.* **2021**, *25*, 121–128. [CrossRef]

14. Hajisalem, V.; Babaie, S. A hybrid intrusion detection system based on ABC-AFS algorithm for misuse and anomaly detection. *Comput. Netw.* **2018**, *136*, 37–50. [CrossRef]
15. Kabir, E.; Hu, J.; Wang, H.; Zhuo, G. A novel statistical technique for intrusion detection systems. *Future Gener. Comput. Syst.* **2018**, *79*, 303–318. [CrossRef]
16. Sharafaldin, I.; Lashkari, A.H.; Ghorbani, A.A. Toward generating a new intrusion detection dataset and intrusion traffic characterization. *ICISSp* **2018**, *1*, 108–116.
17. Kumar, V.; Sinha, D.; Das, A.K.; Pandey, S.C.; Goswami, R.T. An integrated rule based intrusion detection system: Analysis on UNSW-NB15 data set and the real time online dataset. *Clust. Comput.* **2020**, *23*, 1397–1418. [CrossRef]
18. Seo, E.; Song, H.M.; Kim, H.K. GIDS: GAN based intrusion detection system for in-vehicle network. In Proceedings of the 2018 16th Annual Conference on Privacy, Security and Trust (PST), Belfast, Ireland, 28–30 August 2018; pp. 1–6.
19. Cao, B.; Li, C.; Song, Y.; Qin, Y.; Chen, C. Network Intrusion Detection Model Based on CNN and GRU. *Appl. Sci.* **2022**, *12*, 4184. [CrossRef]
20. Fan, J.; Xu, J.; Ammar, M.H.; Moon, S.B. Prefix-preserving IP address anonymization: measurement-based security evaluation and a new cryptography-based scheme. *Comput. Netw.* **2004**, *46*, 253–272. [CrossRef]
21. Haag, P. NFDUMP-NetFlow Processing Tools. 2011. Available online: http://nfdump.sourceforge.net (accessed on 16 June 2023).
22. Ndichu, S.; Ban, T.; Takahashi, T.; Inoue, D. AI-Assisted Security Alert Data Analysis with Imbalanced Learning Methods. *Appl. Sci.* **2023**, *13*, 1977. [CrossRef]
23. Wang, Z.; She, Q.; Ward, T.E. Generative adversarial networks in computer vision: A survey and taxonomy. *ACM Comput. Surv. (CSUR)* **2021**, *54*, 1–38. [CrossRef]
24. Jiang, W.; Hong, Y.; Zhou, B.; He, X.; Cheng, C. A GAN-based anomaly detection approach for imbalanced industrial time series. *IEEE Access* **2019**, *7*, 143608–143619. [CrossRef]
25. Yang, Y.; Nan, F.; Yang, P.; Meng, Q.; Xie, Y.; Zhang, D.; Muhammad, K. GAN-based semi-supervised learning approach for clinical decision support in health-IoT platform. *IEEE Access* **2019**, *7*, 8048–8057. [CrossRef]
26. Wang, X.; Guo, H.; Hu, S.; Chang, M.C.; Lyu, S. Gan-generated faces detection: A survey and new perspectives. *arXiv* **2022**, arXiv:2202.07145.
27. Xia, X.; Pan, X.; Li, N.; He, X.; Ma, L.; Zhang, X.; Ding, N. GAN-based anomaly detection: a review. *Neurocomputing* **2022**, *493*, 497–535. [CrossRef]
28. Durgadevi, M. Generative Adversarial Network (GAN): A general review on different variants of GAN and applications. In Proceedings of the 2021 6th International Conference on Communication and Electronics Systems (ICCES), Coimbatre, India, 8–10 July 2021; pp. 1–8.
29. Zaidan, M.A.; Surakhi, O.; Fung, P.L.; Hussein, T. Sensitivity Analysis for Predicting Sub-Micron Aerosol Concentrations Based on Meteorological Parameters. *Sensors* **2020**, *20*, 2876. [CrossRef] [PubMed]
30. Surakhi, O.; Serhan, S.; Salah, I. On the ensemble of recurrent neural network for air pollution forecasting: Issues and challenges. *Adv. Sci. Technol. Eng. Syst. J.* **2020**, *5*, 512–526. [CrossRef]

electronics

MDPI

Article

Explainable Ensemble Learning Based Detection of Evasive Malicious PDF Documents

Suleiman Y. Yerima [1,*] and Abul Bashar [2]

[1] Faculty of Computing, Engineering and Media, Cyber Technology Institute, De Montfort University, Leicester LE1 9BH, UK

[2] Department of Computer Engineering, Prince Mohammad bin Fahd University, Khobar 31952, Saudi Arabia; abashar@pmu.edu.sa

* Correspondence: syerima@dmu.ac.uk

Abstract: PDF has become a major attack vector for delivering malware and compromising systems and networks, due to its popularity and widespread usage across platforms. PDF provides a flexible file structure that facilitates the embedding of different types of content such as JavaScript, encoded streams, images, executable files, etc. This enables attackers to embed malicious code as well as to hide their functionalities within seemingly benign non-executable documents. As a result, a large proportion of current automated detection systems are unable to effectively detect PDF files with concealed malicious content. To mitigate this problem, a novel approach is proposed in this paper based on ensemble learning with enhanced static features, which is used to build an explainable and robust malicious PDF document detection system. The proposed system is resilient against reverse mimicry injection attacks compared to the existing state-of-the-art learning-based malicious PDF detection systems. The recently released EvasivePDFMal2022 dataset was used to investigate the efficacy of the proposed system. Based on this dataset, an overall classification accuracy greater than 98% was observed with five ensemble learning classifiers. Furthermore, the proposed system, which employs new anomaly-based features, was evaluated on a reverse mimicry attack dataset containing three different types of content injection attacks, i.e., embedded JavaScript, embedded malicious PDF, and embedded malicious EXE. The experiments conducted on the reverse mimicry dataset showed that the Random Committee ensemble learning model achieved 100% detection rates for embedded EXE and embedded JavaScript, and 98% detection rate for embedded PDF, based on our enhanced feature set.

Keywords: malicious PDF detection; PDF malware; feature engineering; reverse mimicry attack; malicious content injection; shapely additive explanation; ensemble learning; explainable machine learning

Citation: Yerima, S.Y.; Bashar, A. Explainable Ensemble Learning Based Detection of Evasive Malicious PDF Documents. *Electronics* **2023**, *12*, 3148. https://doi.org/10.3390/electronics12143148

Academic Editor: Tomasz Rak

Received: 12 June 2023
Revised: 10 July 2023
Accepted: 12 July 2023
Published: 20 July 2023

1. Introduction

Malicious documents have been one of the growing methods used by attackers to propagate malware. This has been made possible due to the growing numbers of unsuspecting document users and failure of detection by modern antivirus software [1]. Portable Document Format (PDF) has become a major attack vector because of its flexibility, cross-platform widespread usage, and the ease of embedding different types of content such as encoded streams, JavaScript code, executable files, etc. Since PDF files are not perceived to be dangerous like EXE files, they are usually treated with less caution by users. Thus, they can be used as an effective means to launch social engineering attacks, for example to convey ransomware. In [2], it was reported that Sophos Labs discovered a spam campaign where a variant of the Locky Ransomware was launched by a VBA macro hidden in Word Document that is deeply nested inside a PDF file. The malicious PDF file was spread by email as an attachment. Such types of malicious components can be embedded in PDF files using tools such as Metasploit. Furthermore, PDF files pose a higher risk compared to Portable Executables since the embedded content can be encrypted or encoded [3]. PDF documents have also been used in targeted attacks and advance persistent threat (APT) campaigns to accomplish one or more stages of a multi-stage attack, for instance, the MiniDuke APT campaign [4], where infected PDF files that targeted an Adobe Reader vulnerability (i.e., CVE-2013-0640) was used for the first stage of the attack.

The detection of malicious PDF documents is made more challenging by the fact that its format is complex, and it is susceptible to a wide range of attacks, many of which take advantage of legitimate PDF functionality, e.g., the embedding and encoding of a wide variety of content types. Several static and dynamic analysis tools are available to facilitate manual analysis of PDF documents for potentially malicious content. Examples of such tools include PDFiD [5], PeePDF [6], PhoneyPDF [7], and PDF Walker [8]. However, the volume of malicious PDF files that are constantly emerging makes it infeasible for the security community to rely on manual analysis alone. While signatures can be utilized to facilitate automated analysis to detect malicious files, this also comes with its own set of challenges, including susceptibility to obfuscation and aging of signatures against the appearance of new types of attacks.

To overcome these limitations, learning-based systems have been proposed by researchers based on different types of features. Two popular kinds of features used in the current learning-based PDF malware detection systems include JavaScript-based features and structural features. Learning-based systems that utilize JavaScript-based features extract them by analyzing embedded Javascript code to detect malicious behaviour, for example in PJScan [9] or Lux0R [10]. Such systems, however, are only effective for detecting malicious PDF files that contain JavaScript code. Examples of proposed learning-based systems that rely on structural features of PDF files include PDF Slayer [11], Hidost [12], and PDFRate [13]. The use of structural features with machine learning became more widespread because it enables fast automated detection of a wide variety of attacks including newly appearing variants. Recently, detection systems that employ visualization-based features are also being proposed. For example, ref. [14] proposed a system where PDF files are first converted to grayscale images before extracting visualization-based features for machine learning.

According to [15], one of the problems with machine learning-based classifiers in the PDF malware detection domain is that mimicry attacks and reverse mimicry attacks are quite effective against them. A reverse mimicry attack involves injecting or embedding malicious content into benign PDF files such that the features of the benign file will effectively mask the presence of the embedded malicious content from being detected by detection systems. It is a form of evasive adversarial attack that can be performed on a large scale using automated tools. Existing machine learning-based solutions such as PDF Slayer [11], Hidost [12], and PDFRate [13] have been shown to have limited robustness against reverse mimicry attacks. Even the more recent attempt at utilizing visualization techniques for high accuracy PDF malware detection presented in [14] did not show substantial resilience in the reverse mimicry attack experiments.

Hence, despite the advances that have been made with learning-based malware PDF detection, their resilience against evasive or adversarial attacks remains a significant challenge. In order to mitigate the problem, this paper proposes a novel approach that uses an enhanced feature set which extends existing structural features with anomaly-based ones, and utilizes the power of ensemble learning to provide a high accuracy malicious PDF detection system that is also resilient against injection-based adversarial attacks. The main contributions of this paper are as follows:

- The paper proposes an ensemble learning-based system that employs an enhanced feature set comprising structural and anomaly-based features. This feature set is a unique one that is designed to enable robust and effective detection of malicious PDF files including those that employ evasive techniques.
- The novel anomaly-based features that enable robust maldoc detection are described, discussing their impact on the performance of the learning-based detection system, as well as its resilience to reverse mimicry attacks.

- An extensive performance evaluation of the proposed system for malicious PDF detection is undertaken, using the recently released Evasive-PDFMal2022 dataset. The results showed that the ensemble learners demonstrated high accuracy with the enhanced feature set.
- Furthermore, several experiments are performed using a publicly available reverse mimicry attack dataset consisting of three types of injection attacks. A comparative analysis with several existing systems is presented to demonstrate the robustness of our proposed approach against reverse mimicry attacks. We also present explanations of the models prediction in each attack scenario using the SHapely Additive exPlanation (SHAP) approach.

This paper is organized as follows: after the Introduction in Section 1, Section 2 gives an overview of PDF file format and is followed by related works in Section 3. Section 4 presents the development of the proposed system, and describes the new anomaly-based features that are incorporated with structural features to enable more robust malicious PDF file detection. The experiments and results are discussed in Sections 5 and 6. Finally the paper is concluded in Section 7 with recommendations for future work.

2. Structure of a PDF File

PDF was created as a versatile format to enable sharing of text, rich media, images, etc. independent of hardware or software platforms and in a consistent way. It was invented by Adobe in 1993 and has now become one of the most widely used standards for sharing documents. The PDF format was standardized into an ISO 32000-1:2008 [16] open standard. The typical structure of a PDF document is shown in Figure 1 and consists of four parts:

- **The header**: contains PDF file version information according to the ISO standard.
- **The body**: This section typically contains the contents that are displayed to the user. It shows the number of objects that define the operations to be performed by the file. The body section also contains the embedded data such as text, images, scripting code, etc. which are also presented as objects. Within an object, operations such as decompression of data or decryption are defined if needed and will typically take place during the rendering of the file.
- **The cross-reference (x-ref) table**: This contains a list of the offsets of each object that are to be rendered within the file by the reader application. The offsets within the x-ref table makes it possible to randomly access any of the objects in the file. The x-ref table is also the section that enables incremental updates to a document, as allowed by the PDF standard. Thus, when a document is updated, extra x-ref tables and trailers are appended at the end of the document.
- **The trailer**: The trailer is a special object corresponding to the last section of the file. It points to the object identified by the/Root tag, which is the first object that will be rendered by the document viewer. The offset of the start of the x-ref table is also located in the trailer. The last line of the file, which is the end of file string '%%EOF' is also part of the trailer section.

Basically, When a PDF reader displays a file, it begins from the trailer object and parses each indirect object referenced by the x-ref table, and at the same time decompresses the data so that all pages, texts, images, and other components of the PDF file are progressively rendered. This means that a PDF file is organized as a graph of objects that contain instructions for the PDF reader, which represents the operations to be performed for presenting the file contents to the user [17].

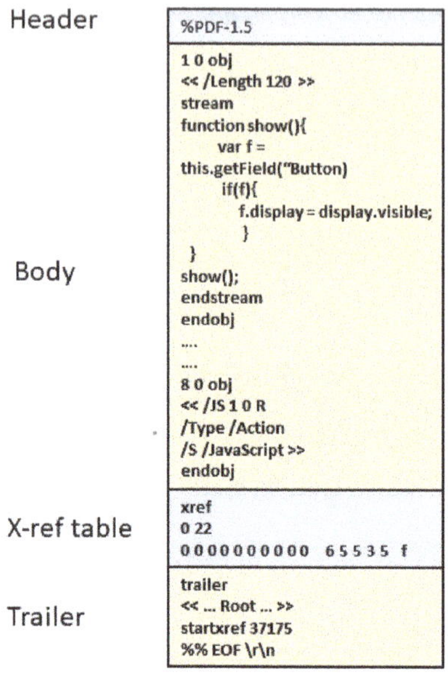

Figure 1. The sections of a typical PDF file.

3. Related Work

Learning-based detection of malware and malicious content in PDF documents have proliferated in recent years due to the drive to create new approaches that will enhance or complement existing anti-malware systems. In [3], the authors proposed an approach to detect malicious content embedded in PDF documents. They focused on data encoded in the 'stream' tag along with other structural information. Their method decrypts encrypted blocks and decodes encoded blocks within the stream tags and also utilizes other structural features. These are given to a decision tree for classification. Although the paper claims that the method is effective against mimicry attacks, no empirical evaluation was presented to support the claim. In [18], a method for detecting and classifying suspicious PDF files based on YARA scan and structural scan is presented. Their system inspects PDF documents to search for features that are important in labelling PDF documents as suspicious. In [10], a system to detect malicious JavaScript embedded in PDF files was presented. The system was called 'Lux 0n discriminant References' (Lux0R). The authors of [9] presented PJScan, a tool which is designed to uncover JavaScript from the malicious file and to extract its lexical properties via a tokenizer. The output, which is a token sequence, is then used to train a machine learning algorithm to detect malicious JavaScript-bearing PDF files.

Jeong, Woo, and Kang [19] presented a convolutional neural network (CNN) designed to take the byte sequence of a stream object contained within a PDF file and predict whether the input sequence contains malicious actions or not. The CNN model achieved superior performance compared to traditional machine learning classifiers including SVM, Decision Tree, Naive Bayes, and Random Forest. Albahar et al. [20] presented two learning-based models for detection of malicious PDFs and experimented on 30,797 infected and benign documents collected from the Contagio dataset and VirusTotal. Their first model was a CNN model that used tree-based PDF file structure as features and yielded 99.33% accuracy; the second model was an ensemble SVM model with different kernels which used n-gram with object content encoding as features and yielded an accuracy of 97.3%. In [21], Bazzi and Onozato used LibSVM to build a classification model which utilizes features extracted from a report generated through dynamic analysis with Cuckoo sandbox. The study used 6000 samples for training and 10,904 samples for testing, obtaining an accuracy of 97.45%.

In [22], a PDF maldoc detection system was proposed based on extracting features with PDFiD and PeePDF. They used both tools to extract keyword and structural features and used malicious document heuristics to derive an additional set of features. Trough feature selection, the top 14 important features were selected, which led to an improved accuracy of up to 97.9% for the ML classifier. Zhang proposed MLPdf in [23] which uses an MLP classifier to detect PDF malware. Their system extracted a group of high quality features from two real world datasets that contained 105,000 malicious and benign PDF documents. The MLP model achieved a detection rate of 95.12% and low false positive rate of 0.08%. Jiang et al. [24] applied semi-supervised learning to the problem of malicious PDF document detection in [24] by extracting structural features together with statistical features based on entropy sequences using wavelet energy spectrum. They then employed a random sub-sampling approach to train multiple sub-classifiers, with their method achieving an accuracy of 94%.

The authors of [25] did a performance comparison of machine learning classifiers to traditional AV solutions by experimenting on PDF documents with embedded JavaScript. They used 995 samples for training, 217 samples for validation, and 500 samples for testing and obtained 92%, 50%, and 96% accuracy with Random Forest, SVM, and MLP, respectively. In [14], the authors applied image visualization techniques of byte plot and Markov plot and extracted various image features from both. They evaluated the performance using the Contagio PDF dataset, obtaining very good results when testing with samples from the same dataset. They also evaluated their models on a reverse mimicry attack dataset, with very limited success but showing slightly improved robustness over the PDF Slayer approach. They experimented with both Markov plot and byte plot visualization methods, applying various image processing techniques used in extracting features to train RF, K-Nearest Neighbor (KNN), and Decision Tree (DT) classifiers. The best method (byte plot + Gabor Filter + Random Forest) achieved an F1-score of 99.48%.

Al-Haija, Odeh, and Hazem proposed in [26] a detection system for identifying benign and malicious PDF files. Their proposed system used an optimally-trained AdaBoost decision tree and their experiments were performed using the Evasive-PDFMal2022 dataset [27] (which is also used in this paper). Their system achieved 98.4% prediction accuracy with 98.80% precision, 98.90% sensitivity, and a 98.8% F1-score. In [28], the authors also utilized the Evasive-PDFMal2022 dataset and applied an enhanced structural feature set to investigate the efficacy of the enhanced set. Seven machine learning classifiers were evaluated on the dataset using the enhanced features, and improved classification accuracy was noticed with 5 out of 7 of the classifiers compared to the baseline scenario without the enhanced features.

In [29], a system for detecting evasive PDF malware was proposed based on Stacking ensemble learning. The detection system is based on a set of 28 static features which were divided into 'general' and 'structural' features. Their system was evaluated on the Contagio dataset, yielding an accuracy of 99.89% and F1-score of 99.86%. They also evaluated the system on their newly generated Evasive-PDFmal2022 dataset [27] for which they achieved 98.69% accuracy and a 98.77% F1-score, respectively.

From our review of related work, it is evident that several of the proposed learning-based detectors utilized features extracted only from JavaScript obtained from the PDF files, e.g., [9,10,19]. While such systems may be able to detect PDF files incorporating content injection attacks that involve embedded JavaScript, they may not be effective against other types of content embedding attacks, e.g., those involving embedded PDF, Word, EXE, or other types of content. Other works, such as [22,23,26,29], utilized structural features in their work, but did not evaluate their approach against any type of adversarial attacks. Different from the existing works, this paper aims to improve the robustness of malicious PDF document detection by enhancing structural features with novel anomaly-based features and utilizing the enhanced feature set to train ensemble learning classifiers. Furthermore, we present experiments to demonstrate the resilience of our proposed approach to reverse mimicry injection attacks, enabled by the new anomaly-based features.

4. Methodology

This section presents our proposed approach to automated ensemble learning-based malicious PDF detection, which is based on an enhanced feature set consisting of 35 features (29 structural features and 6 anomaly-based features). These features are extracted from the labeled files that have been set aside as the training set. The instances consisting of the 35 extracted features are fed into ensemble learning classification algorithms to learn the distinguishing characteristics of benign and malicious PDF files, thus enabling the prediction and classification of unlabeled PDF files as benign or malicious, as shown in Figure 2. The methods used in building the proposed PDF classification system are discussed in the following sub-sections.

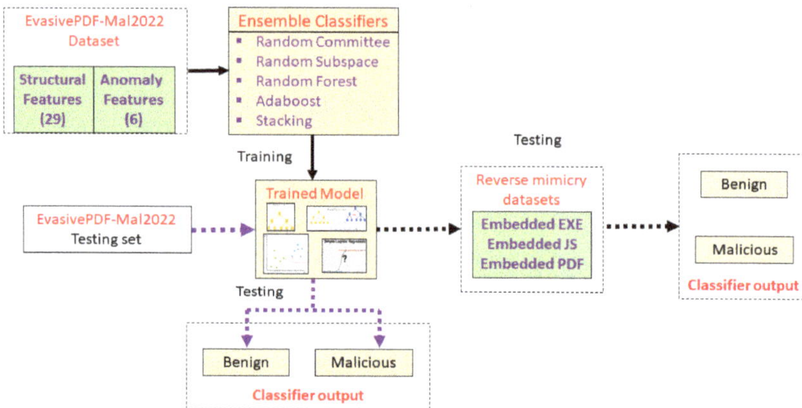

Figure 2. Proposed enhanced features-based approach.

4.1. Datasets

Evasive-PDFMal2022 dataset: The first dataset used for the study in this paper is a recently generated evasive PDF dataset (Evasive-PDFMal2022) [27] which was released by Issakhani et al. [29]. This dataset has been generated as an improved version of the well-known Contagio PDF dataset which has been utilized extensively in previous works. According to [29], the Contagio dataset has several drawbacks which include (a) a high proportion of duplicate samples with very high similarity, which was estimated as 44% of the entire dataset and (b) lack of sufficient diversity of samples within each class of the dataset. Thus, the new dataset aims to address the flaws found with the Contagio dataset and provide a more realistic and representative dataset of the PDF distribution. It consists of 10,025 PDF file samples with no duplicate entries (4468 benign and 5557 malicious).

PDF reverse mimicry dataset: This was the second dataset utilized in our study. It is used to evaluate the robustness of our proposed approach to content injection attacks designed to disguise malicious content by embedding them within benign PDF files. This is known as the reverse mimicry attack, and it is a form of evasive adversarial attack that can be performed on a large scale using automated tools. The reverse mimicry dataset [30] consists of 1500 benign PDF files with embedded malicious components and is available online from the Pattern Recognition and Applications lab (PRAlab), University of Cagliari, Italy. The dataset consists of 500 PDF files containing embedded JavaScript, 500 PDF files containing embedded PDF, and 500 PDF files containing embedded EXE payload. Further details on how these reverse mimicry files were created can be found in [17]. Note that the detection of malicious PDF files created by altering a benign file through such reverse mimicry attacks is a challenging task. This is because the injected file will still retain the characteristics of a benign PDF file, thus making it hard for learning algorithms to discriminate effectively.

4.2. Feature Extraction of Structural Feature Set

Our proposed ensemble learning-based detection system is based on 35 features: 29 structural features and 6 features that are based on anomalies, i.e., properties that are rarely observed from regular harmless files. The structural features we used are similar to those found in previous works, however, the anomaly-based features are novel features aimed at improving the performance of the learning-based detectors. The features were extracted by using our extended version of the the open source PDFMalyzer tool available from [31]. PDFMalyzer is based on PDFiD and PyMuPDF and it enabled the 29 structural features to be extracted. By extending the tool using Python scripts, we were able to extract the new anomaly-based features and combine them with the 29 structural features into a feature vector to represent each of the PDF files being used in our experiments. The initial set of 29 structural features are listed in Table 1.

Table 1. Initial feature set containing 29 structural features (NK: non-keyword based, K: keyword-based).

Feature Name	Type	Description
pdfsize	NK	Size of the PDF file
metadata size	NK	Metadata size
pages	NK	Number of pages in the document (not from keyword)
title characters	NK	Number of characters contained in the title of the file
isEncrypted	NK	Whether or not the file is encrypted (not from keyword)
embedded files	NK	Indicates that an embedded file is present (not from keyword)
images	NK	Indicates whether the document contains images
text	NK	Indicates presence of text within the document
obj	NK	Number of obj tags found
endobj	NK	Number of endObj tags found
stream	NK	Number of stream tags found
endstream	NK	Number of endstream tags found
xref	NK	Number of xref tables in the file
trailer	NK	Number of trailers in the file
startxref	NK	Number of xref start indicators
/Page	NK	Number of pages in the PDF document
/Encrypt	K	Document has DRM or needs a password to be read
/ObjStm	K	Number of object streams that can contain other objects
/JS	K	Number of JS objects
/JavaScript	K	Number of JavaScript objects
/AA	K	Automatic action to be performed upon an event
/OpenAction	K	Automatic action to be performed on viewing document
/Acroform	K	Contains traditional forms authored in Adobe Acrobat
/JBIG2Decode	K	Indicates if the PDF document uses JBIG2 compression
/RichMedia	K	Presence embedded Flash or embedded media
/launch	K	Number of launch actions
/EmbeddedFile	K	Number of EmbeddedFile keywords found
/XFA	K	Keyword for XML Forms Architecture.
/Colors	K	Number of colours present in the file

4.3. Enhancing the Structural Feature Set with New Features

Structural features are related to the characteristics of the name object present in the PDF file [17]. Structural features have the ability to detect the presence of different types of embedded contents such as JavaScript or ActionScript, which can aid in the detection of malicious PDF files. Note that the keywords representing the structural features could be missed if a deliberate attempt has been made to evade their detection, e.g., through obfuscation, or due to errors from the analysis tools being used to extract the features. These uncertainties in feature extraction motivated the derivation of new (anomaly-based) features to improve robustness. The proposed anomaly-based features are described next.

When a user directly modifies an existing PDF file, this creates a new x-ref table and trailer which are added to the file. This means that a manually updated PDF file will typically have more than one trailer and x-ref table. Hence, the feature vector of a benign file should consist of more than one occurrence of /trailer, /xref, and /startxref features. Thus, having only one occurrence of those features in the feature vector should be considered an anomaly. Based on this reasoning, we defined two new features (mal_trait1 and mal_trait2) derived by observing the number of /trailer, /xref, or /startxref occurrences (which are typically the same) together with keyword features that are indicative of possible malicious content. These indicators of malicious content for each of these two new features include (a) presence of JavaScript and (b) the presence of one or more embedded files. The anomaly-based features are explained below:

- **mal_trait1**: This is a new feature being proposed to represent the situation where /xref, /trailer, and /startxref are found only once in the PDF file, but with JavaScript detected within the file as well. This could indicate the injection or embedding of JavaScript code with an automated tool (such as Metasploit), since having only one occurrence the aforementioned three keywords does not suggest user modification.
- **mal_trait2**: This is a new feature being proposed to represent the situation where /xref, /trailer, and /startxref are only found once in the PDF file, but an embedded file is also detected (regardless of whether JavaScript is present or not). This could also indicate that another file was injected or embedded within the PDF file using an automated tool (such as Metasploit), since having only one occurrence of the aforementioned three keywords does not suggest user modification.
- **mal_trait3**: The purpose of this new feature is to search for the presence of both JavaScript code and embedded files within the PDF file. The intuition behind this feature is that the JavaScript code can be used to launch a malicious embedded file.
- **diff_obj**: This feature captures anomalies observed with the opening and closing tags of objects in a PDF file as described in [22]. Each object in the file is expected to begin with an opening tag (obj) and have a corresponding closing tag (endObj). A difference in the occurrences of the opening and closing tags indicates possible file corruption (usually a missing closing tag). This is an obfuscation technique designed to bypass some parsing tools that strictly conform to PDF standards. On the other hand, the file will still be rendered correctly by the PDF readers, thus enabling the intended malicious activity to occur.
- **diff_stream**: This feature also captures anomalies in a similar manner to diff_obj, by recording the occurrences of 'stream' and 'endStream' which are the opening and closing tags of stream objects. According to [22], this evasive technique of omitting a stream object tag is intended to corrupt the file such that parsing tools within detectors will be confused but the file will still be rendered and shown to the user by reader applications.
- **mal_traits_all**: This is a new composite feature that is intended to help with the identification of files that exhibit one or more of the above five anomalous features. The intuition behind this is to create a robust feature that will maintain its relevance even if new techniques evolve to defeat a subset of the new features. For instance, the ability to obfuscate the /trailer, /xref, or /startxref values may produce errors in capturing mal_trait1 and mal_trait2 features or make them obsolete in the future. However, mal_traits_all will still remain relevant in the presence of such obfuscation because it is created as a compound feature. Moreover, the failure of extraction tools could lead to missing or erroneous values

for some of the standard features. Mal_traits_all therefore provides an indicator that has resilience against the occurrence of such errors.

4.4. Ensemble Learning Classifiers

In this section, we provide brief descriptions of the ensemble learners used in our proposed system. The ensemble classifiers are first evaluated using the initial 29 structural features, and then the 36 features, including the anomaly-based ones. The trained ensemble learners are also evaluated on three reverse mimicry datasets. The results of these experiments are presented in Section 5.

4.4.1. Random Committee

This is an ensemble learner that utilizes randomizable base classifiers to build an ensemble. It builds each base classifier using a different random number seed but based on the same data. Hence, a randomizable base classifier must be chosen as it does not accept non-randomizable classifiers such as J48, Simple Logistic, or rule-based classifiers. Random Committee uses the same type of base classifier, e.g., Random Tree. Different seeds are used to generate different random numbers for the underlying base leaner which, although it uses the same mechanism, will result in a different model as a result of being initialized differently. With the Random Committee, since each base learner is built from the same data, diversity of models can only come from random behaviour. The outcomes of these models are averaged to generate a final prediction.

4.4.2. Random Subspace

Random Subspace [32] is an ensemble learner that constructs models in randomly chosen subspaces, with the training data samples in the feature space. The output of the models is then combined by a simple majority vote.

4.4.3. Random Forest

Random Forest [33] combines decision trees with bagging (bootstrap aggregating), and retains many of the benefits of decision trees while being able to handle a large number of features. Each model in the ensemble uses a randomly drawn subset of the training set, and the combined outcome is derived from a majority vote, with each model having equal weight. Random Forest has been widely applied to different classification problems, and generally shows very good performance compared to other non-ensemble learners in many problem domains.

4.4.4. AdaBoost

AdaBoost is based on Boosting, which incrementally builds an ensemble by training each new model instance to emphasize the training instances that were miss-classified in previous iterations. Boosting [34] iteratively builds a succession of models with each one being trained on a dataset with previously miss-classified instances given more weight. All of the models are then weighted according to their success and the outputs are combined by voting or averaging. With AdaBoost, the training set does not need to be large to achieve good results, since the same training set is used iteratively.

4.4.5. Stacking

This is also called Stacked Generalization [35]. It combines multiple base learners by introducing the concept of a meta-learner and can be used to combine models of different types, unlike boosting or bagging-based ensemble learners. The training set is split into two non-overlapping sets and the first part is used to train the base learners while testing them on the second part. Using the prediction/classification outcomes from the test set as inputs, and correct labels as outputs, the meta-learner is trained to derive a final classification outcome.

5. Experiments and Results

In this section, we present the results of the experiments for quantifying the impact of the new features on the performance of ensemble learning classifiers (the core component of our proposed overall approach). In the previous section, we already explained how the features provide resilience against some obfuscation and extraction errors. Ideally, the new features should not have a negative impact on the classification accuracy when incorporated with the existing ones. First, a baseline experiment is performed where we train the five ensemble classifiers using only the original 29 features. Afterwards, a second set of experiments is carried out with the enhanced set containing all the 35 features. The configurations of the ensemble learners are shown in Table 2.

Table 2. Ensemble Classifier Configurations.

	Base Classifier(s)	Configurations
Random Forest	Decision Tree	100 trees
Random Committee	Random Tree	100 iterations; 100 trees
AdaBoost	Random Tree	100 iterations; 100 trees
Random Subspace	Random Tree	100 trees; 100 iterations
Stacking	J48, SVM, Simple Logistic	LR meta classifier

5.1. Original Feature Set Results

Table 3 presents the 10-fold cross validation results of five ensemble classifiers trained using the original 29 structural features extracted from the PDF samples in the dataset. These results are based on the 10,025 samples of the Evasive-PDFMal2022 dataset. From the table, it can be observed that the Random Forest, Random Committee, and Random Subspace models yielded higher overall accuracy >99%. The Stacking and AdaBoost models obtained an overall accuracy of 98.78% and 98.63%, respectively. These results show that the ensemble learners performed well with the 29 baseline structural features since all of the classifiers showed >98% accuracy.

Table 3. Ensemble classifiers results without new features (10-fold CV).

	Precision Mal/Ben	Recall Mal/Ben	F1 Mal/Ben	Accuracy (%)
Random Forest	0.994/0.992	0.993/0.992	0.993/0.992	99.27
Random Committee	0.994/0.994	0.995/0.993	0.995/0.994	99.42
AdaBoost	0.989/0.983	0.986/0.986	0.988/0.985	98.63
Random Subspace	0.994/0.994	0.996/0.992	0.995/0.993	99.40
Stacking	0.987/0.988	0.991/0.984	0.989/0.986	98.78

5.2. Enhanced Feature Set Results

Table 4 presents the 10-fold cross validation results of five ensemble classifiers trained using the enhanced set with 35 features. These results are based on the 10,025 samples of the Evasive-PDFMal2022 dataset. From the table, it can be observed that there is an improvement in the performance of Random Forest with the overall accuracy slightly increased to 99.33%. The AdaBoost and Stacking models also increased their performance with accuracy rising to 98.83% and 98.84%, respectively. On the other hand, the overall accuracy of Random Subspace dropped slightly by 0.06%, while that of Random Committee also dropped by 0.06%. These results show that the introduction of the new features did not have a negative impact on the ensemble classifiers. However, our main goal is to examine whether these features provide resilience by improving the performance of the models in adversarial scenarios. Our next set of experiments on the reverse mimicry attack dataset will underscore the impact of the novel features to the performance of the ensemble learning models.

Table 4. Ensemble classifiers results with the new features (10-fold CV).

	Precision Mal/Ben	Recall Mal/Ben	F1 Mal/Ben	Accuracy (%)
Random Forest	0.994/0.992	0.994/0.993	0.994/0.993	99.33
Random Committee	0.994/0.993	0.995/0.992	0.994/0.993	99.36
AdaBoost	0.990/0.987	0.989/0.987	0.989/0.987	98.83
Random Subspace	0.993/0.994	0.996/0.991	0.994/0.993	99.34
Stacking	0.990/0.988	0.989/0.988	0.990/0.987	98.84

5.3. Investigating the Effect of the New Features against the Reverse Mimicry Attacks

As mentioned earlier, the reverse mimicry dataset consists of three content injection attacks each with 500 samples. They include (a) embedded executable, (b) embedded JavaScript, and (c) embedded PDF. The experiments were conducted by training the ensemble models with all of the Evasive-PDFMal2022 samples and then using each of the 500 samples in the reverse mimicry dataset as the testing set. The first model training was done with only the 29 baseline structural features and then the models were evaluated on the attack samples. The same process was repeated with the full set of 35 features including the new anomaly-based features. The results of these experiments are shown in Tables 5 and 6. The numbers depicted in brackets in the table heading denote the number of samples used in the evaluation (a few of the initial samples failed during the experiments).

Table 5. Reverse mimicry attack dataset—ensemble classifiers results without the new features.

	Embedded EXE (498)	Embedded JS (500)	Embedded PDF (499)
Random Forest	46.78% (233)	41% (205)	5.6% (28)
Random Committee	68.7% (342)	31.8% (159)	5% (25)
AdaBoost	71.2% (355)	45.6% (228)	12.2% (61)
Random Subspace	67.9% (338)	33.4% (167)	5.2% (26)
Stacking	17.7% (88)	58.8% (293)	6.2% (31)

Table 6. Reverse mimicry attack dataset—ensemble classifiers results with the new features.

	Embedded EXE (498)	Embedded JS (500)	Embedded PDF (499)
Random Forest	63.6% (317)	60.2% (301)	10.6% (53)
Random Committee	90.1% (449)	39.6% (198)	11.2% (56)
AdaBoost	60.8% (303)	23.4% (117)	68.9% (344)
Random Subspace	83.7% (417)	43.6% (218)	14.4% (72)
Stacking	93.6% (466)	95.4% (477)	38.9% (194)

From Table 5 (without the new features), it can be seen that the best result for embedded Exe was AdaBoost, with 71.2%, i.e., 355 samples detected. For the embedded JavaScript, the best was the Stacking model which detected 293 samples (58.8%). In the embedded PDF set, the highest was only 12.2% (61 samples) detected. This shows that the embedded PDF was the most challenging attack to detect. One possible reason for this could be the lack of structural features (keywords) that directly indicate when a PDF file is present in the PDF file. In the feature set there are two keywords directly related to JavaScript, which may make it easier to detect embedded JavaScript attacks. Another possible reason could be the way the embedded PDF attack was crafted. The embedded PDF can be used to nest other features which will not appear within the parent benign PDF, thus tricking the classifier into predicting the sample as benign.

From Table 6 (with the new anomaly-based features), there is significant improvement in the detection of the mimicry attacks. Random Committee and Stacking detected 449 (90.1%) and

466 (93.6%) samples of the embedded EXE attack, respectively. For the embedded JavaScript, Stacking also obtained 477 (95%) detected samples, while for the embedded PDF, the highest was AdaBoost, with 68.9% (344 samples). Again, this highlights how challenging it is to detect the PDF embedding attack, for the reasons mentioned earlier. However, there is improvement compared to the results in the previous table; this can be attributed to the new anomaly-based features introduced into the feature set. The significant improvement in the performance of the Stacking learning model highlights the impact of the anomaly-based features introduced into the mix. The new features mal_trait2, mal_trait3, and mal_traits_all are most likely to be responsible for enhancing the ability of the ensemble learners to detect more embedded EXE samples. The new feature mal_trait2 is likely to have had the most impact in improving the models' ability to detect embedded PDF samples. Figures 3–5 visually depict the percentages of detected samples with and without the new features for each of the three types of reverse mimicry attacks investigated.

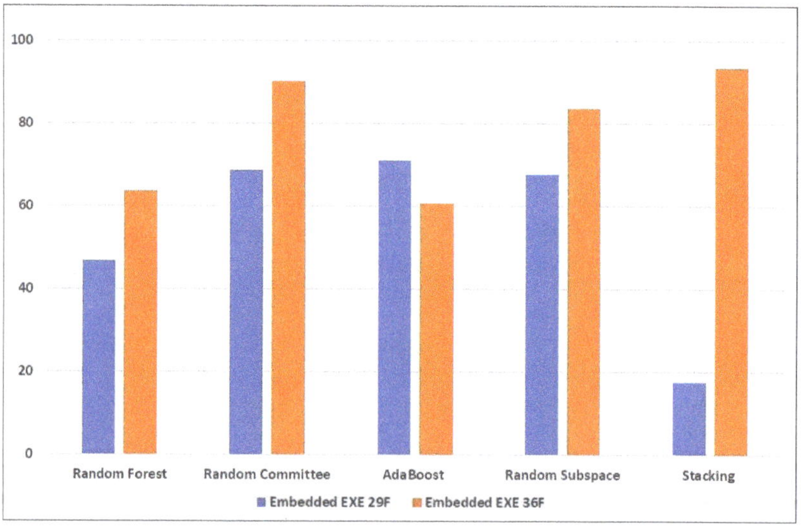

Figure 3. Performance of the ensemble learners on the embedded EXE reverse mimicry samples (with and without the new features).

5.4. *Experimenting with Training Set Augmentation*

At the initial stage of our investigation, we hypothesized that the detection of adversarial samples could be facilitated by augmenting the training set with some examples from the attack dataset. This is expected to enable the classifier models to learn the characteristics of the adversarial samples and be equipped to classify new unseen examples correctly. Based on this hypothesis, another set of experiments was performed, where 10% of the samples from each type of content injection attack set was taken and used to augment the training set. The results of the experiments are shown in Tables 7 and 8.

From Table 7, the results of the ensemble learners' performance when trained without the new features seem to confirm our hypothesis in the case of embedded EXE and embedded JavaScript detection of reverse mimicry attack detection. Random Forest and Random Committee models detected all the samples from both attacks. However, they still performed poorly when tested with the embedded PDF sample set, despite having augmented the training set with 50 samples from the embedded PDF set. Data augmentation of the training set with adversarial examples clearly made the detection of embedded EXE and embedded JavaScript mimicry attacks much easier to detect, even without the new features.

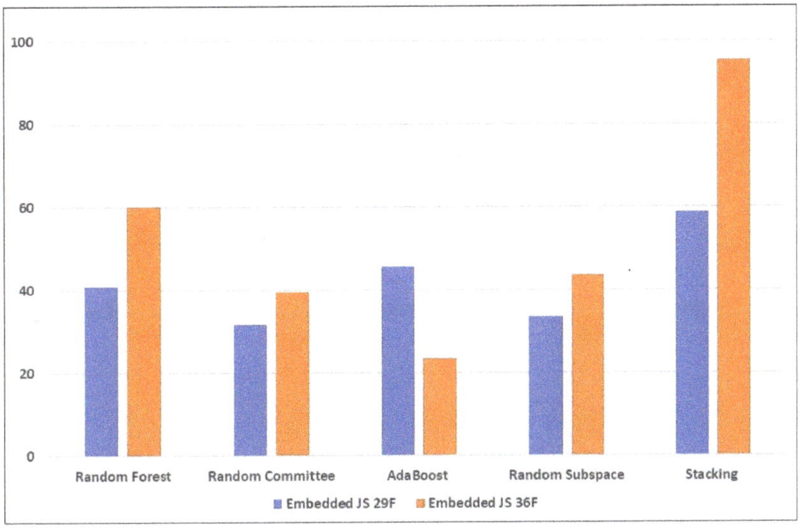

Figure 4. Performance of the ensemble learners on the embedded JS reverse mimicry samples (with and without the new features).

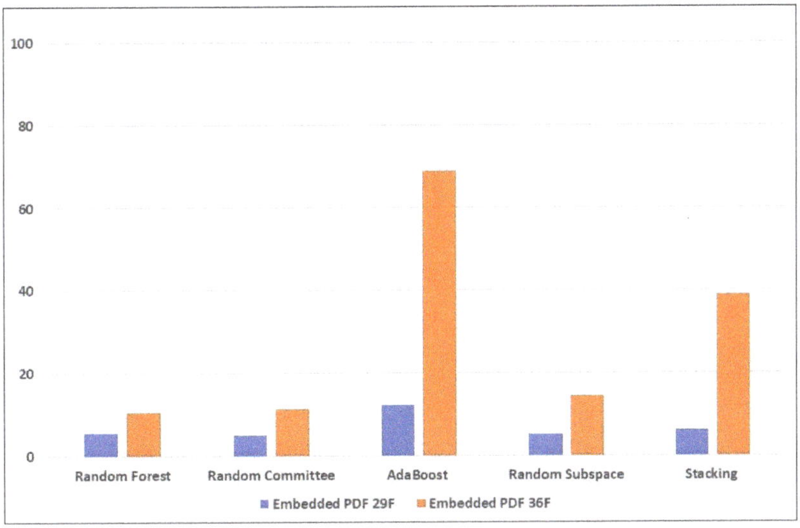

Figure 5. Performance of the ensemble learners on the embedded PDF reverse mimicry samples (with and without the new features).

In Table 8, the results show dramatic improvement when the new anomaly-based features were utilized in the training and testing sets (after augmenting the training set with adversarial samples). Figures 6–8 visually depict the percentages of detected samples with and without the new features, and with data augmentation for each of the three types of reverse mimicry attacks investigated. It can be seen that Random Committee detected 98% of the embedded PDF attacks compared to only 43.2% without the new features. These results show that data augmentation as a means to improve detection of adversarial samples would be more effective only if we have the right feature set. The possible reason for significant improvement in embedded PDF detection due to the new features can be explained as follows: it is highly likely that the combination of the anomaly-based features with other features produced new patterns that were learned by the ensemble models, and these patterns were present in the samples that the training set was augmented with. In a nutshell, we can conclude that the new anomaly-based features significantly enhanced the robustness of the ensemble learning models against reverse mimicry attacks via content injection.

Table 7. Reverse mimicry attack dataset—ensemble classifiers results with training set augmentation but without the new features.

	Embedded EXE (448)	Embedded JS (450)	Embedded PDF (449)
Random Forest	100% (448)	100% (450)	20.49% (92)
Random Committee	100% (448)	100% (450)	43.2% (194)
AdaBoost	92% (412)	98.7% (444)	27.39% (123)
Random Subspace	99.8% (447)	100% (450)	35.9% (161)
Stacking	97.1% (435)	100% (450)	12.9% (58)

Table 8. Reverse mimicry attack dataset—ensemble classifiers results with training set augmentation and the new features included.

	Embedded EXE (448)	Embedded JS (450)	Embedded PDF (449)
Random Forest	100% (448)	100% (450)	90.64% (407)
Random Committee	100% (448)	100% (450)	98% (440)
AdaBoost	99.3% (445)	99.8% (449)	97.6% (438)
Random Subspace	100% (448)	100% (450)	95.1% (427)
Stacking	96% (430)	100% (450)	85.7% (385)

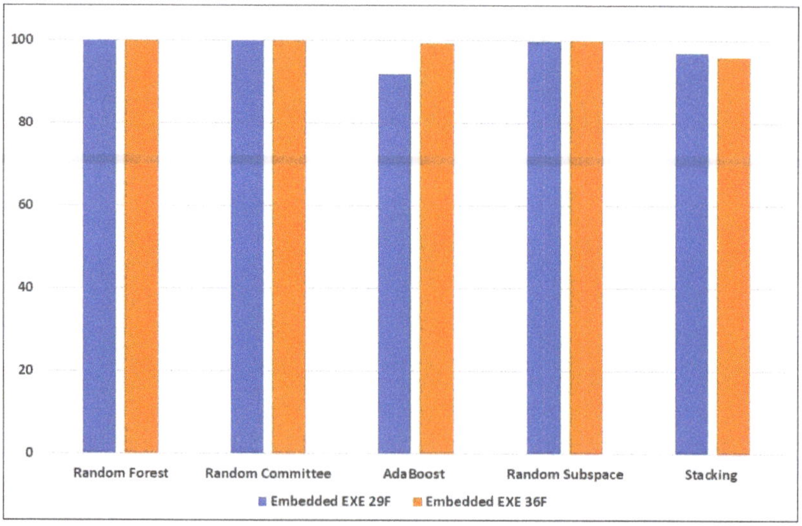

Figure 6. Performance of the ensemble learners on the embedded EXE reverse mimicry samples (with and without the new features), using training set augmented with attack samples.

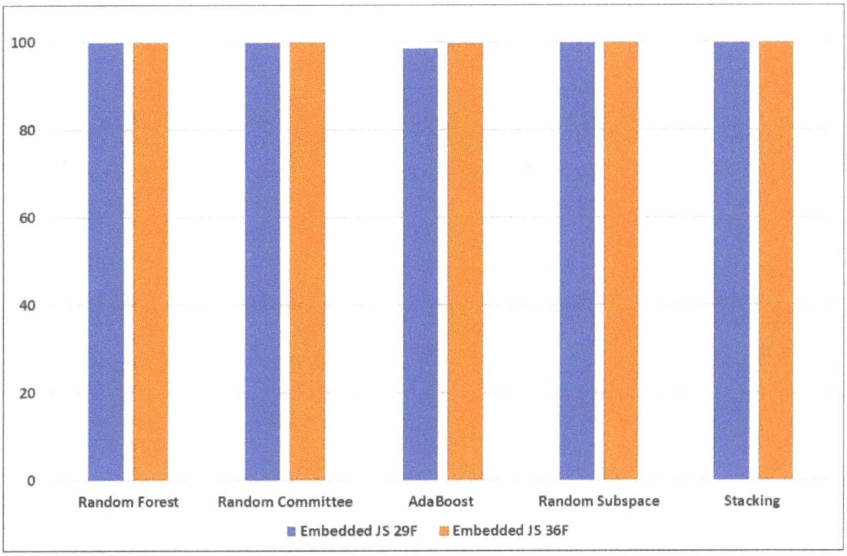

Figure 7. Performance of the ensemble learners on the embedded JS reverse mimicry samples (with and without the new features), using training set augmented with attack samples.

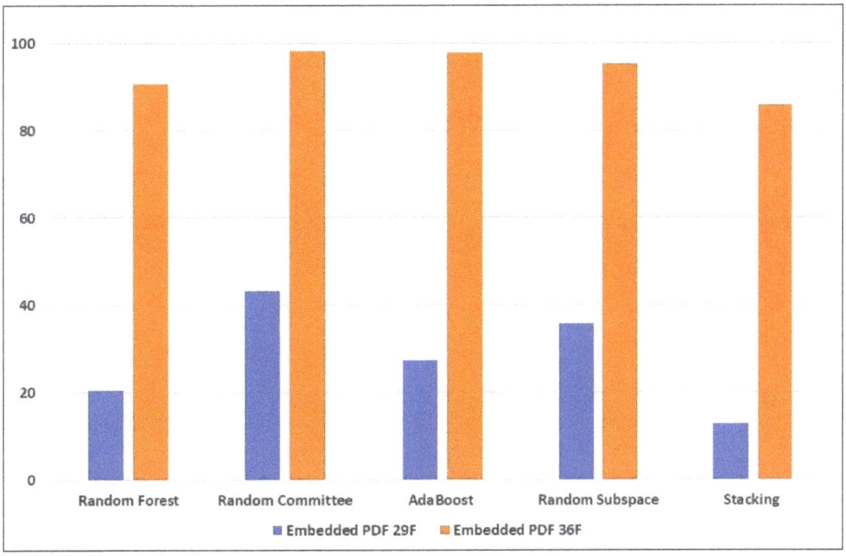

Figure 8. Performance of the ensemble learners on the embedded PDF reverse mimicry samples (with and without the new features), using training set augmented with attack samples.

5.5. Explaining and Interpreting the Ensemble Model Using SHapely Additive exPlanation

In this section we will explain the ensemble model for evasive malicious PDF detection using SHapely Additive exPlanation (SHAP). SHAP was introduced by Lundberg and Lee in 2017 [36] as a model-agnostic method of explaining machine learning models based on Shapley values taken from game theory. SHAP determines the impact of each feature by calculating the difference between the model's performance with and without the feature. Thus, it provides an understanding of how much each feature contributes to the prediction. In Figure 9 a SHAP summary chart can be seen from which we can visualize the importance of the features and their impact on predictions. This plot was generated from building an ensemble model with 80% of the dataset and testing on the remaining. The features are sorted in descending order of SHAP value magnitudes over all testing samples. The SHAP values are also used to show the distribution of the impacts each feature has on the prediction. The colour represents the feature value, with red indicating high while blue indicates low.

From Figure 9, we can see that metadata size, JavaScript, and mal_traits_all were the top three that had the most impact, according to the SHAP values. It can also explain that when the metadata size is low (blue) the model predicts positively, i.e., as a malicious PDF in most cases. However, when the metadata size is large (red), that impacts on the prediction by making the model classify documents as benign. We can also see that in most cases when Javascript or JS is present (red), the model predicts malicious PDF, while it predicts benign PDF if it is absent (blue). When there is mal_traits present (red), then malicious PDF is predicted, and when it is not present (blue), in many instances that led to a prediction of benign PDF. The same is true for mal_trait2. The plot also shows us that for many test samples, when text, images, and number of streams are low (blue endstream, stream) or number of objects are low (blue obj and endobj), then the PDF is likely to be predicted as malicious. For text, high values (red) indicate benign PDF in many cases; which makes sense because those will be genuine documents as opposed to crafted PDF that have been manipulated for nefarious purposes. The plot also shows us that the model predicts malicious PDF for many instances where XFA, OpenAction, and EmbeddedFile were present (red). Note that these plots only relate to the particular test set that was used and will be different from another test set which will have a different distribution of the features.

In Figure 10, the SHAP summary chart depicts the impacts of the top 10 features on an ensemble model's prediction on the embedded exe reverse mimicry test set. It shows the the presence of Acroform (which is indicative of potential manual input into the document) has a negative impact on the prediction (i.e., benign is predicted) while the opposite is true. This also happens when metadata size is large (red) or there is a large number of objects or pages in the PDF document. The presence of mal_traits (i.e., any of the new anomaly features) leads to a positive prediction, and so does the presence of embedded files.

The SHAP summary chart in Figure 11 depicts the impacts of the top 10 features on an ensemble model's prediction on the embedded pdf reverse mimicry test set. It shows that high number of streams (endstream and stream being red) indicates malicious PDF while in some cases low number of streams does also indicate malicious PDF. This could mean that a combination with other features influences the prediction, or some of these instances could be incorrectly classified. The figure also shows us that positive predictions (i.e., malicious PDF) are made when metadata size is low, title characters are absent when PDF size is large (which can be an indicator for embedded PDF) and when OpenAction (which could be used to manipulate the embedded PDF) is present (red).

In Figure 12, the impacts of the top 10 features on an ensemble model's prediction on the embedded JavaScript reverse mimicry test set is shown. The model's positive (malicious PDF) prediction can be explained by seeing low metadata size, smaller number of objects, fewer title characters, fewer streams, and the absence of Acroform. The presence of Acroform, JS keyword, and AA feature seem to be indicators of negative (benign PDF) prediction amongst the samples of embedded JavaScript from this reverse mimicry dataset used to analyze the model with SHAP. In a nutshell, these SHAP summary charts demonstrate the explainability of the models which is crucial in increasing the trust of our proposed approach while giving us insight into the models' decision-making.

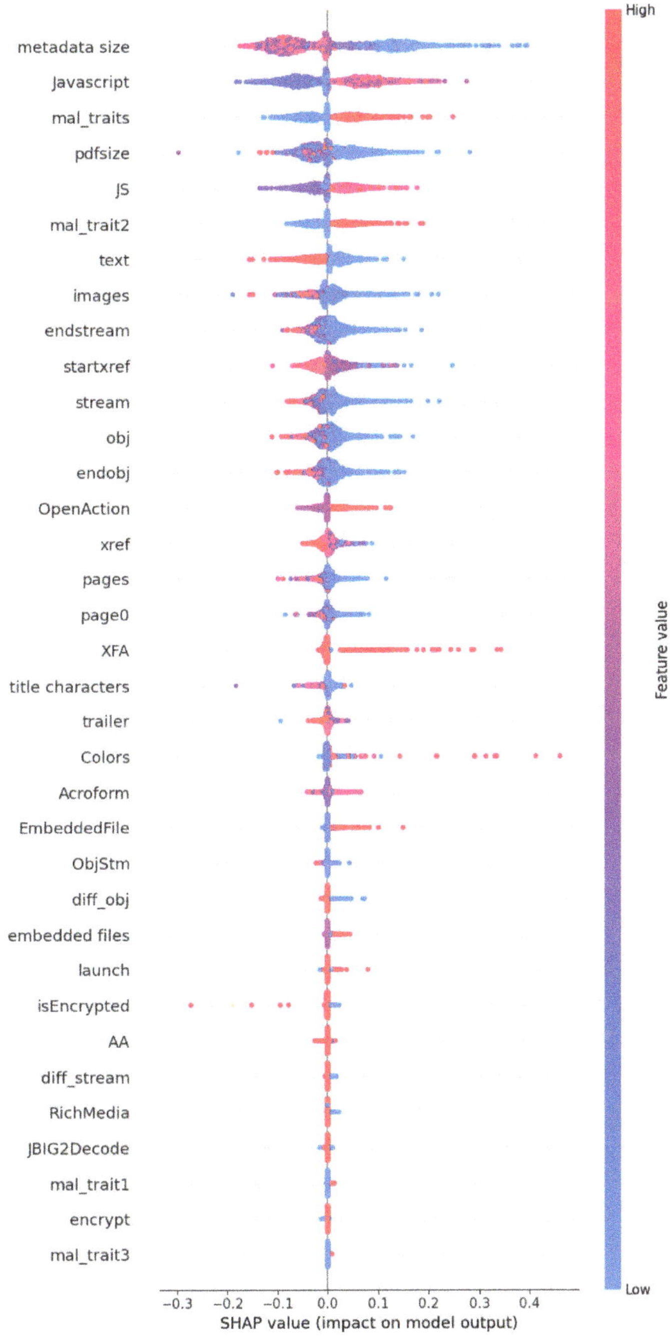

Figure 9. Each features impact on model's predictions as determined by SHAP, for the ensemble model's prediction on test samples that do not contain reverse mimicry content injection.

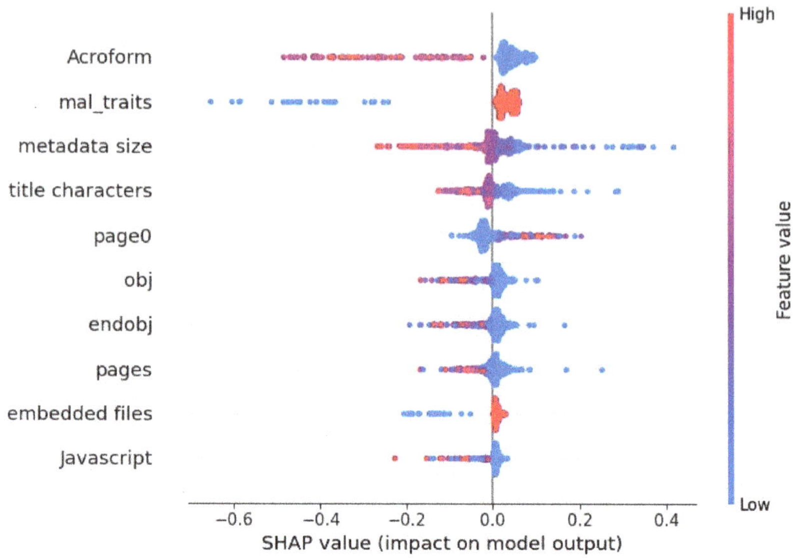

Figure 10. Each features impact on model's predictions as determined by SHAP, for the ensemble model's prediction on test samples that consist of embedded exe within the pdf files.

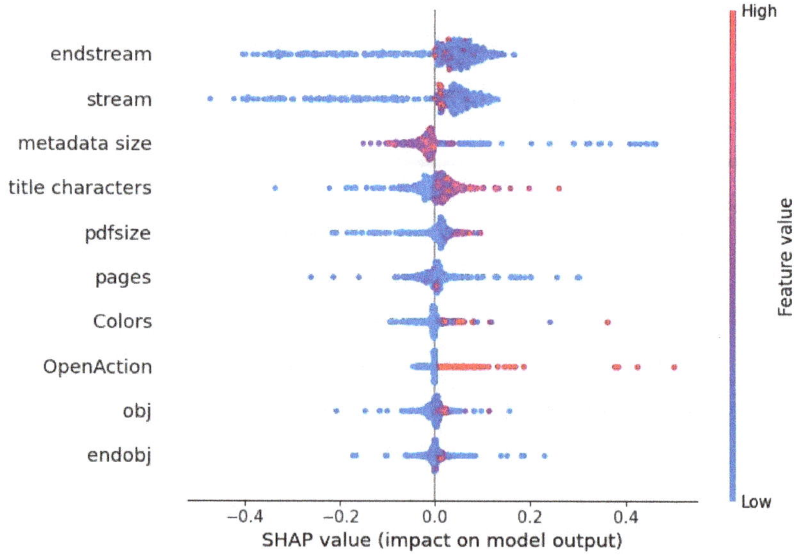

Figure 11. Each features impact on model's predictions as determined by SHAP, for the ensemble model's prediction on test samples that consist of embedded pdf within the pdf files.

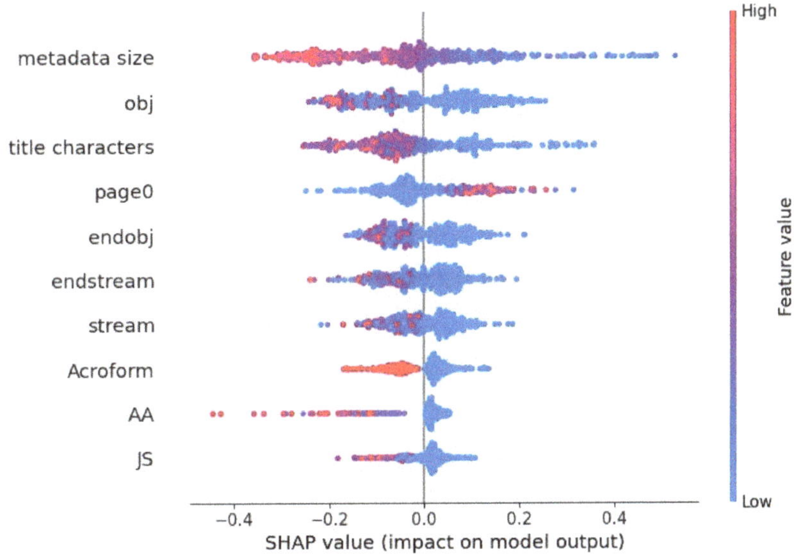

Figure 12. Each features impact on model's predictions as determined by SHAP, for the ensemble model's prediction on test samples that consist of embedded JavaScript within the pdf files.

5.6. Comparing Our Results with Existing Works

In this section we present a comparison of our approach to other works in the literature using the reverse mimicry dataset. In a reverse mimicry attack, malicious content is injected into a benign file. This type of attack does not exploit any specific knowledge of the attacked system [17]. The dataset of 1500 evasive PDF files with injected content was used in [17] to evaluate several existing PDF detectors (Hidost, PJScan, PDFRate, Slayer Neo—Keyword and Full versions).

Corum et al. [30] employed visualization techniques for PDF malware detection. The results of their model on the reverse mimicry dataset is shown in Table 9. The best result from that paper was obtained from using Byte plot + Gabor + Random Forest classifier, which detected only 95 out of about 500 samples for EXE embedding; 176 out of 500 for JS embedding; and 111 out of 500 for PDF embedding. In Table 9 and Figure 13 this approach is named as Corum-BGR. Their Byte plot + Local entropy + Random Forest approach detected only 70 out of 500 for EXE embedding; 162 out of 500 for JS embedding; and 85 out of 500 for PDF embedding. In Table 9 and Figure 13 this approach is named as Corum-BLR.

From Table 9, the results of the first two rows indicate that the visualization techniques which were reported to have achieved high accuracies on the Contagio PDF dataset performed poorly when tested with the reverse mimicry attack samples. On the other hand, even though Slayer Neo struggled to detect embedded EXE and embedded JavaScript, it was quite effective in detecting embedded PDF attacks. This is because of the way the system was designed. Moreover, note that PDFRate was the only system that detected a high percentage of Embedded JavaScript. This is because it was created specifically to detect JavaScript-bearing malicious PDF files. Note that these tools and approaches constitute the stat-of-the-art in the domain of malicious PDF document detection. Table 9 and Figure 13 both illustrate that the approach proposed in this paper has outperformed these state-of-the-art methods in terms of resilience to reverse mimicry content injection attacks.

Table 9. Reverse mimicry attack dataset—comparison with existing works.

	Embedded EXE	Embedded JS	Embedded PDF
Corum-BGR [14]	19% (95)	35.2% (176)	22.2% (111)
Corum-BLR [14]	14% (70)	32.4% (162)	17% (85)
Hidost	69%	40.8%	1%
PJScan	1%	87.7%	3%
PDFRate	95.2%	28.8%	1.2%
Slayer Neo (Keywords) [17]	59.8%	17.8%	94.8%
Slayer Neo (Full) [17]	9.4%	35.9%	96%
Stacking with our enhanced feature set	93.6% (466)	95.4% (477)	38.9% (194)
Random Committee with our enhanced feature set and training set augmentation	100% (448)	100% (450)	98% (440)

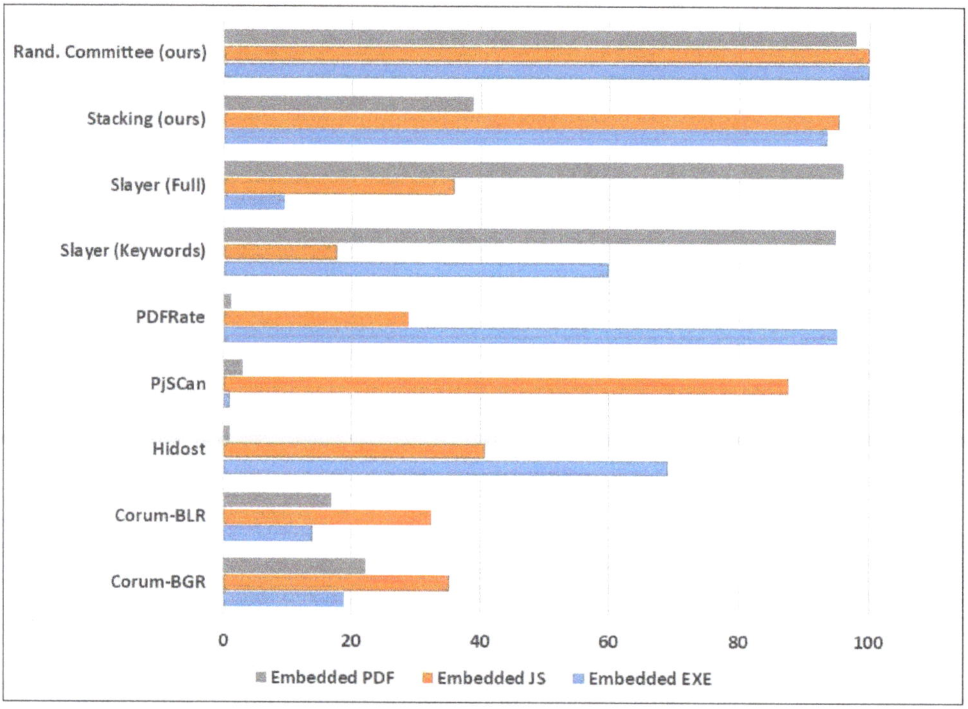

Figure 13. Performance of the ensemble learners on the reverse mimicry samples, with training set augmented with attack samples.

6. Limitations of Our Proposed Approach

In this section we discuss the limitations of the proposed approach presented in this paper. The first one is that the extraction of the features is reliant upon existing static analysis tools (i.e., PDFMaLyzer, which in turn utilizes PDFiD). This means that the approach is prone to the limitations of these tools as well. Hence, if a feature can be hidden from those tools, it would affect extraction in our proposed system as well. However, the extended anomaly features set is designed to counteract the tools' failure to some extent. A direction for future improvement is to make the underlying feature extraction tools more robust or to extract a hybrid of complementary features that the system can use to make it resilient to such failures. Another limitation of our proposed approach is that it could be susceptible to obfuscation, whereby some of the features could be masked. A possible countermeasure for this is to perform content analysis rather than relying solely on extraction of such features from structural keywords. The content analysis-based features could also provide a hybrid composite features approach when combined with the structural and anomaly-based features.

7. Conclusions and Recommendation for Future Work

In this paper we presented a malicious PDF detection system based on ensemble learning with an enhanced feature set. The enhanced feature set consists of 6 new anomaly-based features which we have added to 29 structural features derived from existing PDF static analysis tools. In the first part of our experiments, the results have shown that the introduction of the new features did not diminish performance after testing five ensemble learning algorithms using the Evasive-PDFMal2022 dataset. The second part of our experiments performed on the PDF reverse mimicry dataset showed the robustness of the new features against content injection attacks designed to disguise malicious content by embedding them within benign PDF files. By comparing our results with existing approaches including Hidost, PJScan, PDFrate, Slayer Noe, and other approaches, there was a significant improvement in detection rates by our proposed approach. The experiments conducted on the reverse mimicry dataset showed that the Random Committee ensemble learning model achieved 100% detection rates for embedded EXE and embedded JavaScript, and 98% detection rate for embedded PDF, based on our enhanced feature set. The experiments also showed that data augmentation will not enhance the detection of adversarial samples unless accompanied by effective feature engineering, which our system incorporates through the new anomaly-based features. For future work, we recommend investigating how to improve the resilience of other types of existing PDF detection systems, e.g., those that utilize visualization approaches, to incorporate more resilience against reverse mimicry attacks. Another recommendation for future work is on how to extend the system proposed in this paper with content-based features.

Author Contributions: Conceptualization, S.Y.Y.; Methodology, S.Y.Y. and A.B.; Software, S.Y.Y. and A.B.; Validation, A.B.; Formal analysis, A.B.; Resources, A.B.; Writing—original draft, S.Y.Y. and A.B.; Writing—review & editing, A.B. All authors have read and agreed to the published version of the manuscript.

Funding: This research received no external funding.

Data Availability Statement: The original datasets used in this research work were taken from the public domain. The pre-processed datasets are available on request from the author.

Acknowledgments: This work is supported in part by the 2022 Cybersecurity research grant number PCC-Grant-202228, from the Cybersecurity Center at Prince Mohammad Bin Fahd University, Al-Khobar, Saudi Arabia.

Conflicts of Interest: The authors declare no conflict of interest.

References

1. Singh, P.; Tapaswi, S.; Gupta, S. Malware detection in PDF and office documents: A survey. *Inf. Secur. Glob. Perspect.* **2020**, *29*, 134–153. [CrossRef]
2. Goud, N. Cyber Attack with Ransomware Hidden Inside PDF Documents. Available online: https://www.cybersecurity-insiders.com/cyber-attack-with-ransomware-hidden-inside-pdf-documents/ (accessed on 31 October 2022).
3. Nath, H.V.; Mehtre, B. Ensemble learning for detection of malicious content embedded in pdf documents. In Proceedings of the 2015 IEEE International Conference on Signal Processing, Informatics, Communication and Energy Systems (SPICES), Kozhikode, India, 19–21 February 2015; pp. 1–5.
4. Mimoso, M. MiniDuke Espionage Malware Hits Governments in Europe Using Adobe Exploits. Available online: https://threatpost.com/miniduke-espionage-malware-hits-governments-europe-using-adobe-exploits-022713/77569/ (accessed on 31 October 2022).
5. Stevens, D. PDF Tools. Available online: https://blog.didierstevens.com/programs/pdf-tools/ (accessed on 25 September 2022).
6. Stevens, D. Peepdf—PDF Analysis Tool. Available online: https://eternal-todo.com/tools/peepdf-pdf-analysis-tool (accessed on 25 September 2022).

7. Bandla, K. PhoneyPDF: A Virtual PDF Analysis Framework. Available online: https://github.com/kbandla/phoneypdf (accessed on 8 November 2022).
8. Gdelugre, G. PDF Walker: Frontend to Explore the Internals of a PDF Document with Origami. Available online: https://github.com/gdelugre/pdfwalker (accessed on 25 September 2022).
9. Laskov, P.; Srndic, N. Static Detection of Malicious JavaScript-Bearing PDF Documents. In Proceedings of the 27th Annual Computer Security Applications Conference (ACSAC 2011), Orlando, FL, USA, 5–9 December 2011; pp. 373–382.
10. Corona, I.; Maiorca, D.; Ariu, D.; Giacinto, G. Detection of malicious pdf-embedded javascript code through discriminant analysis of api references. In Proceedings of the 2014 Workshop on Artificial Intelligent and Security Workshop, Scottsdale, AZ, USA, 7 November 2014; pp. 47–57.
11. Maiorca, D.; Ariu, D.; Corona, I. A pattern recognition system for malicious pdf files detection. In *Machine Learning and Data Mining in Pattern Recognition*; Springer: Berlin/Heidelberg, Germany, 2012; pp. 510–524.
12. Srndic, N.; Laskov, P. Hidost: A Static Machine-learning-based Detector of Malicious Files. *Eurasip J. Inf. Secur.* **2016**, *2016*, 22. [CrossRef]
13. Smutz, C.; Stavrou, A. Malicious PDF detection using meta-data and structural features. In Proceedings of the 28th Annual Computer Security Applications Conference (ACSAC 2012), Orlando, FL, USA, 3–7 December 2012; pp. 239–248.
14. Corum, A.; Jenkins, D.; Zheng, J. Robust PDF Malware Detection with Image Visualization and Processing Techniques. In Proceedings of the 2019 2nd International Conference on Data Intelligence and Security (ICDIS), South Padre Island, TX, USA, 28–30 June 2019; pp. 1–5.
15. Kang, H.; Yuefei, Z.; Yubo, H.; Long, L.; Bin, L.; Wei, L. Detection of Malicious PDF Files Using a Two-Stage Machine Learning Algorithm. *Chin. J. Electron.* **2020**, *28*, 1165–1177.
16. *ISO 32000-1:2008*; Document Management—Portable Document Format—Part 1: PDF 1.7. ISO: Geneva, Switzerland, 2008. Available online: https://www.iso.org/standard/51502.html (accessed on 30 October 2022).
17. Maiorca, D.; Biggio, B. Digital Investigation of PDF Files: Unveiling Traces of Embedded Malware. *IEEE Secur. Priv. Mag. Spec. Issue Digit. Forensics* **2017**, *17*, 63–71. [CrossRef]
18. Khitan, S.J.; Hadi, A.; Atoum, J. PDF Forensic Analysis System using YARA. *Int. J. Comput. Sci. Netw. Secur.* **2017**, *17*, 77–85.
19. Jeong, Y.; Woo, J.; Kang, A. Malware detection on byte streams of pdf files using convolutional neural networks. *Secur. Commun. Netw.* **2019**, *2019*, 8485365. [CrossRef]
20. Albahar, M.; Thanoon, M.; Alzilai, M.; Alrehily, A.; Alfaar, M.; Alghamdi, M.; Alassaf, N. Toward Robust Classifiers for PDF Malware Detection. *Comput. Mater. Contin.* **2021**, *69*. [CrossRef]
21. Bazzi, A.; Yoshikuni, O. Automatic Detection of Malicious PDF Files Using Dynamic Analysis. In Proceedings of the JSST 2013 International Conference on Simulation Technology, Tokyo, Japan, 11 September 2013; pp. 3–4.
22. Falah, A.; Pan, L.; Huda, S.; Pokhrel, S.R.; Anwar, A. Improving Malicious PDF Classifier with Feature Engineering: A Data-Driven Approach. *Future Gener. Comput. Syst.* **2021**, *115*, 314–326. [CrossRef]
23. Jason, Z. MLPdf: An Effective Machine Learning Based Approach for PDF Malware Detection. *arXiv* **2018**, arXiv:1808.06991.
24. Jiang, J.; Song, N.; Yu, M.; Liu, C.; Huang, W. Detecting malicious pdf documents using semi-supervised machine learning. In *Advances in Digital Forensics XVII. Digital Forensics 2021. IFIP Advances in Information and Communication Technology*; Peterson, G., Shenoi, S., Eds.; Springer: Berlin/Heidelberg, Germany, 2021; Volume 612. [CrossRef]
25. Torres, J.; Santos, S. Malicious PDF Documents Detection using Machine Learning Techniques—A Practical Approach with Cloud Computing Applications. In Proceedings of the 4th International Conference on Information Systems Security and Privacy (ICISSP 2018), Funchal, Portugal, 22–24 January 2018; pp. 337–344. [CrossRef]
26. Abu Al-Haija, Q.; Odeh, A.; Qattous, H. PDF Malware Detection Based on Optimizable Decision Trees. *Electronics* **2022**, *11*, 3142. [CrossRef]
27. CIC. PDF Dataset: CIC-Evasive-PDFMal2022. Available online: https://www.unb.ca/cic/datasets/PDFMal-2022.html (accessed on 25 September 2022).
28. Yerima, S.Y.; Bashar, A.; Latif, G. Malicious PDF detection Based on Machine Learning with Enhanced Feature Set. In Proceedings of the 14th International Conference on Computational Intelligence and Communication Networks, Al-Khobar, Saudi Arabia, 4–6 December 2022.
29. Issakhani, M.; Victor, P.; Tekeoglu, A.; Lashkari, A.H. PDF Malware Detection based on Stacking Learning. In Proceedings of the 8th International Conference on Information Systems Security and Privacy (ICISSP 2022), Online, 9–11 February 2022; pp. 562–570.
30. PRALAB. PDF Reverse Mimicry Dataset. Available online: https://pralab.diee.unica.it/en/pdf-reverse-mimicry/ (accessed on 31 October 2022).
31. Lashkari, A.H. PDFMALyzer. Available online: https://github.com/ahlashkari/PDFMalLyzer (accessed on 25 September 2022).
32. Ho, T.K. The Random Subspace Method for Constructing Decision Forests. *IEEE Trans. Pattern Anal. Mach. Intell.* **1998**, *20*, 832–844.
33. Breiman, L. Random forests. *Mach. Learn.* **2001**, *45*, 5–32. [CrossRef]
34. Freund, Y.; Schapire, R.E. Experiments with a new boosting algorithm. In Proceedings of the Thirteenth International Conference on Machine Learning, Bari, Italy, 3–6 July 1996; pp. 148–156.

35. Wolpert, D.H. Stacked generalization. *Neural Netw.* **1992**, *45*, 214–259. [CrossRef]
36. Lundberg, S.M.; Lee, S.I. A Unified Approach to interpreting model predictions. In Proceedings of the 31st International Conference on Neural Information Processing Systems, Long Beach, CA, USA, 4–9 December 2017; Volume 30.

electronics

MDPI

Article

SPE-ACGAN: A Resampling Approach for Class Imbalance Problem in Network Intrusion Detection Systems

Hao Yang [1], Jinyan Xu [2], Yongcai Xiao [1] and Lei Hu [2,*]

[1] State Grid Jiangxi Electric Power Research Institute, Nanchang 330096, China; birdeye@163.com (H.Y.); xiaoyongcaixidian@163.com (Y.X.)
[2] School of Computer and Information Engineering, Jiangxi Normal University, Nanchang 330022, China; x_jinyan@163.com
* Correspondence: hulei@jxnu.edu.cn

Abstract: Network Intrusion Detection Systems (NIDSs) play a vital role in detecting and stopping network attacks. However, the prevalent imbalance of training samples in network traffic interferes with NIDS detection performance. This paper proposes a resampling method based on Self-Paced Ensemble and Auxiliary Classifier Generative Adversarial Networks (SPE-ACGAN) to address the imbalance problem of sample classes. To deal with the class imbalance problem, SPE-ACGAN oversamples the minority class samples by ACGAN and undersamples the majority class samples by SPE. In addition, we merged the CICIDS-2017 dataset and the CICIDS-2018 dataset into a more imbalanced dataset named CICIDS-17-18 and validated the effectiveness of the proposed method using the three datasets mentioned above. SPE-ACGAN is more effective than other resampling methods in improving NIDS detection performance. In particular, SPE-ACGAN improved the F1-score of Random Forest, CNN, GoogLeNet, and CNN + WDLSTM by 5.59%, 3.75%, 3.60%, and 3.56% after resampling.

Keywords: network intrusion detection system; imbalanced network traffic; resampling method

Citation: Yang, H.; Xu, J.; Xiao, Y.; Hu, L. SPE-ACGAN: A Resampling Approach for Class Imbalance Problem in Network Intrusion Detection Systems. *Electronics* **2023**, *12*, 3323. https://doi.org/10.3390/electronics12153323

Academic Editors: Dariusz Rzońca and Tomasz Rak

Received: 18 June 2023
Revised: 31 July 2023
Accepted: 1 August 2023
Published: 3 August 2023

1. Introduction

Since 2020, Corona Virus Disease 2019 (COVID-19) has spread worldwide, dramatically changing people's lifestyles, and forcing people to shift their learning, work, and entertainment activities from offline to online. However, with the continuous development of the network, a series of constantly evolving network attack means, such as worms and buffer overflow, threaten the transportation, energy, education, medical and other industries. Many companies and organizations lack the experience and skills of synchronized confrontation with network attacks, so it is difficult to detect these network attacks. NIDS is a monitoring system for network traffic, which can detect suspicious network attack activities from network traffic and respond to alerts in a timely manner to protect the network before hackers intrude.

There are three main types of NIDS: misuse-based, anomaly-based and hybrid NIDS [1]. Misuse-based NIDS [2,3] uses a library of features or fingerprints of known attacks to match each traffic feature or fingerprint, and if the match is successful, the traffic is determined to be malicious. Anomaly-based NIDS [4,5] models normal behavior and does not require attacks to be explicitly identified in the training data. The model describes traffic activity in the normal state of the protected system, and any network traffic that does not match the behavior described by the model is captured and reported. Hybrid NIDS [6,7] uses both misuse and anomaly-based, which allows the NIDS to have a lower false alarm rate and higher accuracy than the above two methods alone. The use of deep learning to implement hybrid NIDS is the dominant approach today, and by learning the characteristics within the network traffic, attack signatures can be obtained and anomalous behavior can also be identified. The use of hybrid NIDS can compensate for the shortcomings of misuse-based NIDS and anomaly-based Anomaly NIDS. Deep learning models such as Convolutional Neural Networks (CNNs) [8], Long Short-Term Memory Networks (LSTM) [9] and GoogLeNet [10] have been proven to be effective in detecting attacks.

However, the extremely unbalanced distribution of network traffic [11,12] greatly hinders further development of deep learning-based intrusion detection research: the behavior of Internet users is almost normal, with only a small number of malicious attacks. As a result, a large sample of traffic is generated in cyberspace, but only a small percentage is malicious, and the distribution between different malicious traffic is also uneven. Deep learning network models require a large number of samples for training, and they have more robust performance when a large amount of data are available for training, but the performance of deep learning algorithms also degrades significantly when learning unbalanced data [13–15]. In general, the lack of a certain class of network traffic may cause NIDS to favor the majority class of samples and neglect learning from the minority class. Therefore, balancing network traffic is necessary for NIDS to fully learn the features of each class of samples.

Many studies have resampled network traffic when training the NIDS model: oversampling the minority class of samples or undersampling the majority class of samples, such as the Synthetic Minority Over-Sampling Technique (SMOTE) [16], Random Undersampling (RUS) [17] and Generative Adversarial Networks (GANs) [18]. SMOTE requires traversing each minority class sample and selecting one sample to calculate its distance from neighboring samples. This is very resource intensive in a data set with a large volume of data. In addition, the samples generated via interpolation increase the possibility of overlapping samples of each class and the possibility of overfitting of the classification mode. RUS removes most class samples in a random way, so it may remove samples that are on the classification boundary, which may result in information loss. The samples generated by GAN are random in nature and cannot be generated on demand, so most of the samples generated have difficulty fitting the features of the minority class of samples.

In this work, we introduced a novel resampling method, SPE-ACGAN, based on the combination of Auxiliary Classifier GAN (ACGAN) [19] and Self-Paced Ensemble (SPE) [20] to deal with the problem of imbalanced network traffic. The imbalanced datasets are divided into a minority class subset and a majority class subset, and then the minority class subset is fed into ACGAN to generate the specified number of samples, and the majority class subset is fed into SPE to remove the majority class samples until the number of samples reaches the specified value.

The main contributions of this work are described as follows:

- For NIDS, a resampling method SPE-ACGAN based on the combination of SPE and ACGAN is proposed to alleviate the data imbalance problem, which is able to reduce the majority class samples and increase the minority class samples to make the training set more balanced.
- We merge the CICIDS-2017 dataset and the CICIDS-2018 dataset into a new dataset, named CICIDS-17-18. The CICIDS-17-18 dataset is a more imbalanced dataset with a larger amount of data to show the effectiveness of SPE-ACGAN.
- Our proposed method is experimented on the above three datasets and compared with some existing resampling methods. The performance metrics of some typical NIDS models are improved after applying our proposed method.

The next section, Section 2, discusses the existing methodology. Section 3 presents the proposed method. Section 4 compares and analyses the performance of the proposed method and existing methods. Finally, Section 5 concludes the whole paper.

2. Related Work

Nowadays, for NIDS, the deep learning-based method is the essential classification model to identify different types of attacks. An imbalance of training samples can lead to overfitting of the classification model and affect its generalization ability. We introduce the related work from the deep learning-based method and sample resampling.

Deep learning-based network intrusion detection is mainly based on training models to learn potential features of data samples for classification and prediction purposes, which can be divided into supervised and unsupervised learning. Supervised learning includes Long and Short-Term Memory networks (LSTM) [9], Convolutional Neural Networks (CNNs) [8], etc. Yang et al. [21] proposed a Gradient-Boosting Decision Tree (GBDT)–parallel quadratic ensemble learning method for intrusion detection systems with a Gated Recurrent Unit (GRU) model and special modification to network traffic to handle temporal data. Experimental results based on the CICIDS2017 dataset show that the advanced temporal intrusion detection system based on integrated learning achieves better accuracy, recall, precision and F1 scores compared to existing methods. Unsupervised learning mainly consists of Auto Encoder (AE) [22,23] and Self-supervised Learning (SSL), which can learn from a large number of unlabeled samples and also effectively learn the features of different classes of traffic data [24]. Vaiyapuri et al. [25] proposed an unsupervised IDS model that uses deep autoencoder (DAE) to learn traffic features and then uses one class support vector machine (OCSVM) to segment the decision hyperplane, using the NSL-KDD dataset and UNSW-NB15 dataset. The proposed model was verified as having good performance.

Considering the impact of data imbalance, the minority class of samples will tend to be overfitted during training, and the model prediction will be more biased towards the majority of samples, which is less accurate in identifying malicious attacks. Therefore, many scholars have started to study how to solve the problem of extremely unbalanced data distribution of network traffic. Yan et al. [26] proposed an improved locally adaptive composite minority sampling algorithm (LA-SMOTE) to deal with network traffic imbalance and then detected network traffic anomalies based on a deep learning GRU neural network. Abdulhammed et al. [27] used data oversampling and undersampling methods to deal with the imbalanced dataset CIDDS-001 and used a deep neural network, random Forests and variational autoencoder classifiers to evaluate the dataset. Ga et al. [28] used ACGAN for minority class sample synthesis on the CICIDS-2017 dataset and then used CNN for classification to achieve the final OA and F1-score of 99.48% and 98.71%. Park et al. [29] used GAN for data synthesis on the minority class attack data in the training phase and then used Auto Encoder (AE) to optimize the generated data, and experimental results on NSL-KDD, UNSW-NB15 and IoT datasets show that reasonably increasing data can improve the performance of existing deep learning-based NIDS by solving the data imbalance problem. Table 1 provides a comparison of typical resampling methods.

Table 1. The typical resampling methods.

Method	Oversampling	Undersampling
SMOTE	√	
RUS		√
GAN	√	
SPE-ACGAN (our method)	√	√

In this paper, for NIDS models, we propose SPE-ACGAN, a resampling method based on a combination of supervised learning ACGAN and SPE, to resample unbalanced network traffic in order to solve the problem of unbalanced network work traffic. The ACGAN network adds to the GAN network the ability to generate a specified class, which can generate the minority class samples of a specified category, and its discriminator continuously improves the quality of the data it generates. SPE efficiently reduces the number of majority class samples and is able to retain most of the samples that are on the classification boundary.

3. Methods and Materials

In this section, the general structure of SPE-ACGAN proposed in this paper and the working principle of each module are first introduced. Then, the datasets used in the proposed algorithm and the implementation details of the algorithm, are presented.

3.1. SPE-ACGAN

Considering the unbalance of training samples of the network traffic in NIDS, we resample the training samples from two dimensions. Using SPE to decrease the number of samples in the majority class and using ACGAN to increase the number of samples in the minority class.

3.1.1. SPE

SPE is a framework for unbalanced classification [20], the core idea of which is to propose a concept of classification hardness and to coordinate data hardness by undersampling self-paced to generate a new undersampled dataset. The process of SPE is shown in Figure 1.

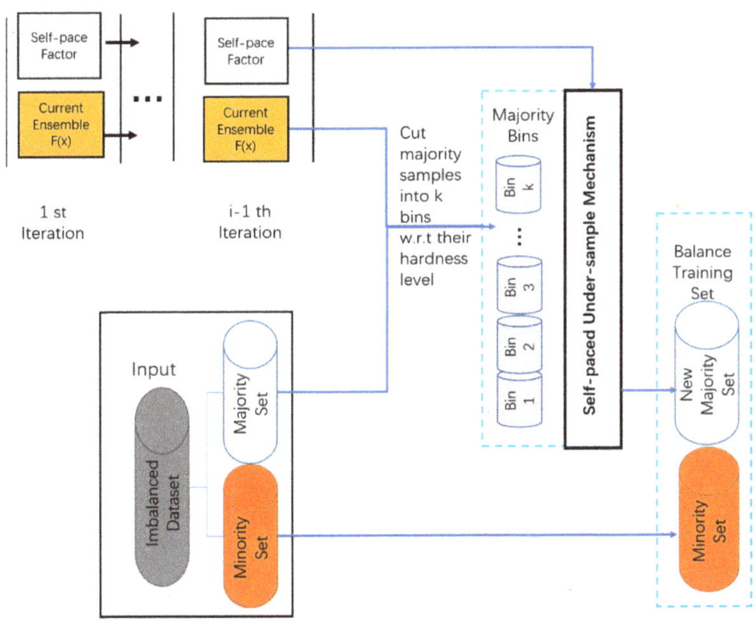

Figure 1. The process of SPE.

SPE first divides the input, the imbalanced dataset, into a majority set N and a minority set P. Then, each sample in N is randomly placed into k bins according to the categorical hardness of each sample, and each bin has a total categorical hardness. The above steps are repeated until the number of majority class samples equals the number of minority class samples or the specified number to complete the resampling. To obtain a balanced dataset, SPE keeps the total categorical hardness of each bin as the same. The hardness value is derived from a hardness function, H, which is a "categorical hardness function", such as Absolute Error, MSE and Cross Entropy. For a given model $F(x)$, the categorical hardness of the sample (x, y) is given by Equation (1):

$$\mathcal{H}_x = \mathcal{H}(x, y, F) \tag{1}$$

The hardness grade is given by Equation (2), where B_ℓ is the hardness grade of the ℓ-th box:

$$B_\ell = \left\{ (x,y) \middle| \frac{\ell-1}{k} \leq \mathcal{H}_x = \mathcal{H}(x,y,F) \leq \frac{\ell}{k} \right\} \mathcal{H}(\cdot) \in [0,1] \quad (2)$$

3.1.2. ACGAN

ACGAN is mainly composed of Generation (G) and Discrimination (D), and its structure is shown in Figure 2. ACGAN works as follows: ACGAN works by the network generating a random set of noise values z. According to the input of the specified category, the generator G modifies the noise values z into X_{fake} of the corresponding category, and the discriminator D trained by X_{real} to identify whether the generated X_{fake} it real data, and if it is virtual data, what is the probability of belonging to each category, respectively, and the error is found by the loss function. The generator G is instructed to update the parameters.

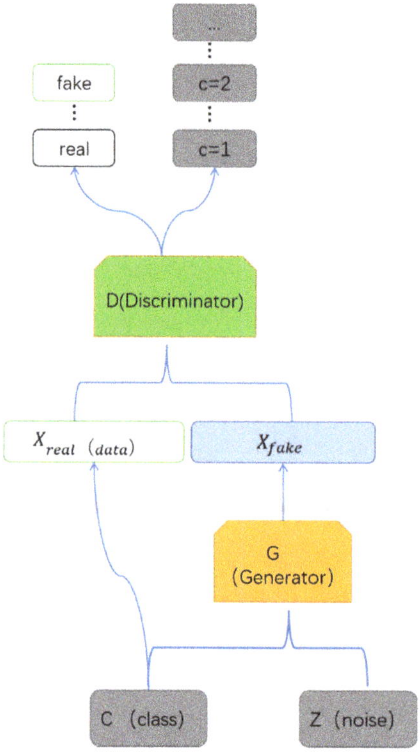

Figure 2. The architecture of ACGAN.

3.1.3. Overall Model Architecture

The SPE-ACGAN resampling method works in two steps, which are performed by ACGAN and SPE, respectively. The first step is the oversampling of the minority class samples, which is fed into ACGAN to increase the number of minority class samples; the second step is the undersampling of the majority class samples, which is fed into SPE to reduce the number of majority class samples. The workflow of the SPE-ACGAN resampling method is shown in Figure 3.

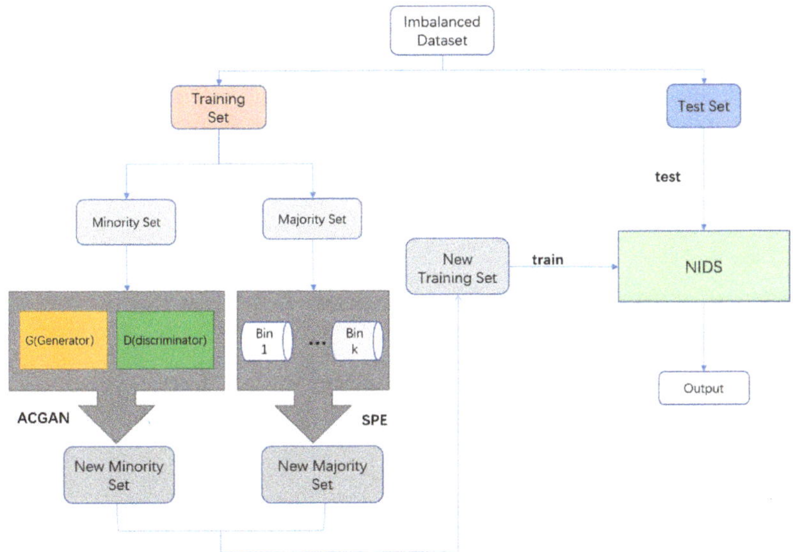

Figure 3. The architecture of the proposed resampling model.

3.2. Details of the SPE-ACGAN

3.2.1. Dataset

CICIDS-2017 and CICIDS-2018 [30] are network intrusion detection datasets published by the Canadian Institute for Cybersecurity (CIC) on Amazon Web Services (AWS) in 2017 and 2018. The two datasets mentioned above have as many as 14 types of attacks, such as DDOS, XSS, Heartbleed and Brute force, each over 100 GB in size, which make them the richest datasets of all publicly available datasets in terms of category.

In addition, there is an imbalance in the samples for each category in both of these datasets. In the CICIDS-2017 dataset, the lowest number of Heartbleed, Infiltration and Web Attack-Sql Injection is only 11, 36 and 21, while the highest number of Benign is 2,273,097. In the CICIDS-2018 dataset, the number of Heartbleed and Port Scan is 0, while Benign has 6,376,223. To exacerbate this imbalance and test our proposed resampling method, we merge the CICIDS-2017 dataset and the CICIDS-2018 dataset to form the CICIDS-17-18 dataset. The distribution of data before and after the merging of the CICIDS-17-18 dataset is shown in Table 2. After the CICID-17-18 dataset is merged, the proportions of FTP-Patator, SSH-Patator, Bot, etc., are increased, and the gap between them and Heartbleed, Infiltration, Web Attack-Sql Injection becomes wider and wider. This aggravates the imbalance and makes it more difficult for NIDS to learn the features of Heartbleed, Infiltration and Web Attack-Sql Injection.

3.2.2. Dataset Resampling

We randomly divided the above data set into a training set and a test set by 8:2. SPE-ACGAN resampled the training sets for the CICIDS-17-18 dataset, the CICIDS-2017 dataset, and the CICIDS-2018 dataset with the aim of moderating the extreme imbalances in these datasets. ACGAN performed data synthesis for several categories of samples, and SPE censored most of the categories so that benign traffic and all malicious traffic are close to each other, and the number of malicious flows is close to each other in proportion.

The number of resamples is an important factor in the quality of resampling. After some experiments, like the threshold of the minority category with 5000, 10,000 and 20,000, in the resampling process, we define categories with less than 10,000 to be the minority category, which need to be oversampled to increase their number by 10,000, while categories with more than 20,000 and less than 50,000 are reduced in number by 50% and rounded down. For categories with more than 50,000, we reduce the number by 70% and round down. Finally, based on the total number of all malicious traffic, the number of benign traffic is adjusted to be approximately equal to the number of malicious traffic.

Table 2. Quantity distribution of CICDS-17-18 dataset before and after consolidation.

Class	Samples of CICIDS-2017	Composition (%)	Samples of CICIDS-2018	Composition (%)	Samples of CICIDS-17-18	Composition (%)
Benign	2,273,097	80.301	6,376,223	76.041	8,649,320	76.023
FTP-Patator	7938	0.281	193,353	2.306	201,291	1.771
SSH-Patator	5897	0.209	187,588	2.237	193,485	1.702
Bot	1966	0.070	285,289	3.402	287,255	2.523
DDos	128,027	4.523	687,840	8.203	815,867	7.176
Dos GoldenEye	10,293	0.364	461,911	5.509	472,204	4.155
Dos Hulk	231,073	8.163	41,507	0.495	272,580	2.401
Dos Slowhttptest	5499	0.195	139,889	1.668	145,388	1.282
Dos Slowloris	5796	0.205	10,989	0.131	16,785	0.148
Heartbleed	11	0.001	0	0	11	0.001
Infiltration	36	0.001	161,095	1.921	161,131	1.416
Port Scan	158,930	5.615	0	0	158,930	1.397
Web Attack-Brute Force	1507	0.054	610	0.001	2117	0.001
Web Attack-Sql Injection	21	0.001	86	0.001	107	0.001
Web Attack-XSS	652	0.024	229	0.001	881	0.001

When ACGAN generates an analogous sample, it can generate 10 samples of a specified type per round, and we can achieve the required number of samples by having it generate multiple batches of samples of a specified type. When SPE undersamples, we use the hardness level of classification as the hardness index and calculate the hardness index of each sample. The samples are reordered after each training session, and when the number of iterations is satisfied, the samples are selected from the front to the back according to the number of undersamples. Tables 3–5 show the data distribution of the above three datasets before and after resampling by SPE-ACGAN, respectively.

Table 3. CICIDS-2017 distribution of the number of training sets before and after resampling.

Class	Before Resampling	Composition (%)	After Resampling	Composition (%)
Benign	1,818,477	80.301	300,000	52.685
FTP-Patator	7938	0.281	17,938	3.153
SSH-Patator	6350	0.209	16,350	2.874
Bot	1572	0.07	11,572	2.034
DDos	102,421	4.523	30,726	5.401
Dos GoldenEye	8234	0.364	18,234	3.205
Dos Hulk	184,858	8.163	55,457	9.748
Dos Slowhttptest	4399	0.195	15,499	2.724
Dos Slowloris	4636	0.205	14,636	2.573
Heartbleed	8	0.001	10,008	1.759
Infiltration	28	0.001	10,028	1.763
Port Scan	127,144	5.615	38,143	6.705
Web Attack-Brute Force	1295	0.054	11,295	1.810
Web Attack-Sql Injection	16	0.001	10,016	1.761
Web Attack-XSS	521	0.023	10,521	1.847

Table 4. CICIDS-2018 distribution of the number of training sets before and after resampling.

Class	Before Resampling	Composition (%)	After Resampling	Composition (%)
Benign	509,778	76.041	509,778	49.708
FTP-Patator	154,682	2.306	46,404	4.569
SSH-Patator	150,070	2.237	45,021	6.742
Bot	228,231	3.402	68,469	8.427
DDos	550,272	8.203	165,081	16.255
Dos GoldenEye	369,528	5.509	11,858	1.671
Dos Hulk	33,205	0.495	16,602	11.257
Dos Slowhttptest	111,911	1.668	33,573	1.635
Dos Slowloris	8791	0.131	18,791	1.850
Heartbleed	0	0	0	0
Infiltration	128,876	1.921	38,662	3.807
Port Scan	0	0	0	0
Web Attack-Brute Force	488	0.001	10,488	1.033
Web Attack-Sql Injection	68	0.001	10,068	0.991
Web Attack-XSS	183	0.001	10,183	0.992

Table 5. CICIDS-17-18 distribution of the number of training sets before and after resampling.

Class	Before Resampling	Composition (%)	After Resampling	Composition (%)
Benign	6,919,456	76.023	700,000	49.793
FTP-Patator	161,032	1.771	48,309	3.463
SSH-Patator	154,788	1.702	46,436	3.328
Bot	229,804	2.523	68,941	4.942
DDos	652,693	7.176	195,807	14.035
Dos GoldenEye	377,763	4.155	113,328	8.123
Dos Hulk	218,064	2.401	65,419	4.689
Dos Slowhttptest	116,311	1.282	34,893	2.501
Dos Slowloris	13,428	0.148	13,428	0.963
Heartbleed	8	0.001	10,008	0.713
Infiltration	128,904	1.416	38,671	0.646
Port Scan	127,144	1.397	38,143	2.771
Web Attack-Brute Force	1693	0.001	11,634	0.834
Web Attack-Sql Injection	86	0.001	10,086	0.723
Web Attack-XSS	704	0.001	10,704	0.761

4. Experimentation and Result Analysis

In this section, we detail the experimental setup and evaluation metrics and present the experimental results to demonstrate the validity of the proposed method.

4.1. Experimental Setup

In this work, the settings on all experimental environments are as follows: the deep learning framework is the Tensorflow 2.4 open source framework, the operating system is the Windows 10 Professional operating system, the processor is an Intel(R) Core (TM) i5 10400F CPU @ 2.90 GHz, the memory size is 32 GB, the graphics card uses a single NVIDIA GeForce GTX 1080Ti, the development environment is PyCharm and Anaconda3, and the development language is Python.

A machine learning NIDS, Random Forest [31], and three deep learning network intrusion detection models, CNN + WDLSTM (weight-dropped LSTM) [9], CNN [32] and GoogLeNet [10], are used as the validation models to verify the effectiveness of the SPE-ACGAN resampling method proposed in this paper.

The CICIDS-2017 dataset, CICIDS-2018 dataset and CICIDS-17-18 dataset are used as datasets for validating the SPE-ACGAN resampling method, and RUS [17], SMOTE [16], ACGAN and SPE are used as the resampling methods for comparison.

4.2. Performance Metrics

In network intrusion detection, there are many evaluation metrics that can be referred to. In this paper, we would like to use Precision (P), Recall (R) and F1-Score (F1) as the criteria to evaluate the performance of the model.

P: The proportion of attack samples correctly predicted by the classifier to all samples predicted as attacks, whose formula is shown in (3):

$$P = \frac{TP}{TP + FP} \tag{3}$$

R: The ratio of all samples correctly classified by the classifier as attacks to all samples actually attacked, with the formula shown in (4):

$$R = \frac{TP}{TP + FN} \tag{4}$$

F1: The summed average of precision and recall to check the stability of the system by considering the precision and recall of the system with the formula shown in (5):

$$F1 = \frac{2 \times P \times R}{P + R} \tag{5}$$

True Positive (TP) means that the classifier correctly predicts a positive sample as a positive sample; True Negative (TN) means that the classifier correctly predicts a negative sample as a negative sample; False Positive (FP) means that the classifier incorrectly predicts a negative sample as a positive sample; and False Negative (FN) is a false negative, meaning that the classifier incorrectly predicts a positive sample as a negative sample.

Table 6 summarizes the performance changes of each NIDS after the resampling of the CICIDS-2017 and CICIDS-2018 datasets by SPE-ACGAN. After the resampling of the CICIDS-2017 dataset by SPE-ACGAN, Random Forest achieved 93.03%, 94.93% and 93.97% in the Precision, Recall and F1-score metrics, 0.86%, 1.14% and 1% higher than before resampling. CNN + WDLSTM achieved 98.68%, 98.88% and 98.78% in the Precision, Recall and F1-score metrics, 0.61%, 0.46% and 0.54% higher than before resampling. CNN achieved 96.85%, 98.11% and 97.48% in the Precision, Recall and F1-score metrics, 0.17%, 0.06% and 0.12% higher than before resampling.

Table 6. The outcome of the proposed method before and after resampling.

Method	CICIDS-2017			CICIDS-2018		
	P (%)	**R (%)**	**F1 (%)**	**P (%)**	**R (%)**	**F1 (%)**
Random Forest	92.17	93.79	92.97	91.68	89.65	90.65
GoogLeNet	92.88	94.53	93.69	92.94	91.39	91.71
CNN	96.68	98.05	97.36	93.62	92.10	92.34
CNN + WDLSTM	98.07	98.42	98.24	94.97	94.88	94.63
Our Proposed + Random Forest	93.03	94.93	93.97	92.70	90.64	91.66
Our Proposed + GoogLeNet	93.34	94.10	93.72	93.17	92.43	92.80
Our Proposed + CNN	96.85	98.11	97.48	94.71	93.33	94.01
Our Proposed + CNN + WDLSTM	98.68	98.88	98.78	95.92	96.13	96.02

In addition, after the CICIDS-2018 dataset was resampled by SPE-ACGAN, Random Forest achieved 92.70%, 90.64% and 91.66% in the Precision, Recall and F1-score metrics, 1.02%, 0.99% and 1.01% higher than before resampling. CNN + WDLSTM achieved 95.92%, 96.13% and 96.02% in the Precision, Recall and F1-score metrics, 0.95%, 1.25%

and 1.39% higher than before resampling. CNN achieved 94.71%, 93.33% and 94.01% in Precision, Recall and F1-score metrics, 1.09%,1.23% and 1.67% higher than before resampling. GoogLeNet achieved 93.17%, 92.43% and 92.80% in the Precision, Recall and F1-score metrics, 0.23%, 1.04% and 1.09% higher than before resampling. The above experimental results all show that the SPE-ACGAN resampling method can moderate the network traffic imbalance problem.

Table 7 summarizes the comparison of SPE-ACGAN with other methods in the CICIDS-17-18 dataset experiments. After resampling by SPE-ACGAN, Random Forest achieved 75.63%, 77.14% and 76.38% in Precision, Recall and F1-score, 2.02%, 2.77% and 5.59% higher than before resampling. CNN + WDLSTM achieved 82.23%, 82.54% and 82.38% in Precision, Recall and F1-score, 2.77%, 3.3% and 3.56% higher than before resampling. CNN achieved 83.94%, 82.78% and 81.66% in Precision, Recall and F1-score,5.41%,5.48% and 3.75% higher than before resampling. GoogLeNet achieved 77.57%, 80.20% and 78.86% in Precision, Recall and F1-score, 3.41%, 3.80% and 3.60% higher than before resampling. The performance of each NIDS on the three metrics of Precision, Recall and F1-score before and after ACGAN resampling is shown in Figures 4–6.

Table 7. Comparison of the proposed method and different methods.

Method	CICIDS-17-18		
	P (%)	R (%)	F1 (%)
Random Forest	73.34	74.37	70.79
GoogLeNet	74.16	76.40	75.26
CNN	78.53	77.30	77.91
CNN + WDLSTM	79.46	79.24	78.82
RUS + Random Forest	75.13	75.15	75.14
RUS + GoogLeNet	76.58	79.36	77.94
RUS + CNN	78.38	78.23	75.03
RUS + CNN + WDLSTM	77.68	79.06	78.36
SMOTE + Random Forest	74.76	79.87	76.23
SMOTE + GoogLeNet	77.82	80.14	78.96
SMOTE + CNN	78.09	81.96	75.32
SMOTE + CNN + WDLSTM	80.55	81.67	81.11
SPE + Random Forest	75.58	75.22	75.40
SPE + GoogLeNet	75.69	77.58	76.57
SPE + CNN	80.53	80.12	80.32
SPE + CNN + WDLSTM	81.02	81.82	81.42
ACGAN + Random Forest	74.44	74.55	74.49
ACGAN + GoogLeNet	75.55	77.52	76.52
ACGAN + CNN	78.98	77.21	78.08
ACGAN + CNN + WDLSTM	78.36	77.06	77.70
Our Proposed + Random Forest	75.63	77.14	76.38
Our Proposed + GoogLeNet	77.57	80.20	78.86
Our Proposed + CNN	83.94	82.78	81.66
Our Proposed + CNN + WDLSTM	82.23	82.54	82.38

Furthermore, the SPE-ACGAN method proposed in this paper takes F1-score values of 76.38%, 82.38%, 81.66% and 78.86% in Random Forest, CNN + WDLSTM, CNN and GoogLeNet after resampling. After resampling by the SPE-ACGAN method proposed in this paper, the F1-score in Random Forest, CNN + WDLSTM, CNN and GoogLeNet takes the values of 76.38%, 82.38%, 81.66% and 78.86%. After resampling by the RUS, the F1-score takes the values of 75.14%, 77.94%, 75.03% and 78.36%. After resampling using the SMOTE method, the F1-score takes the values of 76.23%, 78.96%, 75.32% and 81.11% in Random Forest, CNN + WDLSTM, CNN and GoogLeNet. After resampling by SPE method, the F1-score takes the values of 75.40%, 76.57%, 80.32%, and 81.42% in Random

Forest, CNN + WDLSTM, CNN and GoogLeNet. After resampling using the ACGAN method, the F1-score takes the values of 74.49%, 76.52%, 78.08%, and 77.70% in Random Forest, CNN + WDLSTM, CNN and GoogLeNet. By comparing the results with other resampling results, it can be concluded that the resampling method proposed in this paper has the highest performance improvement for each NIDS.

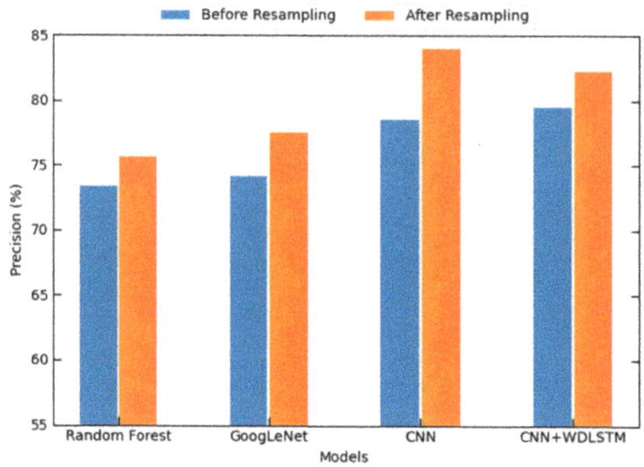

Figure 4. The Performance of Precision before and after resampling.

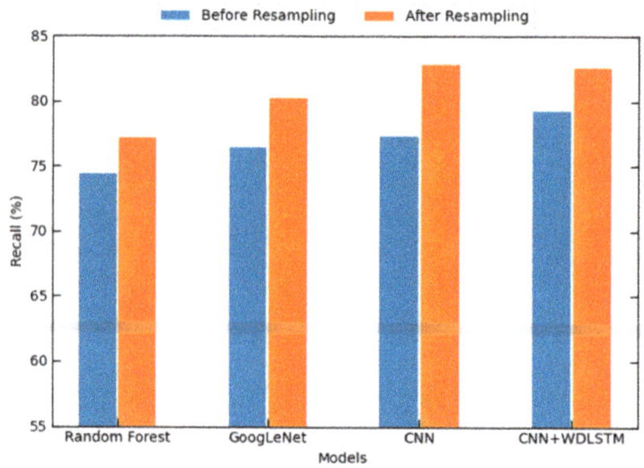

Figure 5. The Performance of Recall before and after resampling.

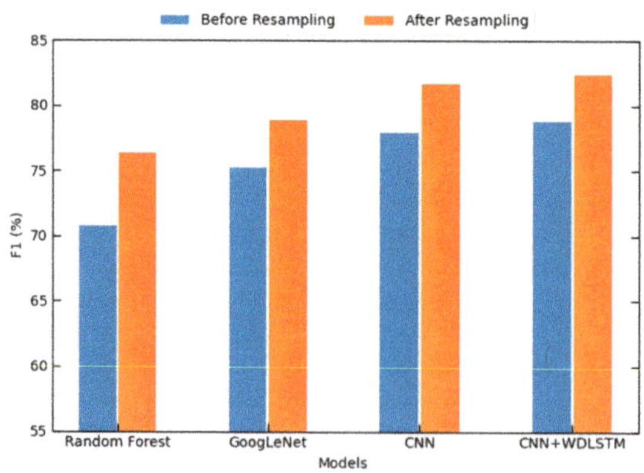

Figure 6. The Performance of F1-score before and after resampling.

5. Conclusions

The sample imbalance of network traffic is one of the important reasons affecting the detection performance of the NIDS classifier. In this paper, the rationale behind our proposed resampling approach is to balance the amount of malicious traffic with the amount of benign traffic and, similarly, to balance the amount of malicious traffic in each category. We propose a resampling method SPE-ACGAN based on the combination of ACGAN and SPE, which balances the network traffic by eliminating majority class samples and generating minority class samples. Compared to existing oversampling methods, ACGAN is able to generate data with specified categories, whereas GAN generates data randomly and needs to be filtered again to find data that match the features of the specified categories. Not only that, ACGAN does not need to traverse the neighboring samples of the minority samples compared to SMOTE, which can greatly improve efficiency. Compared to RUS, SPE is able to retain samples that are on the classification boundary to a great extent rather than randomly removing samples from the majority class. Experimental results show that the resampling method proposed in this paper alleviates the sample imbalance problem of NIDS and not only improves the performance of multi-class NIDS but also achieves a better improvement than other resampling methods.

In NIDS, capturing attack samples is a difficult task, but generating attack samples is more difficult because verifying the effectiveness of generating samples is not an easy task. The processing of small and zero samples will be an important aspect of NIDS.

In addition, scenarios of class imbalance often occur in everyday life, such as the gap between rare disease diagnoses and health cases. The use of resampling techniques enables the model to cope with the imbalance by enabling the features of a small number of class samples during the training process.

Author Contributions: Conceptualization, H.Y., J.X., Y.X. and L.H.; methodology, J.X. and H.Y.; validation, L.H. and J.X.; formal analysis, J.X. and Y.X.; resources, J.X.; writing—original draft preparation, J.X. and L.H.; writing—review and editing, L.H. and Y.X. All authors have read and agreed to the published version of the manuscript.

Funding: This research received no external funding.

Data Availability Statement: The datasets utilized in this paper are the CICIDS-2017 dataset (https://www.unb.ca/cic/datasets/ids-2017.html, accessed on 7 June 2022) and the CICIDS-2018 dataset (https://www.unb.ca/cic/datasets/ids-2018.html, accessed on 7 June 2022).

Conflicts of Interest: The authors declare no conflict of interest.

References

1. Molina-Coronado, B.; Mori, U.; Mendiburu, A.; Miguel-Alonso, J. Survey of network intrusion detection methods from the perspective of the knowledge discovery in databases process. *IEEE Trans. Netw. Serv.* **2020**, *4*, 2451–2479.
2. Viegas, E.K.; Santin, A.O.; Oliveira, L.S. Toward a reliable anomaly-based intrusion detection in real-world environments. *Comput. Netw.* **2017**, *11*, 200–216. [CrossRef]
3. Aggarwal, C.C. *Data Mining: The Textbook*; Springer: Berlin/Heidelberg, Germany, 2015.
4. Shyu, M.L.; Chen, S.C.; Sarinnapakorn, K.; Chang, L.W. A novel anomaly detection scheme based on principal component classifier. In Proceedings of the IEEE Foundation and New Direction of Data Mining Workshop, Melbourne, FA, USA, 19–22 November 2003; pp. 172–179.
5. Goodall, J.R.; Ragan, E.D.; Steed, C.A.; Reed, J.W. Situ: Identifying and explaining suspicious behavior in networks. *IEEE Trans. Vis. Comput. Graph.* **2019**, *1*, 204–214. [CrossRef] [PubMed]
6. Depren, O.; Topallar, M.; Anarim, E.; Ciliz, M.K. An intelligent intrusion detection system (IDS) for anomaly and misuse detection in computer networks. *Expert Syst. Appl.* **2005**, *4*, 713–722. [CrossRef]
7. Bhuyan, M.H.; Bhattacharyya, D.K.; Kalita, J.K. A multi-step outlier-based anomaly detection approach to network-wide traffic. *Inf. Sci.* **2016**, *6*, 243–271. [CrossRef]
8. Wu, K.; Chen, Z.; Li, W. A novel intrusion detection model for a massive network using convolutional neural networks. *IEEE Access* **2018**, *9*, 50850–50859. [CrossRef]
9. Hassan, M.M.; Gumaei, A.; Alsanad, A.; Alrubaian, M.; Fortino, G. A hybrid deep learning model for efficient intrusion detection in big data environment. *Inf. Sci.* **2019**, *3*, 386–396. [CrossRef]

10. Li, Z.P.; Qin, Z.; Huang, K.; Yang, X.; Ye, S.X. Intrusion detection using convolutional neural networks for representation learning. In Proceedings of the NIP 2017, Long Beach, CA, USA, 4–9 December2017; pp. 858–866.
11. Bedi, P.; Gupta, N.; Jindal, V. I-SiamIDS: An improved Siam-IDS for handling class imbalance in network-based intrusion detection systems. *Appl. Intell.* **2021**, *2*, 1133–1151. [CrossRef]
12. Bedi, P.; Gupta, N.; Jindal, V. Siam-IDS:Handling class imbalance problem in Intrusion Detection Systems using Siamese Neural Network. In Proceedings of the Third International Conference on Computing and Network Communications, Vellore, India, 30–31 March 2019; Elsevier: Amsterdam, The Netherlands, 2019; pp. 780–789.
13. Apruzzese, G.; Colajanni, M.; Ferretti, L.; Guido, A.; Marchetti, M. On the effectiveness of machine and deep learning for cyber security. In Proceedings of the International Conference on Cyber Conflict, Swissotel Tallinn, Estonia, 29 May–1 June 2018; IEEE: Piscataway, NJ, USA, 2018; pp. 371–390.
14. Dong, B.; Wang, X. Comparison Deep Comparison deep learning method to traditional methods using for network intrusion detection. In Proceedings of the IEEE International Conference on Communication Software & Networks, Beijing, China, 4–6 June 2016; IEEE: Piscataway, NJ, USA, 2016; pp. 581–585.
15. Wang, S.; Liu, W.; Wu, J.; Cao, L.; Meng, Q.; Kennedy, P.J. Training deep neural networks on imbalanced data sets. In Proceedings of the International Joint Conference on Neural Networks, Vancouver, BC, Canada, 24–29 July 2016; IEEE: Piscataway, NJ, USA, 2016; pp. 4368–4374.
16. Ma, X.Y.; Shi, W. Aesmote: Adversarial reinforcement learning with smote for anomaly detection. *IEEE Trans. Netw. Sci. Eng.* **2021**, *2*, 943–956. [CrossRef]
17. Tahir, M.A.; Kittler, J.; Mikolajczyk, K.; Yan, F. A multiple expert approach to the class imbalance problem using inverse random under sampling. In Proceedings of the International Workshop on Multiple Classifier Systems, Reykjavik, Iceland, 10–12 June 2009; Springer: Berlin, Germany, 2009; pp. 82–91.
18. Lee, J.; Park, K. AE-CGAN model based high performance network intrusion detection system. *Appl. Sci.* **2019**, *9*, 4221. [CrossRef]
19. Odena, A.; Olan, C.; Solens, J. Conditional image synthesis with auxiliary classifier GANs. In Proceedings of the 34th International Conference on Machine Learning, Sydney, Australia, 6–11 August 2017; PMLR: New York, NY, USA, 2017; pp. 2642–2651.
20. Liu, Z.; Cao, W.; Gao, Z.; Bian, J.; Chen, H. Self-paced Ensemble for Highly Imbalanced Massive Data Classification. In Proceedings of the 36th IEEE International Conference on Data Engineering, Dallas, TX, USA, 20–24 April 2020; IEEE: Piscataway, NJ, USA, 2020; pp. 841–852.
21. Yang, J.; Sheng, Y.; Wang, J. A GBDT-paralleled quadratic ensemble learning for intrusion detection system. *IEEE Access* **2020**, *8*, 175467–175482. [CrossRef]
22. Lopez-Martin, M.; Carro, B.; Sanchez-Esguevillas, A.; Lloret, J. Conditional variational autoencoder for prediction and feature recovery applied to intrusion detection in IoT. *Sensors* **2017**, *17*, 1967. [CrossRef] [PubMed]
23. Rifai, S.; Vincent, P.; Muller, X.; Glorot, X.; Bengio, Y. Contractive auto-encoders: Explicit invariance during feature extraction. In Proceedings of the ICM 2011, Bellevue, WA, USA, 28 June–2 July 2011; ACM: New York, NY, USA, 2011; pp. 833–840.
24. Wang, Z.; Li, Z.; Wang, J.; Li, D. Network intrusion detection model based on improved BYOL self-supervised learning. *Secur. Commun. Netw.* **2021**, *2021*, 9486949. [CrossRef]
25. Vaiyapuri, T.; Binbusayyis, A. Enhanced deep autoencoder based feature representation learning for intelligent intrusion detection system. *CMC—Comput. Mater. Contin.* **2021**, *3*, 3271–3288. [CrossRef]
26. Yan, B.H.; Han, G.D. LA-GRU: Building combined intrusion detection model based on imbalanced learning and gated recurrent unit neural network. *Secur. Commun. Netw.* **2018**, *2018*, 6026878. [CrossRef]
27. Abdulhammed, R.; Faezipour, M.; Abuzneid, A.; Abumallouh, A. Deep and machine learning approaches for anomaly-based intrusion detection of imbalanced network traffic. *IEEE Sens. Lett.* **2019**, *1*, 7101404. [CrossRef]
28. Andresini, G.; Appice, A.; Rose, L.D.; Malerba, D. GAN augmentation to deal with imbalance in imaging-based intrusion detection. *Futur. Gener. Comp. Syst.* **2021**, *123*, 108–127. [CrossRef]
29. Park, C.; Lee, J.; Kim, Y.; Park, J.-G.; Kim, H.; Hong, D. An enhanced AI-based network intrusion detection system using generative adversarial networks. *IEEE. IoT-J.* **2023**, *10*, 2330–2345. [CrossRef]
30. Sharafaldin, I.; Lashkari, A.H.; Ghorbani, A.A. Toward generating a new intrusion detection dataset and intrusion traffic characterization. In Proceedings of the International Conference on Information Systems Security & Privacy, Funchal, Portugal, 22–24 January 2018; Elsevier: London, UK, 2018; pp. 108–116.
31. Breiman, L. Random forests. *Mach. Learn.* **2001**, *45*, 5–32. [CrossRef]
32. Ho, S.; Jufout, S.A.; Dajani, K.; Mozumdar, M. A novel intrusion detection model for detecting known and innovative cyberattacks using convolutional neural network. *IEEE Open J. Comput. Soc.* **2021**, *2*, 14–25. [CrossRef]

electronics MDPI

Article

Modeling of Improved Sine Cosine Algorithm with Optimal Deep Learning-Enabled Security Solution

Latifah Almuqren [1], Mohammed Maray [2], Sumayh S. Aljameel [3], Randa Allafi [4,*] and Amani A. Alneil [5]

1 Department of Information Systems, College of Computer and Information Sciences, Princess Nourah bint Abdulrahman University, P.O. Box 84428, Riyadh 11671, Saudi Arabia
2 Department of Information Systems, College of Computer Science, King Khalid University, Abha 61471, Saudi Arabia
3 SAUDI ARAMCO Cybersecurity Chair, Computer Science Department, College of Computer Science and Information Technology, Imam Abdulrahman Bin Faisal University, P.O. Box 1982, Dammam 31441, Saudi Arabia
4 Department of Computers and Information Technology, College of Sciences and Arts, Northern Border University, Arar 91431, Saudi Arabia
5 Department of Computer and Self Development, Preparatory Year Deanship, Prince Sattam Bin Abdulaziz University, Al-Kharj 16278, Saudi Arabia
* Correspondence: randa.allafi@nbu.edu.sa

Abstract: Artificial intelligence (AI) acts as a vital part of enhancing network security using intrusion detection and anomaly detection. These AI-driven approaches have become essential components of modern cybersecurity strategies. Conventional IDS is based on predefined signatures of known attacks. AI improves signature-based detection by automating the signature generation and reducing false positives through pattern recognition. It can automate threat detection and response, allowing for faster reaction times and reducing the burden on human analysts. With this motivation, this study introduces an Improved Sine Cosine Algorithm with a Deep Learning-Enabled Security Solution (ISCA-DLESS) technique. The presented ISCA-DLESS technique relies on metaheuristic-based feature selection (FS) and a hyperparameter tuning process. In the presented ISCA-DLESS technique, the FS technique using ISCA is applied. For the detection of anomalous activities or intrusions, the multiplicative long short-term memory (MLSTM) approach is used. For improving the anomaly detection rate of the MLSTM approach, the fruitfly optimization (FFO) algorithm can be utilized for the hyperparameter tuning process. The simulation value of the ISCA-DLESS approach was tested on a benchmark NSL-KDD database. The extensive comparative outcomes demonstrate the enhanced solution of the ISCA-DLESS system with other recent systems with a maximum accuracy of 99.69%.

Keywords: cloud computing; security; feature selection; machine learning; artificial intelligence

Citation: Almuqren, L.; Maray, M.; Aljameel, S.S.; Allafi, R.; Alneil, A.A. Modeling of Improved Sine Cosine Algorithm with Optimal Deep Learning-Enabled Security Solution. *Electronics* **2023**, *12*, 4130. https://doi.org/10.3390/electronics12194130

Academic Editors: Tomasz Rak and Dariusz Rzońca

Received: 7 August 2023
Revised: 13 September 2023
Accepted: 21 September 2023
Published: 3 October 2023

1. Introduction

Recently, cloud computing (CC) has developed as one of the common Internet-based technologies in the information technology (IT) field [1]. There are three levels that compose CC, e.g., the system layer, platform layer, and application layer [2]. Cloud security is considered the main challenge to cloud adoption by most enterprises [3]. The open and entirely dispersed nature of the cloud platform makes it extremely susceptible to vulnerabilities and security attacks [4]. Therefore, intruders have a high potential to carry out threats against cloud-linked devices or the cloud. Alternatively, cloud cyberattacks and services have a negative impact on CC platform performance and QoS requirements [5]. Conventional security regulation, namely, firewalls and antivirus software, could protect cloud infrastructure from complex cyberattacks [6]. By employing machine learning (ML), enterprises can improve their security actions and decrease the risk of data breaches. The possibility of ML to enhance attack detection and response is the major advantage of utilizing it for cloud security. Classical security regulations, namely, antivirus software and firewalls, are responsive and only react to known attacks [7]. Conversely, the ML method can detect patterns in information, which may specify an attack, even when the attack is not yet known. According to previous information, the ML technique is created for identifying designs that point to security vulnerabilities [8].

To overcome these problems and to improve the security of cloud services, providing a deep learning (DL)-based approach will be an adaptable solution [9]. Currently, DL is employed in various organizations due to its excellent prediction power and pattern recognition [10]. As DL utilizes a multi-mode neural network (NN) idea for simulating activities the same as the working model of the brain, it can be promoted and administered in a cloud-based infrastructure [11]. Using DL methods to train massive databases in the cloud platform can perform the overall processes of computing highly efficiently with low latency. Major attacks like trust difficulties, malware identification, data privacy, and network intrusion could be monitored utilizing DL techniques in real time [12]. Different from other standard security enhancers, DL approaches are learned and have intelligent abilities to offer disruptive outcomes in detecting attacks and improving cloud security in the constantly growing competitive world [13].

This study introduces an Improved Sine Cosine Algorithm with a Deep Learning-Enabled Security Solution (ISCA-DLESS) technique for the CC environment. In the presented ISCA-DLESS technique, the selection of features takes place by the ISCA. Additionally, the chosen features are passed into a multiplicative long short-term memory (MLSTM) model for intrusion detection. To improve the anomaly detection rate of the MLSTM approach, the fruitfly optimization (FFO) algorithm can be utilized for the hyperparameter tuning process. The experimental result analysis of the ISCA-DLESS system has been tested on a benchmark database. In short, the key contribution of the paper is summarized as follows.

- Automated anomaly detection using the ISCA-DLESS technique comprising ISCA-based FS, MLSTM-based detection, and FFO-based hyperparameter tuning for CC is presented. To the best of our knowledge, the ISCA-DLESS technique has never existed in the literature.
- The ISCA-DLESS employs an ISCA-based FS technique with the integration of the oppositional-based learning (OBL) concept with SCA, which reduces the data dimensionality and enhances the detection performance.
- Applying MLSTM-based detection, which has the capability of capturing sequential patterns, makes it appropriate to detect anomalies in time-series data.
- Employing the FFO algorithm for hyperparameter tuning of the MLSTM model efficiently searches for optimal hyperparameter configurations.

The rest of the paper is organized as follows. Section 2 provides the related works, and Section 3 offers the proposed model. Then, Section 4 gives the result analysis, and Section 5 concludes the paper.

2. Related Works

Maheswari et al. [14] developed an intrusion detection system (IDS) for web and CC platforms based on hybrid teacher learning-aided DRNN and cluster-based feature optimization. After feature extraction, the study used a Modified Manta-ray Foraging Optimization (MMFO) to select optimum features to detect further. A hybrid Teacher-Learning Enabled DRNN (TL-DRNN) is developed for the classification of web-cloud intrusion. In [15], an Effective Optimum Security Solution for IDS (EOS-IDS) in a CC platform by using a hybrid DL technique was designed. Pre-processing was performed by the improved heap optimization (IHO) method. Next, the authors offer a chaotic red deer optimizer (CRDO) method for optimal feature selections. Later, a deep Kronecker NN (DKNN) is shown for cloud attack and classification and recognition of intrusion. Toldinas et al. [16] devised an innovative technique for network IDS using multi-phrase DL image detection. The feature network was transformed into four-channel (Red, Green,

Blue, and Alpha) images. Then, the images could be utilized for the classification to test and train the pretrained DL mechanism ResNet_50.

Srilatha and Thillaiarasu [17] introduced a Network Intrusion Detection and Prevention Scheme (NIDPS) to prevent and detect a large number of network attacks. The effective IDPS was tested and implemented in a network environment using different ML approaches. In this study, an improved ID3 was developed for identifying abnormalities in network activities and classifying them. The authors in [18] developed an IDSGT-DNN architecture to enhance security in cloud IDSs. The study incorporated defender and attacker systems for attack and normal data processing. In the DNN model, this technique could be implemented with IWA for the recognition of a better solution. Prabhakaran and Kulandasamy [19] suggested a hybrid semantic DL (HSDL) model by incorporating the SVM, LSTM, and CNN frameworks. The semantic data existing in the network traffic were detected utilizing a semantic layer called a Word2Vec embedding layer. The proposed architecture categorized the intrusion existing in the text and its respective attack classes.

Ravi et al. [20] presented a Cauchy GOA with DL for the Cloud-Enabled IDS (CGOA-DLCIDS) method. The proposed approach carried out feature subset selection by CGOA, which improved the recognition speed and decreased the feature subsets. Following this, the method exploited the attention-based LSTM (ALSTM) mechanism for accurate and automatic detection and classification of intrusion. Jisna et al. [21] presented a cloud-based DL LSTM-IDS technique and assessed it to hybrid Stacked Contractive AE (SCAE) along with the SVM-IDS mechanism. DL techniques such as basic ML were constructed to simultaneously perform attack detection and classification.

Alghamdi and Bellaiche [22] introduced an edge-cloud deep IDS technique in the Lambda framework for IoT security to overcome these problems. This approach minimized the time of the training stage by comparing it with standard ML methods and improved the accuracy of true positive-identified attacks. Moreover, the NN-layers' main DL technique attained higher adaptability and performance compared with the standard ML technique. Alzubi et al. [23] proposed an Effective Seeker Optimization algorithm along with an ML-assisted IDS (ESOML-IDS) approach for the FC and EC platforms. The ESOML-IDS algorithm mainly developed an innovative ESO-based FS technique for optimally selecting feature subsets to detect the existence of intrusions in the FC and EC platforms. Ali and Zolkipli's study [24] comprised a brief description of the IDS and presented to the reviewer some basic principles of the IDS task in CC, further developing a novel Fast Learning Network method for functions dependent upon intrusion detection.

Despite the availability of several anomaly and intrusion detection models, it remains a challenging problem. Due to the continuous deepening of the model, the number of parameters in DL models also increases quickly, which results in model overfitting. At the same time, different hyperparameters have a significant impact on the efficiency of the CNN model, particularly the learning rate. It is also needed to modify the learning rate parameter to obtain better performance. Therefore, in this study, we employed the FFO technique for the hyperparameter tuning of the MLSTM model.

3. The Proposed Model

In this manuscript, we have presented a novel ISCA-DLESS system for the effectual identification of anomalies and intrusions in the CC environment. The purpose of the ISCA-DLESS technique is to exploit the metaheuristic algorithms for FS and the hyperparameter tuning process. In the proposed ISCA-DLESS system, three main procedures are contained, such as ISCA-based FS, MLSTM-based classification, and FFO-based hyperparameter tuning. Figure 1 depicts the entire flow of the ISCA-DLESS method.

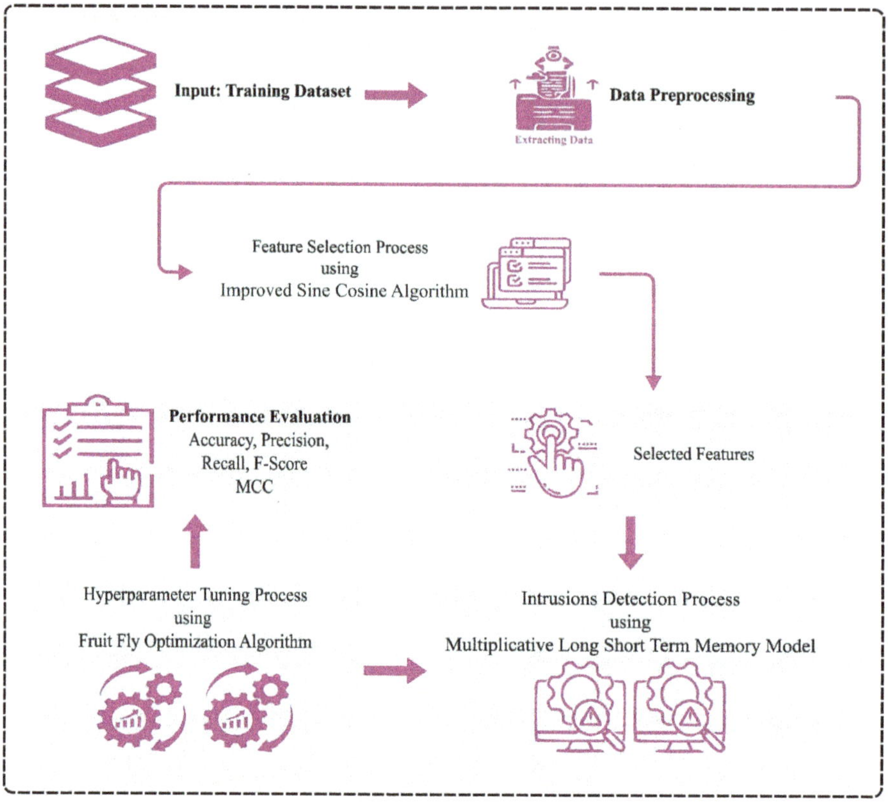

Figure 1. Overall flow of ISCA-DLESS system.

3.1. Stage I: Feature Selection Using ISCA

To elect an optimal set of features, the ISCA was used. SCA is a recent metaheuristic optimization algorithm for resolving global optimization problems [25]. Using SCA, a group of arbitrary populations of candidate performances with standard distribution could be produced to begin the optimizer technique. Then, the locations of candidate performances were upgraded by the following expression:

$$Y_i^{t+1} = Y_i^t + R_1 * \sin(R_2) * \left|R_3 Z_i^t - Y_i^t\right| \tag{1}$$

$$Y_i^{t+1} = Y_i^t + R_1 * \cos(R_2) * \left|R_3 Z_i^t - Y_i^t\right| \tag{2}$$

Now, Y_i^{t+1} and Y_i^t denote the position of ith solution candidate at the t and $t+1$ iterations correspondingly. R_1, R_2, and R_3 show a uniform distribution of random numbers, and Z_i^t shows the target point's position at the ith parameter. The operator || is utilized to define the absolute value:

$$\begin{cases} Y_i^{t+1} = Y_i^t + R_1 * \sin(R_2) * \left|R_3 Z_i^t - Y_i^t\right|, \ R_4 < 0.5 \\ Y_i^{t+1} = Y_i^t + R_1 * \cos(R_2) * \left|R_3 Z_i^{ts} - Y_i^t\right|, \ R_4 \geq 0.5 \end{cases} \tag{3}$$

where R_4 shows the uniformly distributed random value between 0 and 1. R_1 is a randomly generated vector that decides if the solution moves among the search space as well as a better solution. The vector R_2 defines the distance of candidate performances to or in a better solution. The R_3 third parameter describes arbitrary weighted over the better solution to define the micro search ($R_3 < 1$) and macro search ($R_3 > 1$) capabilities of this parameter. Due to this reason, R_3 is highly useful to avoid early convergence. Evolution from cos to sin functions can be assisted by the R_4 random vector. The range of the sin and cos function should be adaptively adjusted to achieve a proper balance between exploitation and exploration, as follows:

$$R_1 = k - Iter\frac{k}{Max\_Iter} \tag{4}$$

In Equation (4), $Iter$ and Max_Iter represent the present and maximal iteration, and k is a constant. The notion of the OBL method relies on an opposite number. Consider that $p \in [x,\ y]$, whereas $y \in \mathcal{R}$, in which $\mathcal{R}$ represents the real number:

$$p_0 = x + y - p \tag{5}$$

Also, this description could be stretched to high dimensions. The opposite number $p_0 = \left(p_o^1,\ p_o^2, \ldots, p_o^d\right)$ for a number p was defined for d-dimensional search space as follows:

$$p_{i,o} = x_i + y_i - p_i \tag{6}$$

The concept of OBL was used for improving the micro search capability of the SCA.

$$Y_{i,o}(Iter) = x_i + y_i - Y(Iter) \tag{7}$$

In the ISCA, the initial population was randomly generated by the uniform distribution, and the fitness of possible solutions was evaluated. Then, the better candidate solution Z was recognized. The OBL method attained a balance among micro as well as macro search capabilities by using the candidate solution. The linear adaptive operator was hybridized with the OBL model. This operator was capable of enhancing the convergence rate by fine-tuning the proper balance among the macro as well as micro search processes. This operator ensured the best exploration and exploitation as the number of generation's problems of varying complexities. OBL was hybridized with linear adaptive (LA) operators to benchmark the function, as shown below:

$$y_{i,0}(Iter) = l_{ac} \times (x_i + y_i - Z_i(Iter)) \tag{8}$$

In Equation (8), $y_{i,0}$ indicates the opposite solution candidate for the ith parameter around the better solution Z_i at the $Iter$ existing iteration. The fitness can be measured after defining the opposite location around a better solution.

The fitness function (FF) of the ISCA is assumed to be the classifier accuracy and FS counts. It minimizes the set dimensional of FSs and maximizes the classifier accuracy. So, the following FF can be employed for measuring separate solutions, as written in Equation (9).

$$Fitness = \alpha * ErrorRate + (1-\alpha) * \frac{\#SF}{\#All\_F} \tag{9}$$

whereas $ErrorRate$ indicates the classifier rate of errors employing the FSs. $ErrorRate$ is measured as the percentage of improper classifiers to the count of classifications made, stated as a value between zero and one. $\#SF$ mentions that the FS counts, and $\#All\_F$ implies the entire attribute counts from the new database. α is employed for controlling the impact of classifier quality and subset length.

3.2. Stage II: MLSTM-Based Classification

In this work, the MLSTM-based classification process could be employed. Classical ANN is constrained in its capability to obtain the sequential data required to handle sequence data in the input [26]. RNN can be used to extract sequential data in the raw information while making predictions, for example, links among the words from the text. An evaluation of RNN future hidden layer (HL) is given in the following: consider the time stamp vector $= (1, \ldots, T)$, a future HL vector $n = (n_1, \ldots, n_T)$, an input $y = (y_1, \cdots, y_T)$, and an output $x = (x_1, x_T)$. Using the following equation, the HL vector is given:

$$n_t = N(W_{yn}y_t + W_{nn}n_{t-1} + b_n), \tag{10}$$

For $x_t = W_{nn}n_{t-1} + b_n$, N shows the activation function of the HL, and W refers to the weight matrix.

The major problem of classical RNNs is that the backpropagation (BP) stage attenuates the loss function, which makes the number smaller, so it could not grant anything to learning. The gradient disappearing problem takes place once these layers gather a small gradient to enhance its weights and learning factors. The input, forget, and output gates are the gating mechanisms of the LSTM network. The forget gate will forbid or grant information and is estimated as follows:

$$F_t = \sigma\left(W_{xf}x_t + W_{hf}h_{t-1} + W_{cf}C_{t-1} + b_f\right), \tag{11}$$

In Equation (11), W_{xf} represents the weighted vector amongst the input and forget gate; x_t shows the existing data; and W_{hf} indicates the weighted vectors amongst the forget gate and HL. If the accumulation of the variable is run with the activation function, the gate allows it to pass if the value is in the range of $[0, 1]$. Otherwise, it removes the data.

Existing and prior outcomes are forwarded to the sigmoidal function that allows updating the cell state memory. At t time, the input vector was defined by the subsequent formula:

$$i_t = \sigma(W_{xi}x_t + W_{hi}h_{t-1} + W_{ci}C_{t-1} + b_i), \tag{12}$$

In Equation (12), W_{xi} denotes the weighted vector of raw information, and W_{hi} shows the weighted vector amongst current values and input gate. The cell layer introduces the existing cell layer, doubles the forgotten variable with the prior cell layer, and drops the variable if doubled by virtual 0:

$$C_t = F_t C_{t-1} + i_t \tanh\left(W_{xf}x_t + W_{hc}h_{t-1} + b_c\right), \tag{13}$$

The second HL is defined by the output gate. At the t timestamp, the resultant vector can be evaluated as follows:

$$o_t = \sigma(W_{xo}x_t + W_{ho}h_{t-1} + W_{co}C_{t-1} + b_o), \tag{14}$$

Lastly, the hyperbolic activation function is represented as follows:

$$h_t = o_t.\tanh(c_t), \tag{15}$$

MLSTM is different from the typical LSTM frameworks that establish a gating mechanism named multiplicative connections. It can be planned to improve the learning and representation abilities of LSTM networks. It presents a novel gating mechanism named "update gate", which is utilized for modulating the cell layer upgrade. The upgrade gate in an MLSTM is determined as the element-by-element product (Hadamard product) among the output of the forget gate as well as the input of the input gate. It implies that the upgrade gate controls several data in the preceding cell layer (determined by the forget gate), and a novel input (determined by the input gate) can be employed for updating the current cell layer. By utilizing element-by-element multiplication, the upgrade gate permits the LSTM to concentrate on particular sizes of the input and selectively upgrade the cell layer that is useful for sequence modelling tasks.

3.3. Stage III: Hyperparameter Tuning Using FFO Algorithm

To enhance the results of the MLSTM approach, the FFO system can be employed. The FFO algorithm is a new nature-inspired optimization approach [27]. Due to its simple computation operation, FFO is easy to apply and comprehend like other metaheuristic approaches. This technique is an SI approach stimulated by the knowledge of the foraging behavioural patterns of FFs. The FF exceeds other species relating to olfaction and vision, which they mainly depend on—FFs can collect miscellaneous aerial smells, notwithstanding the food source being far away. In the scouring stage, the FF scouts and locates food sources near the swarm and evaluates the odour intensity for the food sources. Once the better position with the high odour intensity is identified, the swarm navigates toward it.

Undeniably, the procedure of effectual teamwork and communication between individual FFs is vital to accomplish the strategies of resolving optimization problems. The algorithm has four different stages:

- Initialization;
- Osphresis foraging;
- Population evaluation;
- Vision.

At first, the parameter is set—the maximal amount of iterations and size of populations. The solution, viz. FFs are randomly initialized as follows:

$$X_{ij} = rand(UB_j - LB_j) + LB_j, \tag{16}$$

In Equation (16), X_{ij} denotes ith solution, and jth indicates the element's location at the ith solution. LB indicates a lower boundary, whereas UB shows an upper boundary, and $rand$ denotes a uniformly distributed random integer.

Next, the location updating of the solution takes place according to the osphresis foraging stage. The solution is randomly distributed from the existing position as follows:

$$X_{ij}^{(t+1)} = X_{ij}^{(t)} \pm rand \tag{17}$$

In Equation (17), $X_{ij}^{(t+1)}$ denotes the new location, $X_{ij}^{(t)}$ indicates the existing solution, $rand() \in [-1, 1]$, whereas t refers to the iteration count. The smell and distance are calculated following the location update. Next, the calculation of odor intensity—the function of smell (FF)—for every solution follows. When the optimal FF of the solution is superior to the prior best, the novel location of the solution with the better FF values replaces each solution's position afterwards. Or else, the older solution position will remain. This procedure signifies the vision foraging stage. The process continues until the ending condition is met and produces better outcomes.

Fitness optimal is a key feature of the FFO system. An encoded outcome can be deployed to assess the goodness solution of candidate outcomes. Presently, the accuracy value is the major condition deployed to design an FF.

$$Fitness = \max(P) \tag{18}$$

$$P = \frac{TP}{TP + FP} \tag{19}$$

In which TP and FP define the true and false positive values.

4. Results and Discussion

The proposed model was simulated using Python 3.6.5 tool (The source code will be made available once the funding project is complete). The proposed model was experimented on PC i5-8600k, GeForce 1050Ti 4 GB, 16 GB RAM, 250 GB SSD, and 1 TB HDD. In this section, the simulation validation of the ISCA-DLESS technique can be tested on the NSL-KDD database (available at https://www.kaggle.com/datasets/hassan06/nslkdd (accessed on 13 July 2023)), containing 125,973 instances with five class labels, as represented in Table 1. The ISCA-DLESS technique selected a total of 29 features from the available 42 features. The confusion matrices of the ISCA-DLESS algorithm on distinct databases are shown in Figure 2. The simulation value implied that the ISCA-DLESS approach accurately recognized various classes proficiently.

Table 1. Description of database.

Class	No. of Instances
Dos	45,927
R2l	995
Probe	11,656
U2r	52
Normal	67,343
Total No. of instances	125,973

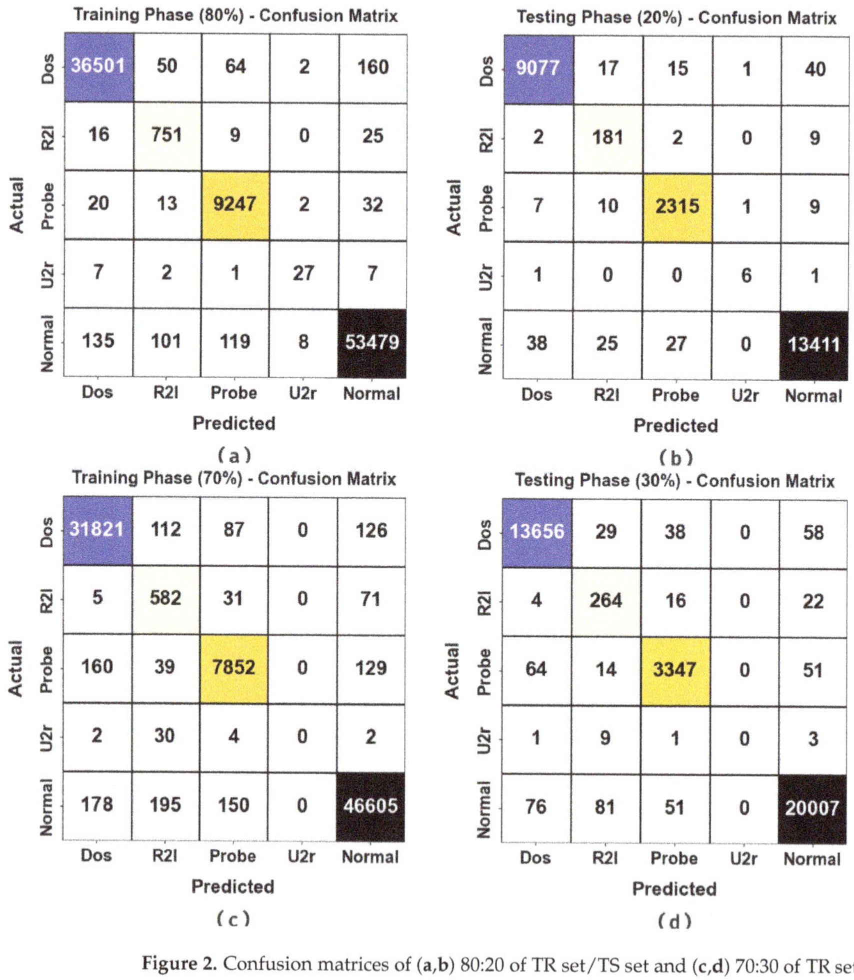

Figure 2. Confusion matrices of (**a**,**b**) 80:20 of TR set/TS set and (**c**,**d**) 70:30 of TR set/TS set.

In Table 2 and Figure 3, the overall outcome of the ISCA-DLESS system with 80:20 of the TR set/TS set is portrayed. The results suggested that the ISCA-DLESS technique reached enhanced performance in all classes.

Table 2. Classifier outcome of ISCA-DLESS algorithm on 80:20 of TR set/TS set.

Class	$Accu_y$	$Prec_n$	$Reca_l$	F_{Score}	MCC
		TR set (80%)			
Dos	99.55	99.51	99.25	99.38	99.03
R2l	99.79	81.90	93.76	87.43	87.52
Probe	99.74	97.96	99.28	98.61	98.47
U2r	99.97	69.23	61.36	65.06	65.16
Normal	99.42	99.58	99.33	99.45	98.83
Average	99.69	89.64	90.60	89.99	89.80
TS set (20%)					
Dos	99.52	99.47	99.20	99.34	98.96
R2l	99.74	77.68	93.30	84.78	85.01
Probe	99.72	98.13	98.85	98.49	98.34
U2r	99.98	75.00	75.00	75.00	74.99
Normal	99.41	99.56	99.33	99.45	98.81
Average	99.67	89.97	93.14	91.41	91.22

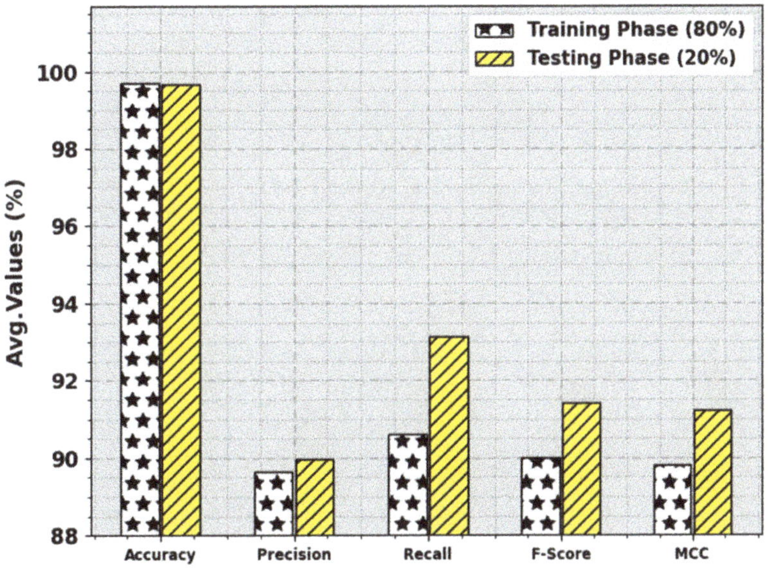

Figure 3. Average of ISCA-DLESS algorithm on 80:20 of TR set/TS set.

With 80% of the TR set, the ISCA-DLESS technique attained average $accu_y$, $prec_n$, $reca_l$, F_{score}, and MCC values of 99.69%, 89.64%, 90.60%, 89.99%, and 89.80%, respectively. Also, with 20% of the TS set, the ISCA-DLESS methodology accomplished average $accu_y$, $prec_n$, $reca_l$, F_{score}, and MCC values of 99.67%, 89.97%, 93.14%, 91.41%, and 91.22% correspondingly.

In Table 3 and Figure 4, the overall outcome of the ISCA-DLESS methodology with 70:30 of the TR set/TS set is portrayed. The results suggested that the ISCA-DLESS system attained greater performance under all classes. With 70% of TR set, the ISCA-DLESS approach obtains average $accu_y$, $prec_n$, $reca_l$, F_{score}, and MCC values of 99.40%, 71.13%, 75.67%, 73.01%, and 72.75% correspondingly. Then, with 30% of TS set, the ISCA-DLESS method gained average $accu_y$, $prec_n$, $reca_l$, F_{score}, and MCC values of 99.45%, 72.34%, 76.13%, 73.98%, and 73.69%, respectively.

Table 3. Classifier outcome of ISCA-DLESS algorithm on 70:30 of TR set/TS set.

Class	$Accu_y$	$Prec_n$	$Reca_l$	F_{Score}	MCC
TR set (70%)					
Dos	99.24	98.93	98.99	98.96	98.36
R2l	99.45	60.75	84.47	70.67	71.38
Probe	99.32	96.65	95.99	96.32	95.95
U2r	99.96	00.00	00.00	00.00	00.00
Normal	99.03	99.30	98.89	99.10	98.06
Average	99.40	71.13	75.67	73.01	72.75
TS set (30%)					
Dos	99.29	98.95	99.09	99.02	98.46
R2l	99.54	66.50	86.27	75.11	75.52
Probe	99.38	96.93	96.29	96.61	96.27
U2r	99.96	00.00	00.00	00.00	00.00
Normal	99.10	99.33	98.97	99.15	98.18
Average	99.45	72.34	76.13	73.98	73.69

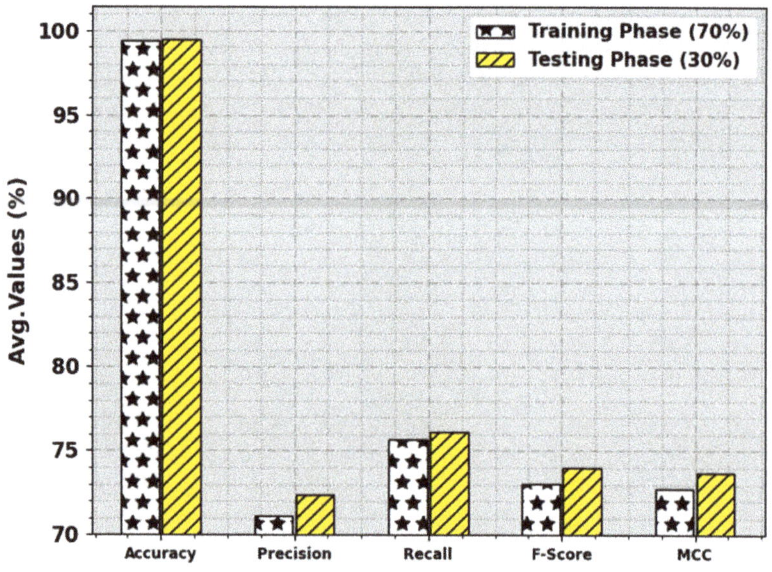

Figure 4. Average of ISCA-DLESS algorithm on 70:30 of TR set/TS set.

Figure 5 demonstrates the training accuracy TR_accu_y and VL_accu_y of the ISCA-DLESS system on 80:20 of the TR set/TS set. The TL_accu_y was defined by the assessment of the ISCA-DLESS technique on the TR dataset, whereas the VL_accu_y was calculated by estimating the performance on a separate testing dataset. The outcomes exhibited that TR_accu_y and VL_accu_y increased with an upsurge in epochs. As a result, the performance of the ISCA-DLESS system improved on the TR and TS datasets with a rise in the number of epochs.

Figure 5. $Accu_y$ curve of ISCA-DLESS algorithm on 80:20 of TR set/TS set.

In Figure 6, the TR_loss and VR_loss curve of the ISCA-DLESS system on 80:20 of the TR set/TS set is depicted. The TR_loss defines the error among the predictive outcome and original values on the TR data. The VR_loss signifies the measure of the solution of the ISCA-DLESS technique on individual validation data. The results stated that the TR_loss and VR_loss tended to be lesser with rising epochs. It depicted the enhanced performance of the ISCA-DLESS technique and its ability to create an accurate classification. The reduced value of TR_loss and VR_loss established the greater performance of the ISCA-DLESS method in capturing patterns and relationships.

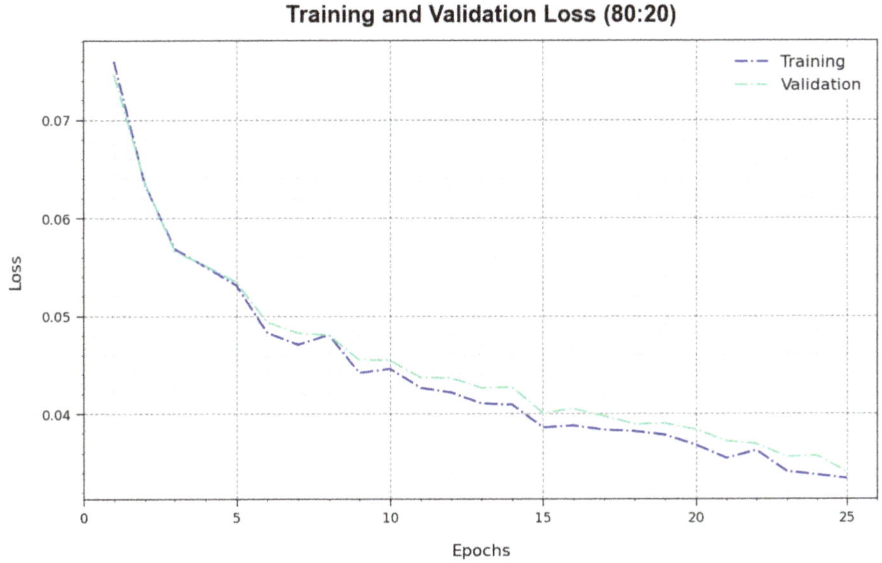

Figure 6. Loss curve of ISCA-DLESS algorithm on 80:20 of TR set/TS set.

A comprehensive precision–recall (PR) analysis of the ISCA-DLESS system is displayed on 80:20 of the TR set/TS set in Figure 7. The simulation value defined the ISCA-DLESS approach solution in greater PR values. Afterwards, it could be clear that the ISCA-DLESS algorithm attained superior performances of PR in five classes.

In Figure 8, a ROC analysis of the ISCA-DLESS algorithm is defined on 80:20 of the TR set/TS set. The simulation value determined that the ISCA-DLESS approach led to maximal values of ROC. Next, the ISCA-DLESS system achieved greater outcomes in ROC in five classes.

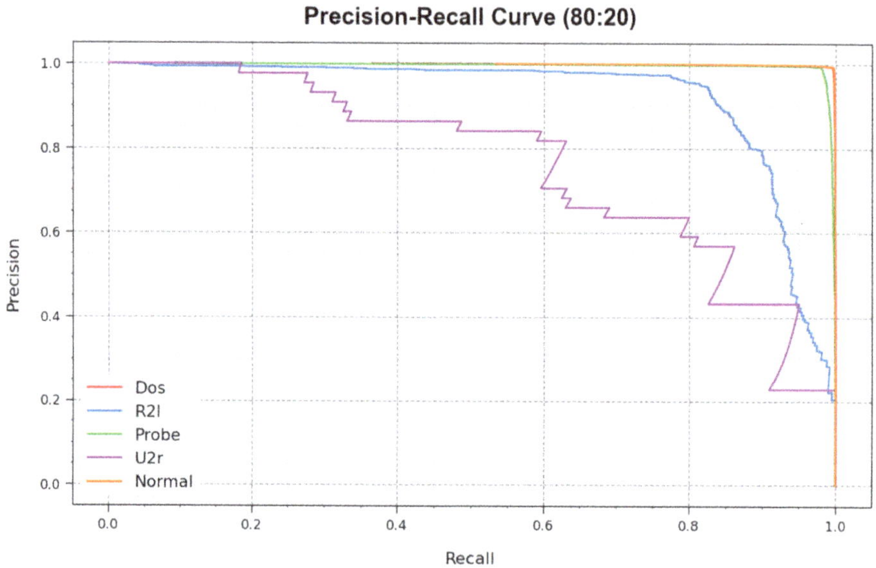

Figure 7. PR curve of ISCA-DLESS algorithm on 80:20 of TR set/TS set.

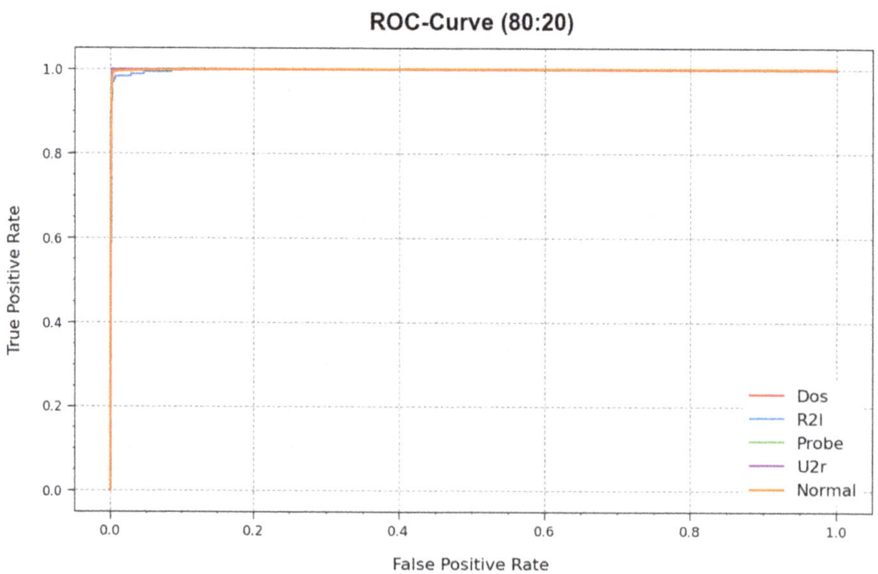

Figure 8. ROC of ISCA-DLESS algorithm on 80:20 of TR set/TS set.

In Table 4, a detailed comparative result of the ISCA-DLESS methodology with recent systems is made [28]. Figure 9 depicts the $accu_y$ and F_{score} outcomes of the ISCA-DLESS approach with other approaches. The obtained values inferred that the LKM-OFLS and PCA-NN models reached poor performance. At the same time, the K-means-OFLS, MLP, and FCM-OFLS models reported moderately improved results. Meanwhile, the IMFL-IDSCS technique attained considerable performance. Finally, the ISCA-DLESS technique showcased better performance, with a maximum $accu_y$ of 99.69% and *an* F_{score} of 89.99%.

Table 4. Comparative outcome of ISCA-DLESS system with recent techniques [28].

Methods	$Accu_y$	$Prec_n$	$Reca_l$	F_{Score}
ISCA-DLESS	99.69	89.64	90.60	89.99
IMFL-IDSCS	99.44	86.15	78.36	81.92
LKM-OFLS	89.47	84.76	74.80	78.38
K-means-OFLS	91.55	85.87	75.63	78.46
MLP Algorithm	91.59	86.72	76.89	75.13
PCA-NN	90.22	84.68	76.19	77.68
FCM-OFLS	93.50	82.85	74.54	75.80

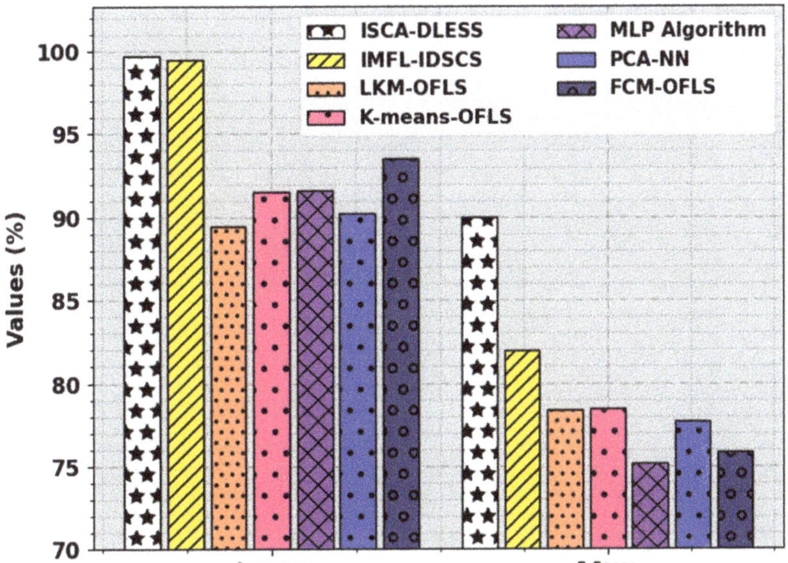

Figure 9. $Accu_y$ and F_{score} outcome of ISCA-DLESS approach with recent systems [28].

Figure 10 represents the $prec_n$ and $reca_l$ analysis of the ISCA-DLESS system with other methods. The simulation values implied that the LKM-OFLS and PCA-NN approaches attained worse outcomes. Then, the K-means-OFLS, MLP, and FCM-OFLS methods reported moderately enhanced performance. In the meantime, the IMFL-IDSCS system attained considerable outcomes. At last, the ISCA-DLESS system demonstrated optimum performance with maximal $prec_n$ of 89.64% and $reca_l$ of 90.60%. Therefore, the ISCA-DLESS technique could be utilized for enhanced cloud security.

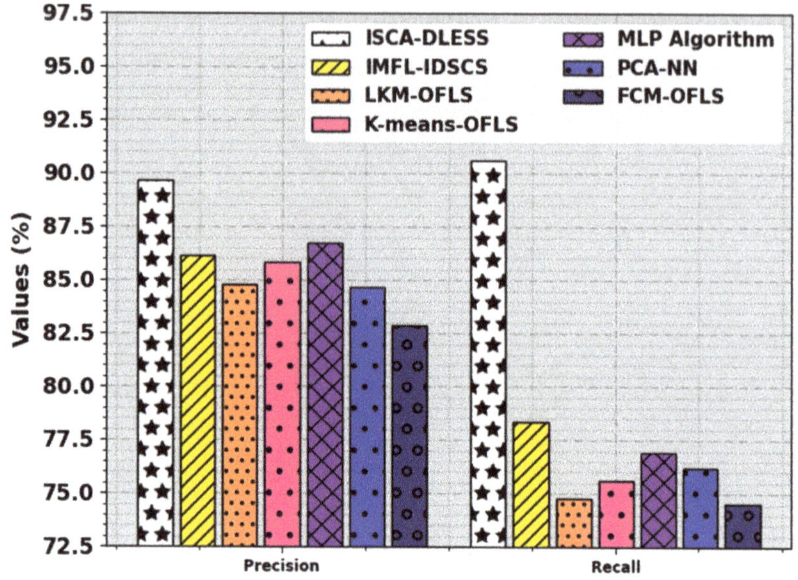

Figure 10. $Prec_n$ and $reca_l$ outcome of ISCA-DLESS approach with recent systems [28].

5. Conclusions

In this study, we derived a novel ISCA-DLESS algorithm for effectual identification of anomalies and intrusions in the CC environment. The ISCA-DLESS technique applied the FS process with a hyperparameter-tuned classification model for anomaly detection. In the proposed ISCA-DLESS system, the three main procedures comprised ISCA-based FS, MLSTM-based classification, and FFO-based hyperparameter tuning. The application of the ISCA-based FS helped in reducing the high dimensionality problem and enhanced the classification performance. Moreover, the use of the FFO algorithm for the hyperparameter tuning of the MLSTM model aided in accomplishing an improved detection rate. The comprehensive analysis demonstrated an enhanced solution in the ISCA-DLESS technique with other recent approaches, with a maximum accuracy of 99.69%. Thus, the ISCA-DLESS technique could be applied for automated anomaly detection in the CC environment. In future, the proposed model could be extended to address cloud-specific threats, such as misconfigurations, data exposure, and supply chain attacks, in the context of anomaly detection. In addition, the proposed model could operate seamlessly across multiple cloud providers and hybrid cloud environments. This includes ensuring interoperability and consistent threat monitoring.

Author Contributions: Conceptualization, L.A.; Methodology, L.A., M.M. and S.S.A.; Software, A.A.A.; Validation, M.M., R.A. and A.A.A.; Investigation, L.A.; Data curation, S.S.A.; Writing—original draft, L.A., M.M., S.S.A. and R.A.; Writing—review & editing, M.M., S.S.A., R.A. and A.A.A.; Visualization, A.A.A.; Project administration, R.A.; Funding acquisition, L.A. All authors have read and agreed to the published version of the manuscript.

Funding: The authors extend their appreciation to the Deanship of Scientific Research at King Khalid University for funding this work through large group Research Project under grant number (RGP2/235/44). This research was funded by Princess Nourah bint Abdulrahman University Researchers Supporting Project number (PNURSP2023R349), Princess Nourah bint Abdulrahman University, Riyadh, Saudi Arabia. The authors extend their appreciation to the Deanship of Scientific Research at Northern Border University, Arar, KSA for funding this research work through the project number "NBU-FFR-2023-0114. We Would like to thank SAUDI ARAMCO Cybersecurity

Chair for funding this project. This study is supported via funding from Prince Sattam bin Abdulaziz University project number (PSAU/2023/R/1444).

Data Availability Statement: Data sharing does not apply to this article, as no datasets were generated during the current study.

Conflicts of Interest: The authors declare no conflict of interest.

References

1. Jain, D.K.; Ding, W.; Kotecha, K. Training a fuzzy deep neural network with honey badger algorithm for intrusion detection in the cloud environment. *Int. J. Mach. Learn. Cybern.* **2023**, *14*, 2221–2237. [CrossRef]
2. Sathiyadhas, S.S.; Soosai Antony, M.C.V. A network intrusion detection system in a cloud computing environment using dragonfly improved invasive weed optimization integrated Shepard convolutional neural network. *Int. J. Adapt. Control Signal Process.* **2022**, *36*, 1060–1076. [CrossRef]
3. Yi, L.; Yin, M.; Darbandi, M. A deep and systematic review of the intrusion detection systems in the fog environment. *Trans. Emerg. Telecommun. Technol.* **2023**, *34*, e4632. [CrossRef]
4. Goyal, S.B.; Bedi, P.; Kumar, S.; Kumar, J.; Karahroudi, N.R. Application of Deep Learning in Honeypot Network for Cloud Intrusion Detection. In Proceedings of the International Conference on Computational Intelligence and Data Engineering: ICCIDE 2021; Springer: Singapore, 2022; pp. 251–266.
5. Asaolu, O.S. Leveraging Deep Learning-Enabled Intrusion Detection Systems for a Cloud Environment. Ph.D. Thesis, Morgan State University, Baltimore, MD, USA, 2023.
6. Cheikhrouhou, O.; Mahmud, R.; Zouari, R.; Ibrahim, M.; Zaguia, A.; Gia, T.N. One-dimensional CNN approach for ECG arrhythmia analysis in fog-cloud environments. *IEEE Access* **2021**, *9*, 103513–103523. [CrossRef]
7. Hussain, M.; Cifci, M.A.; Sehar, T.; Nabi, S.; Cheikhrouhou, O.; Maqsood, H.; Ibrahim, M.; Mohammad, F. Machine learning-based efficient prediction of positive cases of waterborne diseases. *BMC Med. Inform. Decis. Mak.* **2023**, *23*, 11. [CrossRef] [PubMed]
8. Mubeen, A.; Ibrahim, M.; Bibi, N.; Baz, M.; Hamam, H.; Cheikhrouhou, O. Alts: An adaptive load-balanced task scheduling approach for cloud computing. *Processes* **2021**, *9*, 1514. [CrossRef]
9. Salvakkam, D.B.; Saravanan, V.; Jain, P.K.; Pamula, R. Enhanced Quantum-Secure Ensemble Intrusion Detection Techniques for Cloud Based on Deep Learning. *Cogn. Comput.* **2023**, *15*, 1593–1612. [CrossRef]
10. Chakravarthi, S.S.; Kannan, R.J.; Natarajan, V.A.; Gao, X.Z. Deep Learning Based Intrusion Detection in Cloud Services for Resilience Management. *Comput. Mater. Contin.* **2022**, *71*, 3.
11. Priya, S.; Ponmagal, R.S. Network Intrusion Detection System Based Security System for Cloud Services Using Novel Recurrent Neural Network-Autoencoder (NRNN-AE) and Genetic. *Adv. Sci. Technol.* **2023**, *124*, 729–737.
12. Jyothsna, V.; Manisha, C.; NanduSri, B.S. Intrusion Detection System for Detection of DDoS Attacks in Cloud Environment. *Res. Sq.* **2023**, *preprint*. [CrossRef]
13. Basahel, A.M.; Yamin, M.; Basahel, S.M.; Lydia, E.L. Enhanced Coyote Optimization with Deep Learning Based Cloud-Intrusion Detection System. *Comput. Mater. Contin.* **2023**, *74*, 4319–4336. [CrossRef]
14. Maheswari, K.G.; Siva, C.; Nalinipriya, G. Optimal cluster-based feature selection for intrusion detection systems in web and cloud computing environments using hybrid teacher learning optimization enables deep recurrent neural networks. *Comput. Commun.* **2023**, *202*, 145–153. [CrossRef]
15. Mayuranathan, M.; Saravanan, S.K.; Muthusenthil, B.; Samydurai, A. An efficient optimal security system for intrusion detection in a cloud computing environment using hybrid deep learning technique. *Adv. Eng. Softw.* **2022**, *173*, 103236. [CrossRef]
16. Toldinas, J.; Venčkauskas, A.; Damaševičius, R.; Grigaliūnas, Š.; Morkevičius, N.; Baranauskas, E. A novel approach for network intrusion detection using multistage deep learning image recognition. *Electronics* **2021**, *10*, 1854. [CrossRef]
17. Srilatha, D.; Thillaiarasu, N. Implementation of Intrusion detection and prevention with Deep Learning in Cloud Computing. *J. Inf. Technol. Manag.* **2023**, *15*, 1–18.
18. Balamurugan, E.; Mehbodniya, A.; Kariri, E.; Yadav, K.; Kumar, A.; Haq, M.A. Network optimization using defender system in cloud computing security-based intrusion detection system with game theory deep neural network (IDSGT-DNN). *Pattern Recognit. Lett.* **2022**, *156*, 142–151. [CrossRef]
19. Prabhakaran, V.; Kulandasamy, A. Hybrid semantic deep learning architecture and optimal advanced encryption standard key management scheme for secure cloud storage and intrusion detection. *Neural Comput. Appl.* **2021**, *33*, 14459–14479. [CrossRef]
20. Ravi, C.N.; Karthik, T.S.; Manikandan, K.; Kalaivaani, P.; Chopkar, P.N.; Srivastava, A. Cauchy Grasshopper Optimization Algorithm with Deep Learning Model for Cloud-Enabled Cyber Threat Detection System. In Proceedings of the 2023 7th International Conference on Intelligent Computing and Control Systems (ICICCS), Madurai, India, 17–19 May 2023; pp. 202–208.
21. Jisna, P.; Jarin, T.; Praveen, P.N. Advanced intrusion detection using deep learning-LSTM network on cloud environment. In Proceedings of the 2021 Fourth International Conference on Microelectronics, Signals & Systems (ICMSS), Kollam, India, 18–19 November 2021; pp. 1–6.
22. Alghamdi, R.; Bellaiche, M. A deep intrusion detection system in lambda architecture based on edge cloud computing for IoT. In Proceedings of the 2021 4th International Conference on Artificial Intelligence and Big Data (ICAIBD), Chengdu, China, 28–31 May 2021; pp. 561–566.

23. Alzubi, O.A.; Alzubi, J.A.; Alazab, M.; Alrabea, A.; Awajan, A.; Qiqieh, I. Optimized machine learning-based intrusion detection system for fog and edge computing environment. *Electronics* **2022**, *11*, 3007. [CrossRef]
24. Ali, M.H.; Zolkipli, M.F. Intrusion-detection system based on fast learning network in cloud computing. *Adv. Sci. Lett.* **2018**, *24*, 7360–7363. [CrossRef]
25. Sharma, P.; Dinkar, S.K. A linearly adaptive Sine–cosine algorithm with application in a deep neural network for feature optimization in arrhythmia classification using ECG signals. *Knowl.-Based Syst.* **2022**, *242*, 108411. [CrossRef]
26. Anbarasi, A.; Ravi, T.; Manjula, V.S.; Brindha, J.; Saranya, S.; Ramkumar, G.; Rathi, R. A modified deep learning framework for arrhythmia disease analysis in medical imaging using electrocardiogram signal. *BioMed Res. Int.* **2022**, *2022*, 5203401. [CrossRef]
27. Bacanin, N.; Budimirovic, N.; K, V.; Strumberger, I.; Alrasheedi, A.F.; Abouhawwash, M. Novel chaotic oppositional fruit fly optimization algorithm for feature selection applied on COVID-19 patients' health prediction. *PLoS ONE* **2022**, *17*, e0275727. [CrossRef] [PubMed]
28. Alohali, M.A.; Elsadig, M.; Al-Wesabi, F.N.; Al Duhayyim, M.; Mustafa Hilal, A.; Motwakel, A. Enhanced Chimp Optimization-Based Feature Selection with Fuzzy Logic-Based Intrusion Detection System in Cloud Environment. *Appl. Sci.* **2023**, *13*, 2580. [CrossRef]

electronics

MDPI

Article

Semi-Supervised Alert Filtering for Network Security

Hyeon gy Shon, Yoonho Lee and MyungKeun Yoon *

Department of Computer Science, Kookmin University, 77, Jeongneung-ro, Seongbuk-gu, Seoul 02707, Republic of Korea; forever2331@kookmin.ac.kr (H.g.S.); yhya0904@kookmin.ac.kr (Y.L.)
* Correspondence: mkyoon@kookmin.ac.kr; Tel.: +82-2-910-4806

Abstract: Network-based intrusion detection systems play a pivotal role in cybersecurity, but they generate a significant number of alerts. This leads to alert fatigue, a phenomenon where security analysts may miss true alerts hidden among false ones. To address alert fatigue, practical detection systems enable administrators to divide alerts into multiple groups by the alert name and the related Internet Protocol (IP) address. Then, some groups are deliberately ignored to conserve human resources for further analysis. However, the drawback of this approach is that the filtering basis is so coarse-grained that some true alerts are also ignored, which may cause critical security issues. In this paper, we present a new semi-supervised and fine-grained filtering method that uses not only alert names and IP addresses but also semi-supervised clustering results from the alerts. We evaluate our scheme with both a private dataset from a security operations center and a public dataset from the Internet. The experimental results demonstrate that the new filtering scheme achieves higher accuracy and saves more human resources compared to the current state-of-the-art method.

Keywords: intrusion detection; false positive; cyber security; alert fatigue; semi-supervised learning; prototype clustering

Citation: Shon, H.g.; Lee, Y.; Yoon, M. Semi-Supervised Alert Filtering for Network Security. *Electronics* **2023**, *12*, 4755. https://doi.org/10.3390/electronics12234755

Academic Editors: Tomasz Rak and Dariusz Rzońca

Received: 26 October 2023
Revised: 17 November 2023
Accepted: 18 November 2023
Published: 23 November 2023

1. Introduction

Network-based intrusion detection and prevention systems (IDPS) have played a pivotal role in cybersecurity [1–3]. Located at network gateways or critical points in enterprise networks, they inspect every packet to find suspicious activities or cyberattacks. For decades, the IDPS has evolved to include more than thousands of detection rules, or signatures, most of which are represented as strings or regular expressions. Whenever an IDPS finds any signature from a packet, it triggers an alert. If the packet is really related to cyberattacks, the alert becomes a *true alert*; otherwise, a *false alert* occurs, also called a false positive.

Unfortunately, IDPSs have been notorious for their false alerts, resulting in a phenomenon called *alert fatigue*, a huge number of false alerts overwhelming security analysts to ignore or fail to respond to a small number of true alerts [4,5]. Because writing intrusion detection rules is a challenging task, finding the right balance between an overly specific rule and an overly general one is hard to determine [1]. Most rules are developed to catch general attacks or vulnerabilities because IDPSs can be deployed in any environment. This means that IDPSs would generate alerts when a packet includes a suspicious string, and therefore a tremendous volume of false alerts is inevitable.

The sheer number of alerts has consistently overwhelmed human resources, or security analysts, leading to the pervasive issue of alert fatigue. Reducing alert fatigue is a challenging yet essential task for the security industry, particularly in security operations centers (SOC) where alerts are gathered from numerous monitoring sensors and a limited number of analysts work around the clock, 365 days a year [6,7]. Despite decades of efforts by researchers and industries to address the alert fatigue problem in IDPSs [4–6,8–10], a comprehensive solution has not been achieved yet.

Existing Solutions. Security analysts can usually analyze only a small portion of alerts. Recent studies show that one analyst can investigate 76 alerts per day, but the number of daily alerts is greater than 10,000 [4]. Therefore, some true alerts of real threats should be selected first with a high priority whereas false alerts should be filtered out automatically. In this sense, a number of solutions have been proposed, which can be classified into three categories as follows:

First, additional context information very specific to a particular site can be used to automatically filter out some alerts. For example, a security analyst may ignore any alert about Windows Internet Information Services (IIS) if he/she knows the site information that the victim is an Apache web server [4]. However, very detailed context information should be timely updated, which is often impossible in practice because of scarce human resources or poor cooperation between security and operation teams.

Second, extra data or Cyber Threat Intelligence (CTI) information from other sources can be used. For example, additional server logs, alerts from other IDPSs, or blacklists of IP addresses of well-known attackers, can give a hint as to whether the alert is true or false [5,6,8]. However, extra cost and effort are required to purchase and fully utilize CTI information.

Third, alerts can be classified into multiple groups by security analysts and some groups are filtered out together automatically. This heuristic is widely used in the industry because many false alerts can be suppressed at once. The current grouping method is generally based on only the alert name and IP address, which are available from most of the IDPS alerts [9,10]. This approach can be implemented easily because security administrators can write filtering rules consisting of alert names and IP addresses. A group of alerts satisfying the same rule can be automatically ignored together. We call it the state-of-the-art (SOTA) method to mitigate the alert fatigue problem. However, a new security hole appears with SOTA; any detection of real attacks to the same IP address and alert name is automatically filtered out, and the possible detection would be evaded because of the coarse-grained filtering rule. We call this new problem of SOTA as *coarse-grained alert-filtering problem*.

Proposed Solution. We observe that the coarse-grained alert-filtering problem arises because the current filtering practice lacks precision, relying solely on alert names and IP addresses. In this paper, we present a new fine-grained filtering scheme that can more precisely identify groups of false alerts by considering the alert name, the IP address, and the clustering results from the packet payloads of the IDPS alerts via semi-supervised learning. The new scheme identifies a number of large clusters, and only a few samples from each cluster are manually analyzed to determine if the cluster consists of only false alerts, true alerts, or a mix of both. If the cluster includes either false or true alerts only, the alerts of the same cluster can be processed the same way without further manual analysis; otherwise, the alerts of the cluster can be handed over to security analysts for individual analysis. We refer to our scheme as *Semi-supervised Alert Filtering (SAFI)*. The contribution of this paper can be summarized as follows:

- We introduce the important and practical coarse-grained alert-filtering problem and reveal the limitations of SOTA.
- We introduce SAFI, a new alert-filtering scheme, not only to process alerts more precisely but also to save security analysts' time and efforts. This improvement comes from the new features extracted from packet payloads embedded in most IDPS alerts and their semi-supervised clustering results.
- Both public and private datasets are used in experiments to show that SAFI outperforms SOTA in terms of the accuracy of data analysis as well as the analysis cost.

Encrypted traffic. Most alerts are generated by IDPSs when network packets include suspicious strings or patterns. If the packets are encrypted and the IDPS is not entrusted with decryption keys, some alerts cannot be triggered, whereas a separate decryption box can be deployed to convert encrypted packets into plain ones for IDPSs in enterprise networks [11]; this incurs extra cost and privacy issues. The fundamental limitations of

IDPSs, specifically network-based monitoring devices, are beyond the scope of this paper. Here, we focus on an IDPS alert with a packet payload, which amounts to a large volume of data overwhelming security analysts.

The rest of the paper is organized as follows. We introduce the problem and motivation in Section 3. The new scheme is presented in Section 4, and the experimental results are discussed in Section 5. Sections 2 and 6 cover related work and conclusions, respectively.

2. Related work

Alert Fatigue and SOC. Alert fatigue is a common problem for SOCs where threat detection systems such as IDPS and SIEM generate high rates of false alerts [4,5,12,13]. Commercial products adopt general rules to cover more threats than specific exploits. These general rules cause a significant number of false alerts [1,13,14]. Recent threat detection products provide a tuning method to reduce false alerts [9,10], or SOCs have their own heuristics for determining and fixing false alerts [15]. However, over-tuning causes missing real attacks [9,10,15]. In this paper, we present the first semi-automatic tuning method to reduce alert fatigue by introducing new features from alerts.

Network Intrusion Detection and Prevention. A network-based intrusion detection system (IDS) inspects packets to find cyberattacks and suspicious activities. An intrusion prevention system (IPS) not only identifies threats but also blocks the threat [16]. Although these systems have played a pivotal role in cybersecurity in recent decades [1–3,14,17], there are two issues; first, as more network packets are encrypted, IDPSs cannot look up attack signatures. To tackle this problem, a decryption box can be deployed to obtain plain packets [11], or anomaly detection can be used [18,19]. Second, too many false alerts are generated, resulting in alert fatigue [4,5], the main topic of this paper.

Machine Learning for Network Intrusion Detection. Given the huge numbers of alerts from IDPSs, machine learning has become an alternative to expensive manual analysis [20–25]. Although there are many machine learning approaches for network IDPSs, RAID is the first to use new features from packet payloads of alerts and provide clustering-based tuning to reduce alert fatigue.

We emphasize that there have been a lot of research attempts to reduce false alerts, but to the best of our knowledge, this paper is the first that filters out false alerts on the content of an IDPS packet.

3. Problem and Motivation

Alert fatigue, or threat alert fatigue, is an information overload problem in which security analysts miss true attack alerts hidden in the noise of false alarms [5]. Alerts are generated not only by network-based IDPSs [1], but also by other security devices such as host-based IDPSs [26], Endpoint/Network Detection and Response (EDR/NDR), Web Application Firewalls (WAF), and Security Information and Event Management (SIEM). This paper focuses on alerts of network-based IDPSs, but the main idea can be applied to other types of security devices.

We assume that IDPSs are deployed at a gateway and critical points in enterprise networks and they inspect every packet to find suspicious activities or attacks as shown in Figure 1. A network tapping device can copy packets to IDPSs, or a mirroring port of a network switch can provide the IDPS with any packet going through the switch. If a packet contains any attack signatures, the IDPS generates an alert or event. In this paper, we use alerts and events interchangeably.

If an alert is triggered by real attacks, we call it a true alert; otherwise, it is called a false alert. In this sense, a true alert is a true positive whereas a false alert is a false positive. An alert generally consists of several fields of time, a source IP address, a destination IP address, a source port number, a destination port number, an alert name, and a packet payload that has triggered the alert. Although a packet payload is optional for an IDPS alert, we observe that most IDPSs provide it to give security analysts additional information about the alert. We denote an alert as *alert:={time, sIP, dIP, sPort, dPort, alert name, payload}*. A maximum packet payload is around 1600 bytes in length.

Figure 1 shows that a security operations center (SOC) collects alerts from IDPSs where security analysts review the alerts for further analysis. The number of alerts always overwhelms the number of analysts or human resources. Therefore, an alert-filtering step is essential for automatically processing insignificant alerts. If the filtering step is not designed carefully, true alerts may also be filtered out or ignored mistakenly. This may cause a serious security hole.

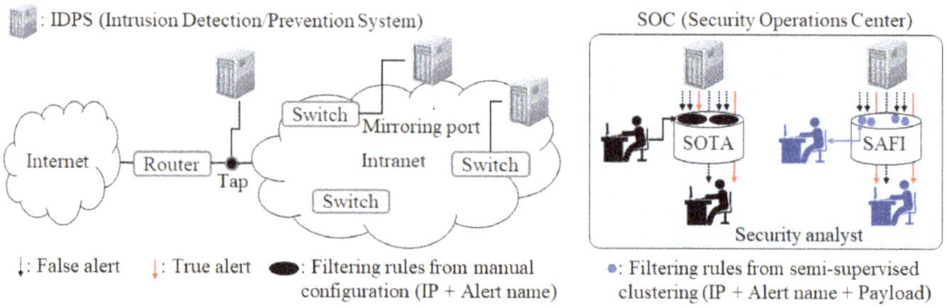

Figure 1. Security monitoring service of SOC on multiple IDPSs deployed in an enterprise network. True alerts can be automatically ignored by SOTA mistakenly. The fine-grained filtering of SAFI can mitigate this problem.

Coarse-grained alert-filtering problem. Latest IDPS and SIEM products provide a simple and practical method for tuning false alerts by establishing a list of exception or filtering rules; the current best practice is that frequent alerts of the same alert name and IP address are grouped together and ignored at once [9,10]. For example, $alert_1$, $alert_2$, and $alert_3$ have the same internal destination IP address and alert name in Figure 2. After manual investigation, a security operator or analyst concludes that these alerts are not real threats and adds a new filtering rule that consists of the IP address and alert name, $[dIP_1||alert\ name_1]$. This filtering rule would suppress or remove any alert of the same group. However, this simple filtering rule may cause a serious problem; a real attack is detected by an IDPS as $alert_4$ later, but this alert would be falsely and automatically ignored by the filtering rule. We call it a *coarse-grained alert-filtering alert* problem that really happened with the Target data breach in 2013 [7]. The problem still exists because a filtering rule still consists of an alert name and an IP address, too coarse-grained to accurately distinguish a true alert and a false alert.

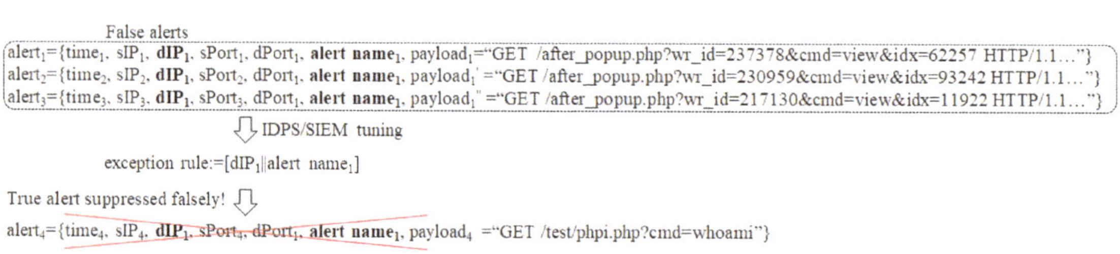

Figure 2. Current best practice for reducing alert fatigue. Exception or filtering rules are based on the alert name and IP address only, which causes the automatic ignoring of a true alert.

Analysis cost and accuracy. We assume that a huge number of m IDPS alerts are given to security analysts. If there were infinite human resources, all of the m alerts could be manually analyzed. If there are no human mistakes, the analysis accuracy would be perfect. Because this is impossible, filtering rules are essential in automatically ignoring as many alerts as possible; then, the remaining alerts are manually analyzed by security analysts. For example, in SOTA, after a few sample alerts are randomly chosen from a group of alerts with the same alert name and IP address, only the samples are manually analyzed. If all of the samples are proven to be false (true) alerts, all alerts from the same group would be considered or predicted as false (true) alerts, and therefore no human resources are spent except the samples. If the samples consist of both true and false alerts, the group is called a mixed group. Ideally, all alerts from the mixed group should be manually analyzed to prevent any security holes. This strategy of SOTA may cause two problems; first, the prediction based on the samples would be wrong. Second, alerts from the mixed groups are too many to be manually analyzed. In practice, only a small portion of these alerts are analyzed, and the others are ignored. In addition, because the critical decision is totally dependent on the heuristics of security analysts, no consistent policy is established in alert filtering.

In this paper, we define the analysis cost for SOTA as the ratio of the number of alerts that security analysts manually analyze to the number of total alerts. For the mixed group, all alerts of the group should be analyzed; otherwise, only sample alerts are analyzed. We also define accuracy as the ratio of the number of correctly estimated alerts on whether they are true or false to the number of total alerts. In this sense, the purpose of SOTA and SAFI is to minimize the analysis cost and to maximize the accuracy.

Motivation. In this paper, we argue that a packet payload embedded inside an alert makes good features for fine-grained filtering rules. The intuition is that frequent false alerts are often caused by similar packet payloads that include the same alert name and the same server IP address. On the contrary, we also observed that the packet payload of a true alert is generally quite different from those of repetitive false alerts as in Figure 2. This motivated us to study a new clustering method that divides a group of alerts of the same name and IP address into multiple subgroups based on the content of a packet payload, which has not been studied in previous work.

4. SAFI: Semi-Supervised Alert Filtering

We present SAFI to mitigate the coarse-grained alert-filtering problem; SAFI distinguishes false alerts and potential true alerts by clustering alerts of the same group into smaller and homogeneous subgroups, or clusters, based on the new features from a packet payload inside an alert. The key idea is to analyze a few samples per cluster rather than per group in a fine-grained way; if all the samples from a cluster are either true alerts or false alerts, all alerts of the cluster are automatically processed the same way to save human resources or analysis costs.

There are two main steps for SAFI to cluster alerts into fine-grained subgroups. Each subgroup has alerts of not only the same alert name and IP address but also similar packet payloads. In this paper, two packets are considered to have similar payloads if their byte sequences are similar to each other. The first step of SAFI is to simply divide alerts into groups of the same alert name and IP address the same as in SOTA [9,10]. Then, SAFI extracts features from packet payloads of alerts, and the features are converted into a fixed-size vector. In this paper, we use Term Frequency-Inverse Document Frequency (TF-IDF) for vectorization (https://scikit-learn.org, accessed on 17 November 2023), but any other schemes can be used instead. In the second step, the fixed-size vectors are used to cluster alerts from the same group into multiple subgroups. Then, n-samples per cluster are selected and manually analyzed. Depending on the analysis results, each cluster is considered as a true-alert cluster, or a false-alert cluster, or a mixed cluster. If the cluster is a true-alert cluster or a false-alert cluster, all alerts of the cluster are considered the same way. We explain each step in detail, and provide how to compare SAFI with SOTA in terms of the accuracy and analysis cost.

4.1. Vectorization and Clustering

In SAFI, each alert is represented as a fixed-size vector. The vector is constructed from a packet payload embedded in an alert. The packet payload of an alert is a byte sequence less than 1600 bytes; we use the TF-IDF vectorization scheme to transform the payload into a fixed-size vector because of its simplicity and efficiency. In this sense, we treat the payload of an alert as a text. We believe that two alerts of very similar packet payloads with the same alert name and IP address would be either true alerts or false alerts. Therefore, a cluster of similar packet payloads can be safely filtered out as either a true-alert cluster or a false-alert cluster.

The soundness of SAFI comes from our observation on real IDPS alert datasets; the same false alerts are repeatedly generated if a certain attack signature or string appears repeatedly. For example, a furniture website may include the string of "drop table" as a furniture type, but this may cause IDPSs to trigger an alert of a SQL injection attack [23]. The problem is that the same alerts repeat whenever that web page is reached by innocent clients. These repeated alerts should often include the same server IP address and alert name. The packet payloads of the alerts probably have very similar contents except a few bytes. The difference can be caused by the packet size or meta information of the packets.

When we apply TF-IDF vectorizer to these packets, the resulting vectors look very similar to each other, and therefore the distance between them is also small. This means that they should form a dense cluster when a clustering algorithm is applied. In this paper, we use a prototype clustering algorithm of [27] because the algorithm is fast and its performance was confirmed against a security dataset. However, any clustering algorithm can be used instead for SAFI.

We explain the prototype clustering in details [27,28]. The quality of clustering results depends on how to properly choose the prototypes. The first prototype alert is randomly selected. Then, for this current prototype vector, we find those vectors that are similar to the prototype; two vectors of v_i and v_j are considered similar only if their cosine similarity is larger than θ, a predefined threshold. All the similar vectors to the current prototype alert make a new cluster. Then, the next prototype alert is selected. According to the heuristic algorithm of [27], we select an alert with the largest distance to the current prototype as the next prototype. Then, the new prototype and its similar vectors make the second new cluster. A vector that already belongs to any cluster is excluded for the next step. This process is repeated until every vector belongs to its cluster. Finally, the clustering algorithm merges two clusters if the similarity of their prototype alerts is beyond θ. The algorithm starts with individual prototypes as singleton clusters before successively merging the two closest clusters. The algorithm terminates when the distance between the closest clusters is larger than $(1 - \theta)$ [27,28]. The default value of θ is set to 0.9.

Figure 3 shows the difference between SOTA and SAFI; only an alert name and an IP address are used to group alerts in SOTA, which in this example are 10.12.1.2 and XSS (Cross-Site Scripting) attack. On the contrary, additional information about the packet payload is also used in SAFI, which can distinguish true alerts and false alerts.

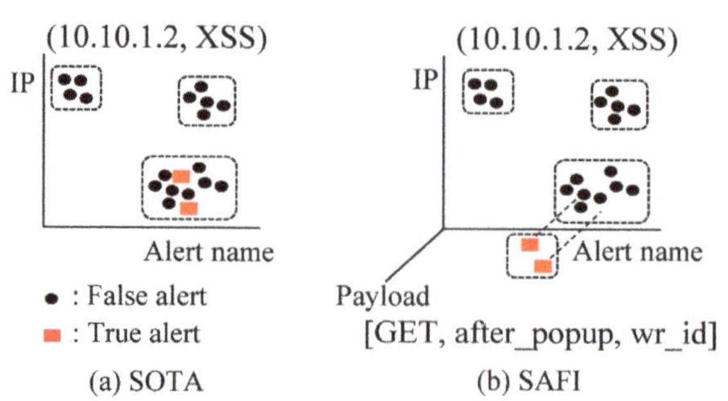

Figure 3. SOTA vs. SAFI in terms of new payload features. Black circles are false alerts and red boxes are true alerts.

4.2. Sampling and Filtering

After clustering on alert vectors is finished, a number of clusters are generated. We define the number of samples per cluster, denoted as n, to randomly select alert samples for each cluster. In SAFI, we conservatively conclude that a cluster is a mixed cluster if at least one sample out of n is different from the others in terms of a true or false alert. If a cluster includes less than n alerts, all alerts of the cluster become samples.

If a cluster is a mixed cluster, we assume that all alerts should be manually analyzed to prevent any security holes. When n becomes larger, we decide the type of cluster more carefully. For example, suppose that n is set to 10. Then, only when 10 sample alerts are all true (false) alerts, the cluster is determined as a true (false) cluster and the automatic filtering can be applied to the other alerts from the cluster. On the contrary, the cluster type is determined only with one sample when n is set to one. It is interesting that the accuracy of SAFI still remains high even with $n = 1$ because there are many dense clusters of similar packet payloads.

4.3. Comparison of SAFI and SOTA

We compare SAFI and SOTA in terms of the analysis cost and accuracy. A combination of an alert name and an IP address makes different alert groups in SOTA [9,10]. In SAFI, each group of SOTA is further divided into subgroups, or clusters, via clustering on the packet payload of an alert. Therefore, the number of clusters of SAFI is bigger than the number of groups of SOTA. We assume that n alert samples are randomly selected from a SAFI cluster and a SOTA group, and these samples are manually analyzed. In this sense, the analysis cost of SAFI might be higher than that of SOTA. This is actually true when $n = 1$. However, the accuracy of SOTA is expected low because many groups from SOTA must be mixed up with true alerts and false alerts.

When n is greater than 1, multiple samples are randomly selected from a SOTA group. Therefore, these multiple samples may reveal that the group is mixed, and in this paper, we leave the alerts of the group to security analysts. This increases the analysis cost of SOTA, and the analysis cost of SAFI becomes lower than that of SOTA.

In this paper, we define the analysis cost as the ratio of the number of alerts that requires manual analysis to the number of total alerts. The analysis cost ranges from 0 to 1; when all alerts are manually analyzed, the cost becomes 1. We assume that a manual analysis correctly determines if the given alert is true or false.

For accuracy, SAFI outperforms SOTA because there are many dense clusters in SAFI. A dense cluster often includes false alerts of similar packet payloads. Therefore, the sample alerts represent the original cluster quite well. The definition of accuracy, precision, recall, and F1 score is as follows: accuracy is the ratio of the number of correctly predicted labels to the total number of alerts. The precision is the ratio of the number of correctly predicted true alerts to the number of alerts that are predicted as true alerts. The recall is the ratio of the number of correctly predicted true alerts to the number of true alerts. It is well known that IDPS alert datasets may include a small number of true alerts but a large number of false alerts. In this case, if SAFI or SOTA predicts all alerts as false, the accuracy and precision would become high but the recall would be low. On the contrary, if SAFI or SOTA predicts all alerts as true, the recall would become high but the precision would be low. Actually, the precision and recall are a trade-off. Therefore, the F1 score is a good choice to measure the quality of label prediction, which is defined as $2 \times \frac{precision \times recall}{precision + recall}$. In this paper, we use the F1 score when comparing SAFI and SOTA.

5. Experiments

We experimentally compare SAFI and SOTA using two datasets, a private dataset from a SOC and a public one from the Internet. We implement both SAFI and SOTA for alert filtering, and measure their analysis cost and F1 score. The experimental results show that SAFI outperforms SOTA by a large degree in both metrics.

5.1. Experimental Setup and Dataset

We use two datasets for experiments; the first dataset is a private dataset of IDPS alerts provided by a SOC in South Korea for academic purposes only. The alerts are related to web attacks, collected over several weeks. All alerts were already manually analyzed and each alert is labeled as either a true or false alert. The IP addresses and alert names are replaced with their hashed values to prevent any information leakage. We use the private dataset to compare SAFI and SOTA.

The second is an open dataset available from the Internet (https://www.isi.csic.es/dataset/, accessed on 17 November 2023), called CSIC2010. The dataset is about normal and anomalous web requests rather than IDPS alerts. The dataset consists of a training part and a test part, and we use only the test part in this paper; we use the normal web requests as false alerts and the anomalous web requests as true alerts because each request is one web-application packet payload. However, the CSIC2010 dataset does not include IP addresses and alert names. Therefore, we use it to confirm the usefulness of SAFI clustering. The value of the second dataset is to guarantee the experiment's reproducibility. The statistics of the two datasets are summarized in Table 1.

Table 1. Statistics for datasets.

Dataset	Alerts	True Alerts	False Alerts
Private	135,852	27,202	108,650
Public	60,668	24,668	36,000

5.2. Experimental Results

We perform two types of experiments with the private and public datasets, respectively. In the first type of experiments, the private IDPS dataset is used to compare SAFI and SOTA in terms of the analysis cost and F1 score. Figure 4a,b show the experimental results. The number of groups with different alert name and IP address is 836 in SOTA. Therefore, when $n = 1$, only 836 samples are manually analyzed, which makes the analysis cost of 836/135,852. As n increases, the analysis cost of SOTA also increases. This is because more groups are identified as mixed from the samples. The serious problem of SOTA is that the F1 score does not improve as n increases. Groups in SOTA are mixed up with true alerts and false alerts because of the coarse-grained basis for grouping, which is shown in Figure 4a. Considering SOTA is a practice in industries, potential security holes may exist in SOCs.

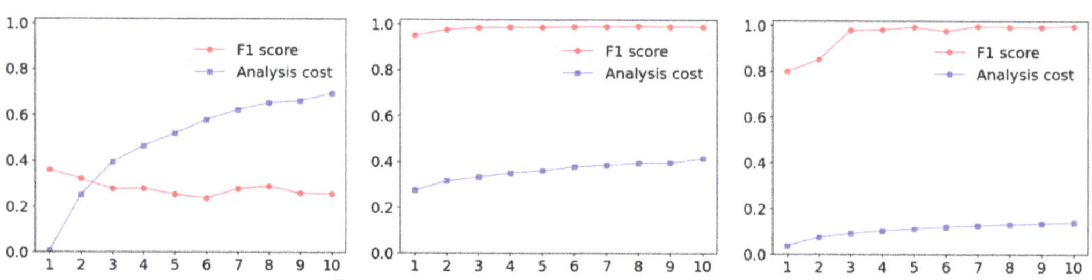

(a) SOTA for private dataset (b) SAFI for private dataset (c) SAFI for public dataset

Figure 4. Comparison of SOTA and SAFI on the private dataset in the left and middle plots. SAFI outperforms SOTA in both F1 score and analysis cost. The right plot shows that high F1 score and low analysis cost are obtained by SAFI clustering for the public dataset.

On the contrary, SAFI can reduce the analysis cost whereas the F1 score can be maintained much higher than SOTA as shown in Figure 4b. It is encouraging that the F1 score is above 0.94 even with $n = 1$. As n increases, both the F1 score and analysis cost also increase. At $n = 2$, the F1 score becomes 0.97 and the analysis cost is 0.33. Therefore, if one third of alerts are manually analyzed, we can determine if any alert is true or false with the F1 score of 0.97.

In the second type of experiment, we use the public dataset to confirm that SAFI is still useful in automatically filtering out alerts, whereas the F1 score is kept high. This dataset can be downloaded from the Internet for reproducible experiments. Because the second dataset does not include IP addresses and alert names, we cannot test SOTA. When $n = 1$, the F1 score of SAFI is around 0.8. When n is not less than 3, the F1 score jumps to higher than 0.97, as in Figure 4c.

Finally, we show alert examples from the second dataset as in Table 2. After SAFI is applied to the dataset, the first and second alerts are taken from a false cluster. The third and fourth alerts are from a true-alert cluster. The similarity between alerts from the same cluster is greater than 0.92 whereas the similarity between alerts from different clusters is less than 0.16. These examples show that alerts from a SAFI cluster are closely related, which leads to a high F1 score and low analysis cost.

Table 2. Alert examples from the public dataset.

No.	Packet Payload	Label
1	GET http://localhost:8080/tienda1/imagenes/1.gif HTTP/1.1	false
2	GET http://localhost:8080/tienda1/imagenes/2.gif HTTP/1.1	false
3	GET http://localhost:8080/tienda1/publico/anadir.jsp? id=2&nombre=Jam%F3n+Ib%E9rico&precio=85&cantidad=%27%3B +DROP+TABLE+usuarios%3B+SELECT+*+FROM+datos+WHERE+nombre+LIKE+%27%25&B1 =A% F1adir+al+carrito HTTP/1.1	true
4	POST http://localhost:8080/tienda1/publico/anadir.jsp HTTP/1.1\nid=2&nombre=Jam%F3n+Ib%E9rico&precio=85&cantidad=%27%3B +DROP+TABLE+usuarios%3B+SELECT+*+FROM+datos+WHERE+nombre+LIKE+%27%25&B1 =A% F1adir+al+carrito	true

6. Conclusions

In this paper, we introduced the alert fatigue problem in network security monitoring and analyzed the limitations of the state-of-the-art method. We presented a new alert-filtering scheme on semi-supervised learning that can process alerts more precisely as well as save security analysts' time and effort much more than the current best method by a large degree. This work was motivated by the observation that current practices may suppress the sheer volume of false alerts, but some true alerts are automatically and incorrectly ignored together because of the coarse-grained alert grouping. We hope that our scheme will be practically deployed at SOCs to mitigate the alert fatigue problem. In this paper, we focused on network IDPS alerts only. Future work would extend to endpoint alerts such as a host, PC, or server, which may include unknown information about potential cyberattacks.

Author Contributions: Conceptualization, H.g.S. and M.Y.; methodology, H.g.S. and M.Y.; software, H.g.S. and Y.L.; validation, H.g.S., Y.L. and M.Y.; formal analysis, M.Y.; investigation, H.g.S. and M.Y.; resources, M.Y.; data curation, H.g.S.; writing—original draft preparation, H.g.S.; writing—review and editing, M.Y.; visualization, H.g.S.; supervision, M.Y.; project administration, M.Y.; funding acquisition, M.Y. All authors have read and agreed to the published version of the manuscript.

Funding: This work was supported by the Institute of Information & Communications Technology Planning & Evaluation (IITP) grant funded by the Korean government (MSIT) (RS-2023-00235509), Development of Security Monitoring Technology.

Data Availability Statement: The dataset includes malware files that may harm computer systems and therefore should not be open to the public. Please contact the corresponding author for the dataset if you need it.

Conflicts of Interest: The authors declare no conflict of interest.

References

1. Snort. Snort—Network Intrusion Detection & Prevention System. Available online: https://www.snort.org/ (accessed on 20 September 2023).
2. Zeek. An Open Source Network Security Monitoring Tool. Available online: https://www.zeek.org/ (accessed on 20 September 2023).
3. Tidjon, L. Intrusion Detection Systems: A Cross-Domain Overview. *IEEE Commun. Surv. Tutor.* **2019**, *21*, 3639–3681. [CrossRef]
4. Automated Incident Response: Respond to Every Alert. Available online: https://swimlane.com/blog/automated-incident-response-respond-every-alert/ (accessed on 20 September 2023).
5. Hassan, W.; Guo, S.; Li, D.; Chen, Z.; Jee, K.; Li, Z.; Bates, A. NoDoze: Combatting Threat Alert Fatigue with Automated Provenance Triage. In Proceedings of the Network and Distributed Systems Security Symposium (NDSS), San Diego, CA, USA, 24–27 February 2019. [CrossRef]
6. Rosso, M.; Campobasso, M.; Gankhuyag, G.; Allodi, L. SAIBERSOC: Synthetic Attack Injection to Benchmark and Evaluate the Performance of Security Operation Centers. In Proceedings of the Annual Computer Security Applications Conference, Austin, TX, USA, 7–11 December 2020; pp. 141–153. [CrossRef]
7. Alahmadi, B.; Axon, L.; Martinovic, I. 99% False Positives: A Qualitative Study of SOC Analysts' Perspectives on Security Alarms. In Proceedings of the 31st USENIX Security Symposium (USENIX Security 22), Boston, MA, USA, 10–12 August 2022; pp. 2783–2800.
8. Zeng, J.; Chua, Z.L.; Chen, Y.; Ji, K.; Liang, Z.; Mao, J. WATSON: Abstracting Behaviors from Audit Logs via Aggregation of Contextual Semantics. In Proceedings of the Network and Distributed Systems Security Symposium (NDSS), Virtual, 21–25 February 2021. [CrossRef]
9. Tuning False Positives. Available online: https://www.ibm.com/docs/en/qsip/7.4?topic=performance-tuning-false-positives (accessed on 20 September 2023).
10. Tuning Intrusion Policies Using Rules. Available online: https://www.cisco.com/c/en/us/td/docs/security/firepower/70/configuration/guide/fpmc-config-guide-v70/tuning_intrusion_policies_using_rules.html (accessed on 20 September 2023).
11. Fan, J.; Guan, C.; Ren, K.; Cui, Y.; Qiao, C. SPABox: Safeguarding Privacy During Deep Packet Inspection at a MiddleBox. *IEEE/Acm Trans. Netw.* **2017**, *25*, 3753–3766. [CrossRef]
12. Axelsson, S. The Base-Rate Fallacy and Its Implications for the Difficulty of Intrusion Detection. In Proceedings of the 6th ACM Conference on Computer and Communications Security, Singapore, 1–4 November 1999; pp. 1–7. [CrossRef]
13. Pietraszek, T. Using Adaptive Alert Classification to Reduce False Positives in Intrusion Detection. In *Recent Advances in Intrusion Detection, Proceedings of the 7th International Symposium, RAID 2004, Sophia Antipolis, France, 15–17 September 2004*; Jonsson, E., Valdes, A., Almgren, M., Eds.; Springer: Berlin/ Heidelberg, Germany, 2004; pp. 102–124. [CrossRef]
14. Paxson, V. Bro: A System for Detecting Network Intruders in Real-Time. *Comput. Netw.* **1999**, *31*, 2435–2463. [CrossRef]
15. Kokulu, F.B.; Soneji, A.; Bao, T.; Shoshitaishvili, Y.; Zhao, Z.; Doupé, A.; Ahn, G.J. Matched and Mismatched SOCs: A Qualitative Study on Security Operations Center Issues. In Proceedings of the 2019 ACM SIGSAC Conference on Computer and Communications Security, London, UK, 11–15 November 2019; pp. 1955–1970. [CrossRef]
16. Intrusion Detection System (IDS) vs. Intrusion Prevention System (IPS). Available online: https://www.checkpoint.com/cyber-hub/network-security/what-is-an-intrusion-detection-system-ids/ids-vs-ips/ (accessed on 20 September 2023).
17. Suricata. Available online: https://suricata.io/ (accessed on 20 September 2023).
18. Areström, E.; Carlsson, N. Early Online Classification of Encrypted Traffic Streams using Multi-fractal Features. In Proceedings of the IEEE INFOCOM 2019—IEEE Conference on Computer Communications Workshops (INFOCOM WKSHPS), Paris, France, 29 April–2 May 2019; pp. 84–89. [CrossRef]
19. van Ede, T.; Bortolameotti, R.; Continella, A.; Ren, J.; Dubois, D.J.; Lindorfer, M.; Choffnes, D.; van Steen, M.; Peter, A. FlowPrint: Semi-Supervised Mobile-App Fingerprinting on Encrypted Network Traffic. In Proceedings of the Network and Distributed System Security Symposium (NDSS), San Diego, CA, USA, 23–26 February 2020; Volume 27. [CrossRef]
20. Ahmad, Z.; Shahid Khan, A.; Wai Shiang, C.; Abdullah, J.; Ahmad, F. Network intrusion detection system: A systematic study of machine learning and deep learning approaches. *Trans. Emerg. Telecommun. Technol.* **2021**, *32*, e4150. [CrossRef]
21. Shen, Y.; Mariconti, E.; Vervier, P.A.; Stringhini, G. Tiresias: Predicting Security Events Through Deep Learning. In Proceedings of the 2018 ACM SIGSAC Conference on Computer and Communications Security, Toronto, ON, Canada, 15–19 October 2018; pp. 592–605. [CrossRef]
22. Tang, R.; Yang, Z.; Li, Z.; Meng, W.; Wang, H.; Li, Q.; Sun, Y.; Pei, D.; Wei, T.; Xu, Y.; et al. ZeroWall: Detecting Zero-Day Web Attacks through Encoder-Decoder Recurrent Neural Networks. In Proceedings of the IEEE INFOCOM 2020—IEEE Conference on Computer Communications, Toronto, ON, Canada, 6–9 July 2020; pp. 2479–2488. [CrossRef]
23. Ede, T.; Aghakhani, H.; Spahn, N.; Bortolameotti, R.; Cova, M.; Continella, A.; Steen, M.; Peter, A.; Kruegel, C.; Vigna, G. DEEPCASE: Semi-Supervised Contextual Analysis of Security Events. In Proceedings of the 2022 IEEE Symposium on Security and Privacy (SP), San Francisco, CA, USA, 23–26 May 2022; pp. 522–539. [CrossRef]
24. Jan, S.T.; Hao, Q.; Hu, T.; Pu, J.; Oswal, S.; Wang1, G.; Viswanath, B. Throwing Darts in the Dark? Detecting Bots with Limited Data using Neural Data Augmentation. In Proceedings of the 2020 IEEE Symposium on Security and Privacy (SP), San Francisco, CA, USA, 18–20 May 2020; pp. 1190–1206. [CrossRef]

25. Mirsky, Y.; Doitshman, T.; Elovici, Y.; Shabtai, A. Kitsune: An Ensemble of Autoencoders for Online Network Intrusion Detection. In Proceedings of the Network and Distributed System Security Symposium (NDSS), San Diego, CA, USA, 18–21 February 2018. [CrossRef]
26. Shen, Y.; Stringhini, G. ATTACK2VEC: Leveraging Temporal Word Embeddings to Understand the Evolution of Cyberattacks. In Proceedings of the 28st USENIX Security Symposium (USENIX Security 19), Santa Clara, CA, USA, 14–16 August 2019; pp. 905–921.
27. Rieck, K.; Trinius, P.; Willems, C.; Holz, T. Automatic analysis of malware behavior using machine learning. *J. Comput. Secur.* **2011**, *19*, 639–668. [CrossRef]
28. Hu, X.; Shin, K.G.; Bhatkar, S.; Griffin, K. MutantX-S: Scalable Malware Clustering Based on Static Features. In Proceedings of the 2013 USENIX Annual Technical Conference (USENIX ATC 13), San Jose, CA, USA, 26–28 June 2013; pp. 187–198.

electronics MDPI

Article

Network Layer Privacy Protection Using Format-Preserving Encryption

Marko Mićović *, Uroš Radenković and Pavle Vuletić

School of Electrical Engineering, University of Belgrade, Bulevar kralja Aleksandra 73, 11000 Belgrade, Serbia; uki@etf.bg.ac.rs (U.R.); pavle.vuletic@etf.bg.ac.rs (P.V.)
* Correspondence: micko@etf.bg.ac.rs

Abstract: Format-Preserving Encryption (FPE) algorithms are symmetric cryptographic algorithms that encrypt an arbitrary-length plaintext into a ciphertext of the same size. Standardisation bodies recognised the first FPE algorithms (FEA-1, FEA-2, FF1 and FF3-1) in the last decade, and they have not been used for network layer privacy protection so far. However, their ability to encrypt arbitrary-length plaintext makes them suitable for encrypting selected packet header fields and replacing their original value with ciphertext of the same size without storing excessive information on the network element. If the encrypted fields carry personally identifiable information, it is possible to protect the privacy of the endpoints in the communication. This paper presents our research on using FPE for network layer privacy protection and describes LISPP, a lightweight, stateless network layer privacy protection system. The system was developed for programmable smart network interface cards (NIC) and thoroughly tested in a real network environment. We have created several implementations ranging from pure P4 to a mix of P4 and C implementations, exploring their performance and the suitability of target-independent P4 language for such processor-intensive applications. Finally, LISPP achieved line rate TCP throughput, up to 4.5 million packets per second, with the penalty of only 30 to 60 microseconds of additional one-way delay, proving that it is adequate for use in production networks. The most efficient implementation was with the FF3-1 algorithm developed in C and carefully adapted to the specific hardware configuration of the NIC.

Keywords: network privacy; format-preserving encryption; programmable networks

Citation: Mićović, M.; Radenković, U.; Vuletić, P. Network Layer Privacy Protection Using Format-Preserving Encryption. *Electronics* **2023**, *12*, 4800. https://doi.org/10.3390/electronics12234800

Academic Editors: Tomasz Rak and Dariusz Rzońca

Received: 17 October 2023
Revised: 19 November 2023
Accepted: 25 November 2023
Published: 27 November 2023

1. Introduction

The impact of communication services and applications on our lives and the reliance on the Internet is increasing daily. This is also followed by the evidence that this usage is being extensively monitored and analysed and that a lot of personal data is gathered for either security or commercial purposes. Therefore, privacy protection and internet usage anonymisation have been important research topics for several decades. Personal data can be gathered from various sources, most directly from operating systems and applications and through analysing network traffic patterns and protocols. This paper focuses on the latter—protecting from network layer profiling and personal data leakage. Assume an Internet user communicates using a permanent public IP address or can be unambiguously linked to some public IP address in a specific period. In that case, adversaries can track the user's behaviour and habits by monitoring the set of visited IP addresses [1]. Even more accurate information can be obtained by tracking users' DNS requests [2,3]. Although IPv6 has some privacy protection mechanisms like prefix rotation and IPv6 privacy extensions, a recent study showed that the privacy of a substantial fraction of end-users is still at risk [4]. Therefore, the European Union's General Data Protection Regulation (GDPR) regulation considers IP addresses as Personally Identifiable Information (PII) [5], and special care must be taken to protect them.

Since web-based applications are the most popular internet services used nowadays, various proxy/VPN services have emerged that enable hiding the original source IP address when visiting a website. However, such systems have a single point of failure and rely on trust in the proxy/VPN provider, which can reside in a foreign legislation environment. The low-latency onion routing system Tor is probably the best-known and widely used network layer anonymisation system [6]. It allows anonymous web page browsing through the onion router circuits, which consist of three servers (called onion routers) and three layers of packet data encryption. It also enables access to hidden web content. By using Tor, a web server or an observer on any single point on the Internet cannot tell which are both endpoints of the web session, preserving user anonymity on a network layer. However, such protection comes with a performance cost. Each packet is split into Tor cells with additional headers, decreasing the useful part of the packet. Each cell processing includes three encryptions and decryptions between the endpoints. The path between the endpoints is (intentionally) not optimal. This additional cost is seen as often slow and annoying web browsing. Also, despite the heavy use of encryption, it is well known that Tor circuits are susceptible to end-to-end timing and rogue Tor router attacks owned by an adversary. Former is an attack in which an adversary monitoring traffic on multiple points in the network can discover the endpoints by correlating traffic patterns. This non-ideal situation inspired new proposals for providing network layer anonymity, which will be more thoroughly described in the next section.

More than 20 years have passed since the first Tor release, and some internet usage patterns have changed since then. Nowadays, almost all web traffic is encrypted [7], raising questions about whether additional data field encryption layers are needed and justified, primarily because they do not provide additional security against timing and rogue onion router attacks. In this paper, we propose a lightweight, stateless system for network layer anonymity which encrypts and obfuscates only the necessary parts of the packet headers to protect the user's privacy. Such an approach using well-known symmetric algorithms (e.g., Advanced Encryption Standard—AES) is not quite feasible because packet header fields are usually shorter and do not align with the block size for block ciphers nor a byte boundary for stream ciphers. For example, if a 12-bit plaintext is encrypted using AES-128, it will have to be padded to the size of the block—128 bits (usually with zeros or a random tweak), and the encryption will produce a 128-bit ciphertext. In order to decrypt back the initial 12-bit plaintext, one has to store the whole 128-bit ciphertext somewhere to perform decryption and later remove the padding (Figure 1a). Suppose a 12-bit plaintext is a packet header field or part of it (e.g., host part of the IP address with mask /20), which is to be encrypted. In that case, storing the encrypted version of that field takes additional space, which implies additional headers, protocols, or storage space. Therefore, we explored using Format-Preserving Encryption (FPE) for network layer privacy protection. FPE enables the encryption of arbitrary-length fields in a manner which allows the replacement of a protocol field with its encrypted version of the same size. This difference between the FPE and block ciphers is shown in Figure 1. Recently, the first such protocols, FEA-1 and FEA-2 [8] in South Korea and FF1 and FF3-1 [9] in the United States, passed the evaluation and adoption by the relevant standardisation bodies. To the best of our knowledge, this is the first use of FPE to protect network packet header fields. We believe its successful and performant implementation demonstrated in this paper through the design and deployment of a LIghtweight Stateless Privacy Protection system (LISPP) will pave the way for further use in networking applications for the privacy protection of all applications, not just the web.

Programmable network devices (e.g., switches, smart network interface cards (NICs) or switches filled with the bump-in-the-wire SmartNICs) became very popular in the last decade among network professionals, with the P4 language as one of the most popular recent innovations. Their programmability and flexibility enabled innovation and boosted research in the field. SmartNICs can be programmed using various programming languages and styles (e.g., P4, C, assembler or a combination of those), and in this way, offload a part of traffic processing from the central server processors and cores. Programmable NICs enable computing tasks execution and traffic processing closer to the data path, shortening processing times and enabling high-speed traffic processing and new applications or packet modifications without sacrificing network traffic performance. We developed the LISPP system and evaluated its performance on Netronome Agilio CX programmable network interface card. LISPP was developed in several implementations ranging from pure P4 to a combination of P4 and Micro-C to explore the performance of portable P4 code and its dependence on the specific hardware configuration of the card.

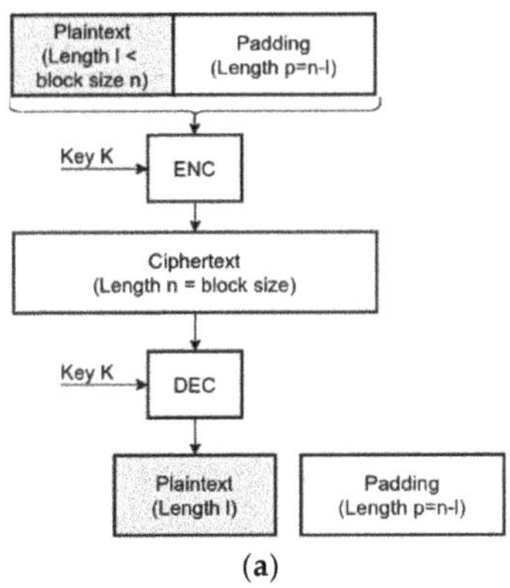

Figure 1. (**a**) Block cipher encryption vs (**b**) Format-preserving encryption of plaintext of length l.

Our research differs from the previous network layer privacy protection proposals in several important points. This is the first study on the use of the FPE for network layer privacy protection. Unlike previous proposals described in Section 2, the implemented LISPP system is fully stateless, which implies low memory requirements, simple multi-homing, fully transparent, and fast line-rate operation on contemporary programmable hardware, as described in the paper. The performance of the proposed system is proven through the experimental evaluation of actual devices within a real network environment. The LISPP achieved line rate TCP throughput, up to 4.5 million packets per second, with the penalty of only 30 to 60 microseconds of additional one-way delay under the conditions described in the remainder of the paper. Such throughput on low-cost SmartNICs proves that FPE is adequate for network layer privacy protection. The LISPP is network layer protocol-independent, ready to protect the privacy of both IPv4 and IPv6 header data, which was not a feature of the previously proposed systems. Finally, we have made three implementations of LISPP using different combinations of P4 and Micro-C code. The analysis of these implementations presented in this paper revealed less than optimal performance of the P4 code on the target system we used and indicated that further work on the P4 compiler optimisation is needed.

The paper is structured as follows: Section 2 gives an overview of the recent research in the field of network layer privacy protection, with special attention on the issues of the proposed solutions, which are a consequence of the use of classic encryption algorithms. Section 3 describes the architecture and principles of privacy protection using LISPP. Section 4 introduces FPE algorithms and presents FPE implementation challenges on network accelerator cards. Section 5 discusses LISPP performance and obtained experimental results, while Section 6 concludes the paper.

2. Related Work

Two recent overview papers of applied research in the field of data plane programming [10,11], among other work, listed the most recent efforts on network layer privacy protection that use novel network programmability mechanisms. Systems like HORNET [12], TARANET [13], PHI [14], LAP [15], or Dovetail [16] aim to provide solutions similar to Tor with multiple cooperating routers/servers along the packet path, usually in conjunction with some Next Generation Internet technology that enhances security. In some cases (e.g., LAP, HORNET), packets are additionally encrypted by the system. In the other (e.g., Dovetail), there is no additional encryption, and the system relies on another protection mechanism. Similarly to Tor, all these systems are vulnerable to timing correlation attacks, but newer systems like HORNET also turned out to have other previously unknown vulnerabilities [17].

A series of papers on network layer privacy from a research group from Princeton University primarily inspired our work. SPINE [18] is a system for IP address, TCP sequence, and acknowledgement numbers obfuscation. It encrypts the source IP address from the original IPv4 packet header and encodes the encrypted data into the newly created IPv6 packet while discarding the original IPv4 packet header. To avoid encrypting one IP address always into the same encrypted value using a single key, SPINE adds a random nonce to the address encryption process. The nonce is different for each packet and randomises the encrypted address values. To achieve reversible decryption, SPINE encodes the encrypted IP address and the used nonce into a newly created IPv6 packet. An IPv6 address that is longer than the IPv4 address can store both the encrypted IPv4 address and the nonce. In order to ensure high-speed operation, SPINE uses a simple XOR-based encryption scheme. SPINE is a VPN-like system in which two or more collaborating autonomous systems are the endpoints of the newly created IPv6 tunnels. Encrypting original IPv4 addresses and storing them in the new IPv6 header hides original communication details from the intermediate autonomous systems. The SPINE system is stateless because it does not need to store the mapping between the original IPv4 and newly created IPv6 addresses—the encryption/decryption process provides the mapping. However, mapping between the destination IPv4 address and the corresponding IPv6 endpoint prefix is needed for the operation, as well as the previous key exchange.

Wang et al. [19], in the P4-based PINOT system for address obfuscation, proposed a scheme in which the source IP address is padded with random padding up to the size of the block of the cryptographic algorithm and then encrypted. The encryption scheme is more complex than in the case of SPINE but still non-standard. PINOT uses a simplified 56 or 64-bit wide two-stage substitution-permutation network to achieve high packet rates. As in SPINE, because the length of the ciphertext is longer than the IPv4 address, and in order to make the process reversible once the packets return from the opposite endpoint, the encrypted data is encoded in the IPv6 packet. The system is stateless for egress source IP addresses for which there is no need for a table lookup—they are just encrypted using a local key. However, the lookup is needed for the destination IPv6 address, which has to be found based on the destination IPv4 address. The PINOT authors assumed that some sort of DNS snooping is used, in which both A and AAAA records are intercepted at the network device and mapped so that the appropriate destination IPv6 address can be created. However, this process does not seem trivial on a network element or without a performance penalty. Cryptographic keys in the PINOT system do not have to be exchanged. They are local to the network element if the egress and ingress points to the network are the same and as long as there are no multiple entries into the network.

Unfortunately, today, privacy-preserving systems like PINOT and SPINE, which use IPv4 to IPv6 translation, do not ensure full Internet connectivity. At the moment of writing this paper, only about one-third of all the autonomous systems on the Internet support IPv6 [20]. Further, once IPv6 becomes fully adopted and the predominant IP protocol on the Internet, applying the same approach for the IPv6 address and transport layer would be challenging, if not impossible. PINOT and SPINE used the fact that IPv6 addresses are longer than IPv4, which enabled storing the ciphertext of the IPv4 address and random nonce in the IPv6 address. However, when a random nonce is padded to an IPv6 address, the resulting ciphertext will be longer than the available space in the IPv6 address. The question is where the excess bits of the ciphertext would be stored—either in a new protocol header or using a stateful operation is required.

Another older IP address mixing system [21] is a stateful system that encrypts the host part of the class B IPv4 address and source port using an RC5-based scheme with the addition of a random number—tweak. Since the output of the RC5 is 128-bit, and the plaintext input, which is replaced with the encrypted value, is only 32 bits long, the solution for the excess ciphertext was to make the system stateful. The system keeps records of all flow to 32-bit encrypted value mappings to perform decryption/replacement operations on the backward packet path. In that case, an encrypted pair (src IP, src port) is used as a key. Although not reported in the paper, the system suffers from the birthday problem and experiences collisions with 64 thousand concurrent flows with a probability of 0.5.

To summarise, the previous research that proposed the encryption of the critical packet header fields using classic block encryption algorithms showed that such an approach successfully hides users' IP addresses. However, block ciphers require a stateful operation or IPv4–IPv6 translation, which poses significant implementation and usage issues, as explained above. This paper explored FPE for packet header field obfuscation and created the LISPP system. This fully stateless system efficiently scrambles packet flow data using FPE, hiding the source IP address from the observers on the Internet and disabling user profiling. The system is built for low-cost programmable SmartNICs for IPv4 and IPv6 and achieves line rate throughput on 10 Gbit/s links, proving that the concept can be used in real networking environments. We have tested the performance of several development and deployment options (pure P4, mixed P4 and C and pure C implementation). LISPP achieved line rate operation on 10 Gbit/s interfaces. Our analysis also showed performance issues in pure P4 deployments. P4 performance, especially for processor-demanding tasks like encryption, still heavily depends on the underlying hardware architecture. The P4 compiler we used does not optimise the executable code most efficiently.

3. LISPP System Architecture

An IP address consists of two inseparable parts: the network part, which determines the host's location on the Internet (the autonomous system) and the host part, which identifies the exact sender or recipient of the packet in that network. Since the information about the location is needed to route the packet properly, IP addresses are usually sent unprotected or unchanged. Some of the previous protection mechanisms use either fully stateful address swapping (e.g., in NAT) or full packet encryption (like in Tor or IPsec). However, such approaches are not always scalable for general Internet usage patterns.

LISPP processes packets at the network boundary. It encrypts the host part of the source IP address and source port in packets that exit the protected network and decrypts them in the opposite direction. In the egress direction, the host part of the original source IP address from the protected network (designated as P—plaintext in Figure 2) and source port are replaced with their encrypted values (designated as C—ciphertext and new port number). The network part of the IP address (Net) remains the same, ensuring proper packet routing back to the protected network. In the ingress direction, decryption using the same key is performed, restoring addresses and ports to their original values.

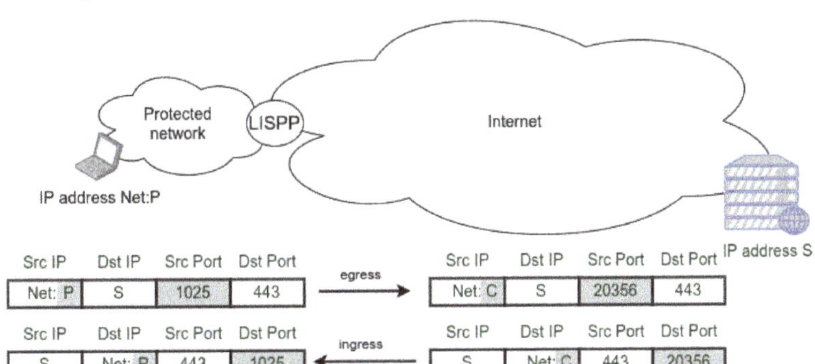

Figure 2. LISPP header field modification at the network boundary.

This way, when the user from the protected network communicates with the external devices, external devices can only know the user's location (network part of the IP address) but not the exact user's original source IP address. The encrypted version of the plaintext changes in every session because the source port takes a new value in subsequent TCP or UDP sessions. Every time a client from the protected network accesses the same external server, the client will appear to have a different IP address with a high probability, ensured by the use of encryption algorithms. Such a behaviour prevents the destination, or an observer in any location between the protected network and the destination, from tracking the behaviour of any specific user in the protected network on a network level because his network sessions will appear to be coming from different IP addresses. LISPP behaviour is similar to Port Address Translation (PAT) because it changes the source IP address and port on the network entry/exit point. However, unlike PAT, which maps a pool of private IP addresses onto a single or a smaller number of public IP addresses, LISPP makes a bijection of a pool of public IP addresses onto that same IP address set. Also, unlike PAT, LISPP is fully stateless, implying that mappings between the original and encrypted pairs of addresses and ports do not have to be stored at the network element because the mapping is performed using encryption/decryption. From this brief description, it is clear that LISPP does not strive to replace or present an alternative to Tor, as it assumes a single point which obfuscates the addresses. However, there are several clear use cases, as described in the remainder of this section, in which LISPP can protect user privacy.

3.1. Packet Processing

In both directions, after packet parsing and checksum verification, LISPP filters packets which will be processed (Figure 3). Since LISPP uses a source port as a part of the plaintext in the egress direction, LISPP can be used to protect any TCP or UDP packet. However, it is possible to program the match filter to push to the encryption phase packets with any specific destination port value (e.g., TCP 443 for TLS or UDP 53 for DNS requests/responses) while all the other packets pass the system unchanged.

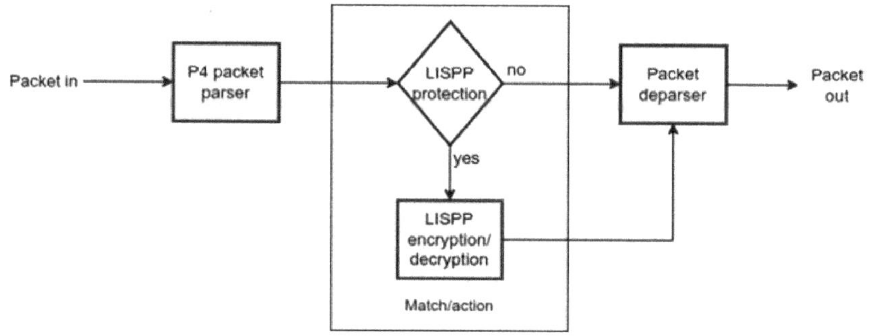

Figure 3. LISPP packet processing diagram.

3.2. Packet Header Field Encryption

Figure 4 shows how LISPP encrypts the packet header elements using FPE. The host part of the source IP address and source port are concatenated and encrypted using a secret key. Since FPE is used, n bits of plaintext are encrypted into exactly n bits of the ciphertext regardless of the number of bits n. In that case, it is possible to obtain a reversible one-to-one mapping between the (src IP, src port) and (enc(src IP), enc(src port)) pairs regardless of the network mask size (and IP version). It is possible to achieve fully transparent and stateless operation in both directions.

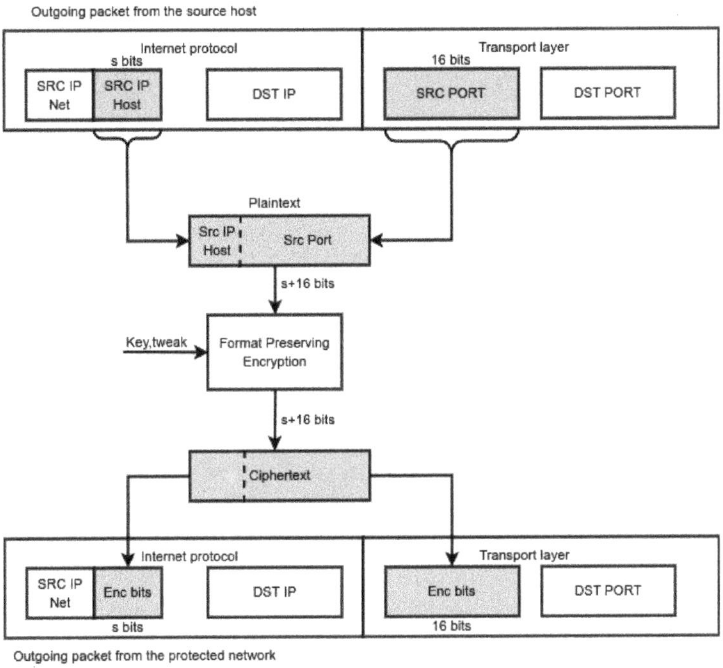

Figure 4. LISPP address and port encryption.

An illustration of LISPP address obfuscation in operation is given in Figure 5. This figure shows the empirical probability of the appearance of encrypted values of host parts of the IP address (enc(src IP)) obtained from a single source IP address with mask /23 and the full range of source ports from 0 to 65,535. Visual inspection shows that LISPP achieves uniform distribution of the encrypted source IP addresses across all possible 512 values of a 9-bit host part of the address. More rigorous randomness testing is presented in Section 5.

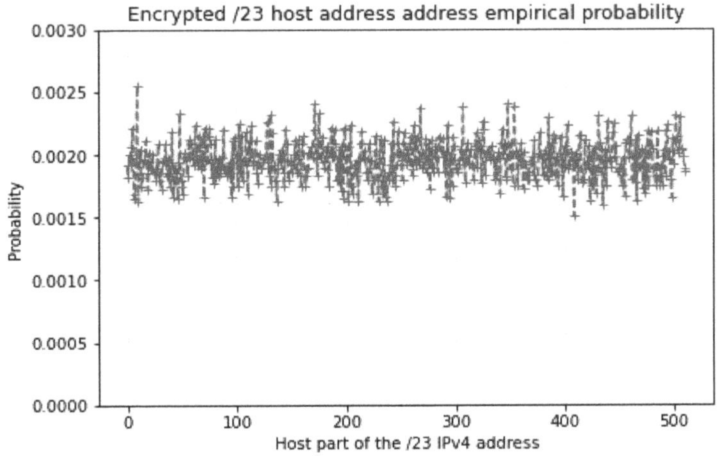

Figure 5. Empirical probability of the appearance of host part values of the IP address obtained from a single source IP address with mask /23 and the full range of source ports using LISPP.

FPE algorithms require a secret cryptographic key and tweak (described in Section 4) to be used to encrypt and decrypt header fields. This cryptographic material can be created directly on a network element using a pseudorandom derivation from a seed defined by the user or taken from some source of randomness on the network element. In the case of a single entry point into the protected network, key and tweak do not have to leave that network element because both encryption and decryption are performed on the same device. However, when a network has multiple entry points and asymmetric ingress and egress flow paths, all devices on the network boundary must use the same cryptographic material. In that case, one boundary network element would create the key and tweak, while the others would receive that material through some secure connection (e.g., TLS or IPsec). In any of these cases, the tweak is not sent along the encrypted header fields, which further strengthens the security of the proposed solution.

Since cryptographic material has a limited operational lifetime, the key and tweak have to be changed periodically (e.g., daily) or after some number of packets are processed. In moments when the key and/or tweak are changed, network flows active in that period will be broken because packet header fields in the egress direction would be encrypted with the old key, while in the ingress would be decrypted with the new, yielding wrong IP address and port numbers for that flow. This transient behaviour, although short, disrupts network operation and has to be planned for quiet periods of network operation. There is a trade-off between stronger data privacy (often key/tweak changes) and reliable network operation.

3.3. Threat Model

LISPP is a network layer privacy protection mechanism that cannot protect privacy at the application level. Like PINOT, it hides the client's IP addresses and flow identifiers from the server side and the intermediate networks while accessing services on the Internet. LISPP assumes trust in the local network operator and does not hide connection/flow details from it or the devices that perform the encryption. An intermediate network between the protected network and the destination or the destination itself can reveal the actual IP address of the client, either by obtaining the encrypted–plaintext mapping or an encryption key through collusion with the client's network provider or by breaking the encryption algorithm, which is computationally hard at the moment of writing this paper.

3.4. LISPP Use Case

LISPP is applicable and is desirable in all cases where the user has a public IP address given by the Internet Service Provider (ISP). ISPs might use mechanisms like stateless, temporary IPv6 address assignment [22], which periodically leases and changes temporary IP addresses. However, by default, this period is one day, giving the adversary a sufficiently large time window to analyse user behaviour and cross-correlate this behaviour with other sources of private data. There is also evidence that despite the use of such mechanisms, there are still substantial privacy leaks [4]. Since LISPP provides per-session address randomisation, it completely breaks any chance of network layer user tracking.

One clear use of LISPP is to mitigate the threat of private information leakage and user profiling by public DNS resolvers (e.g., Google Public DNS: 8.8.8.8, Quad9: 9.9.9.9, Cloudflare: 1.1.1.1 and similar). DNS resolvers receive the set of symbolic names of sites a user visits regardless of how symbolic names are sent to the resolver (encrypted by DNS over TLS or HTTPS or in plaintext). Therefore, for a user behind the specific IP address, DNS resolvers can gather information about the interests and the sites visited. There were some previous attempts to hide the symbolic names from the public resolvers by encrypting and encapsulating them into the regular DNS queries and redirecting them to another resolver [23]. By varying the source IP address for each user's DNS request, LISPP successfully disables such profiling, and the system is significantly simpler than the previous solutions. The ISP can offer LISPP as an additional privacy protection service that prevents third parties (external sites and services) on the Internet from tracking the users and analysing their behaviour on the network layer. An example of LISPP performing DNS request source obfuscation is given in Figure 6. This figure shows eight consecutive DNS requests from a single computer in a /23 network and the set of addresses and ports to which LISPP converted the original data. It is evident that for the DNS server operator, it is difficult, if not impossible, to tell which device it is talking to at any given time. The effect of using LISPP would be the same for any other network protocol (e.g., web, SSH, FTP, etc.).

Original IP	Original source port	LISPP	LISPP IP address	LISPP source port
147.91.1.136	58119	<----------->	147.91.1.2	14728
147.91.1.136	44383	<----------->	147.91.0.13	8922
147.91.1.136	35450	<----------->	147.91.0.72	48789
147.91.1.136	58776	<----------->	147.91.0.107	1313
147.91.1.136	44346	<----------->	147.91.1.92	35997
147.91.1.136	43597	<----------->	147.91.1.249	7244
147.91.1.136	51655	<----------->	147.91.0.28	63909
147.91.1.136	56926	<----------->	147.91.1.211	28101

Figure 6. LISPP translation of 8 consecutive DNS requests coming from a device with the address 147.91.1.136/23.

Furthermore, because of the ever-increasing number of cybersecurity threats and difficulties in identifying the attackers when the attack comes behind the Carrier-grade NAT devices, there are recent incentives to mandate the retention of the metadata that gives the mapping between the user and the IP address [24]. If such regulations are adopted, LISPP can easily comply with them and preserve privacy against third parties. Unlike large logs of NAT mappings, in the case of LISPP, only cryptographic material used by the FPE during the lifetime of that material needs to be kept to reconstruct the actual IP address of the users upon request from the legal authorities.

3.5. LISPP Design Goals

LISPP was designed with the following properties in mind:

- Transparency. Users from the protected network do not have to employ any dedicated application. They are generally unaware of any privacy protection system on the packet path (except for added minimal latency due to the packet processing).
- Stateless operation. The network element does not have to store any state, i.e., mappings between the plaintext and encrypted fields' values or any tables. The stateless operation further brings simple multihoming because there is no need to synchronise states among the entry/exit points.
- Seamless multihoming. Suppose the protected network has multiple entry/exit points, and packet paths are not symmetric in the ingress and egress direction. In that case, LISPP should be deployed on all entry/exit points with the need to exchange only cryptographic material (keys and tweaks, as described in Section 4.1) between the entry/exit points. Deploying LISPP on all entry/exit points is easily achievable using any key exchange mechanism or through already-established cryptographic channels between the endpoints (e.g., IPsec).
- Effortless reconfigurability. LISPP can be configured to protect any TCP or UDP protocol port. Only the appropriate packet filter should be defined to select packets for which the obfuscation will be performed.
- Protocol independence. LISPP works with both IPv4 and IPv6 without any network layer protocol modifications.
- Legal compliance. Operators of the LISPP-protected network can easily reconstruct true packet origins upon legitimate requests from legal authorities.

Another side effect of using LISPP is that port scanning a device in a protected network from the outside is significantly more difficult. Suppose an adversary scans the entire port range for a single destination address in the protected network. In that case, these packets will pass the decryption and be scattered across the whole IP address segment, as shown in

Figure 5, hitting various devices on ports which are not the same as those that an adversary sent, making the analysis significantly more difficult for the external observer. Further, with frequent changes of the cryptographic key and/or tweaks, the scanned footprint will completely change, making the analysis or the attacks even more difficult. LISPP can, in this case, be considered one of the tools and techniques for the Moving Target Defence strategy [25].

4. Format-Preserving Encryption

FPE is a type of encryption that preserves the format (alphabet) and size of the plaintext in the ciphertext. For example, with FPE, the ciphertext of a 16-digit decimal payment card number is also a 16-digit decimal number. The symbol sets and lengths for the plaintext and ciphertext are the same. One of the first algorithms that allowed variable bit size input and the same size output was a Hasty Pudding Cipher (HPC) [26], one of the candidates at the AES algorithm contest. The HPC algorithm did not pass to the later stages of the AES algorithm contest because of its complex and unusual structure. As a result, the cryptographic community did not widely test the HPC, so its resistance to various attacks was not well known.

The first, and so far only, FPE algorithms that passed as a recommendation of a standardisation body are South Korean FEA-1 and FEA-2, as well as FF1 and FF3 from the National Institute of Standards and Technology (NIST). FF3-1 is a revision of FF3 created after finding a security flaw in FF3 [27]. All these FPE algorithms have a very similar Feistel structure with different options for the random function, as will be described in the next section. The cryptanalysis of standardised FPE algorithms showed that the attacks on the FPE algorithms using differential distinguishers are more complex and require more data for the FF3-1 algorithm compared to the FEA standards [28]. In addition, linear cryptanalysis of the FPE algorithms [29] revealed that attacks on FF3-1 are more time-consuming compared to other algorithms in terms of encryption operations, thereby highlighting its enhanced security. Finally, the FF3-1 algorithm's ability to encrypt binary words ranging from 20 to 192 bits in length made it particularly suitable for encrypting specific protocol fields in our system.

FPE specifications, although relatively young compared to the well-known block cipher symmetric cryptographic algorithms, have attracted attention from cryptanalysts. These experts have identified potential vulnerabilities in FPE schemes, particularly highlighting a decrease in the complexity of attacks, especially for shorter plaintext lengths and under specific circumstances, such as when the adversary has knowledge of the tweak parameter [29,30]. In the context of LISPP, these issues are effectively mitigated, as the plaintext lengths utilised exceed those susceptible to such vulnerabilities, and the tweak parameter never leaves the network element, ensuring its confidentiality. Consequently, the attacks reported in the literature are not applicable to LISPP. Nevertheless, we believe that finding such issues in the current algorithms will only improve their future versions and not jeopardise the use of FPE in general.

4.1. FF1 and FF3-1 Algorithm

FF1 and FF3-1 are Feistel-structure tweakable symmetric algorithms. Feistel structure is a well-known primitive block for symmetric algorithm design, known since the time before the Digital Encryption Standard (DES). Tweakable means that the algorithm uses an additional component called tweak as an input to the encryption and decryption process. The tweak does not necessarily have to be kept secret. It is used to increase the input variability because, with the FPE, input strings can be short with a limited set of values. Encrypting as few plaintexts as possible under any given tweak is recommended. However, changing the tweak during network operation can break existing sessions (one endpoint will change), and it should be conducted in carefully defined moments.

The FF3-1 achieves greater throughput because it has eight rounds, two fewer compared to the FF1, while the FF1 supports a wider range of lengths for the plaintext and flexibility in the tweak length [9]. Since per-packet processing time should be as low as possible, we focused on the FF3-1. In the core of each FF3-1 round is an approved block cipher used as a round function (F_k) to create a pseudorandom output. Figure 7 shows two Feistel rounds of FF3-1 encryption and decryption. Plaintext input is divided into two parts (A_i and B_i), which have the same size in case of an even number of plaintext characters or differ in size by one character in case of an odd number of plaintext characters. In our case, one character corresponds to one bit because packet header fields are binary words. The second part (B_i) is copied into the first part of the next round (A_{i+1}), while the first part is added to the output of the F_k round function. The inputs to the round function are one part of the previous block, round number i, random tweak T and the plaintext size n.

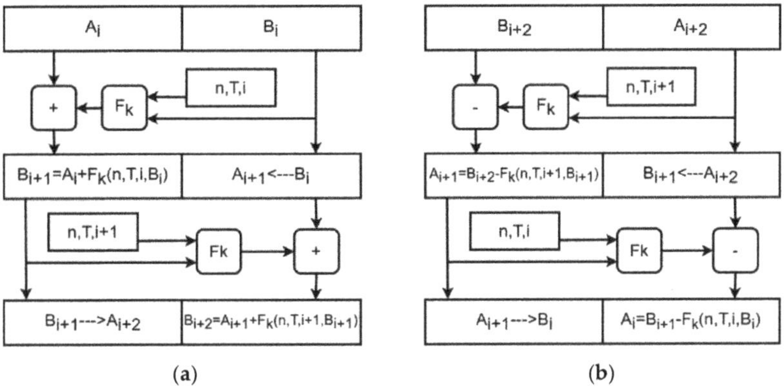

Figure 7. Two Feistel rounds of the FF3-1 (**a**) encryption and (**b**) decryption.

According to the FF3-1 specification, an approved block cipher with secret key K should be used as the F_k, and at this moment, only the AES block cipher fits this profile. However, there are deployments with other lightweight algorithms [31]. Unlike FF1 and FF3-1, FEA-1 and FEA-2 use a modification of the SHARK cryptographic algorithm as a round function. FF3-1 does not use AES-128 to encrypt the data but provides a pseudorandom output truncated to the required number of bits and added to half of the plaintext. Therefore, only AES encryption is used for both FF3-1 encryption and decryption, simplifying the algorithm's implementation. However, as described, FF3-1 consists of 8 rounds in which AES encryption of a 128-bit block is invoked, which means that standard-based FF3-1 implementation is roughly comparable to the encryption of 1024 bits (8 blocks) with AES, and that could present a challenge for the system performance. However, in Section 5, we show that even with the pure software implementation of the cryptographic algorithms, it was possible to achieve line rate performance on SmartNIC.

The FF3-1 algorithm has several parameters which define its behaviour. The base is the number of characters in a plaintext alphabet denoted as the radix. For binary plaintext, radix is 2. For an English plaintext consisting only of letters, the radix is 26. The number of plaintext characters and their base define the domain size of the plaintext as $radix^{length}$. For example, for a 16-digit debit card number, the theoretical domain size is 10^{16}. Still, the actual domain size is somewhat smaller because some fields, like the issuer identification number, are fixed. FF3-1 specifies the minimum domain size of the plaintext to be at least 1 million. Therefore, for binary inputs, any plaintext that is longer or equal to 20 bits complies with the algorithm specification. For LISPP, this limit implies that the longest network prefix that could be used as an input is /28 (4 bits for the host part of the address + 16-bit source port). However, such a small network with only 14 devices might present a different privacy risk to users. Since the number of users using the network is small, side-channel attacks that analyse the user's activity at a certain period are more likely to happen. Better results are obtained for larger networks with shorter prefixes and more users (we consider a mask of at least /24 as recommended). Another FF3-1 limit is the maximum plaintext length, which must be smaller than $2\log_{radix}2^{96}$. For binary inputs, this is 192 bits—enough for almost all uses in packet headers for both IPv4 and IPv6. In the context of LISPP header field encryption, this implies that the source port and the whole IPv6 address can be encrypted using the FF3-1 without reaching the theoretical limit of the algorithm. It is interesting to notice that since the output of the AES algorithm is applied to half of the plaintext, which is of the maximal size of 96 bits, in all cases, regardless of the plaintext size, in each FF3-1 round, only one AES encryption is used, and the algorithm performance will be the same. Our experimental evaluation proved that the system's performance was the same regardless of the IP address mask length.

4.2. FF3-1 Implementation (Target Netronome)

The implementation targets the Netronome Agilio CX 2 × 10GbE NFP-4000 series SmartNIC [32]. The SmartNIC consists of 12 clusters, i.e., islands of different architecture. Islands can be roughly divided into two categories depending on the purpose of the contained Flow Processor Cores (FPC), i.e., Microengines (ME). The first category includes islands containing only multi-threaded MEs for packet processing. There are eight cooperative threads within the ME, with only one thread running at any time; each has its own set of 32 32-bit wide general-purpose registers. Each ME has its Code Store and Local Memory following the Harvard architecture. In addition to its own Local Memory, which stores the data needed for processing every packet, the ME has access to four other kinds of memory. The size and data access time expressed in clock cycles for each kind of memory are given in Table 1 [33]. Cluster Local Scratch is meant to store the data needed to process the majority of packets and smaller tables. Cluster Target Memory stores packet headers and coordinates ME and other subsystems. Internal Memory stores the packet payload and medium-sized tables. External Memory stores large tables.

Table 1. Types of Netronome Agilio CX memories and their access times.

Memory Kind	Size	Access Time (Cycles)
Code Store (CS)	8 K instructions	1
Local Memory (LM)	4 KB	1–3
Cluster Local Scratch (CLS)	64 KB	20–50
Cluster Target Memory (CTM)	256 KB	50–100
Internal Memory (IMEM)	4 MB	150–250
External Memory (EMEM)	3 MB + 2 GB RAM	150–500

The second category includes islands that contain accelerators (ILA, PCIe, Crypto, ARM) and multi-threaded MEs managing those accelerators. The SmartNIC used in the experimental evaluation contained no islands with a cryptographic accelerator. That is why we implemented the FF3-1 algorithm entirely in the software. Netronome Agilio CX cards without crypto accelerators use Linear Feedback Shift Register (LFSR) to generate a pseudorandom number, which can be used by Microengine software as an FF3-1 key and a tweak. It can be initialised using a timestamp and some user-defined value as a pseudorandom seed.

As described above, the FF3-1 algorithm can be used with an arbitrary alphabet or character set. In the case of a binary alphabet whose radix is 2, the following primitive operations of the FF3-1 algorithm were simplified:

- $NUM_{radix}(X)$, the number that the numeral string X represents in base radix when the numerals are valued in decreasing order of significance, has precisely the value X,
- $STR^{m}{}_{radix}(X)$, which is the representation of X as a string of m numerals in base radix, in decreasing order of significance, given a nonnegative integer X less than $radix^{m}$, is X at bit-width m,
- NUM(X) equals the integer that a bit string X represents. When the bits are valued in decreasing order of significance, it is essentially X itself,

- Modulo operation X mod $radix^m$ is implemented by a bit masking as X & ((1 << m)—1).

P4 is a hardware-independent network programming language where users can write the forwarding behaviour of the network devices using the standard forwarding model defined in the P4 architecture [34]. The user does not need to know Network flow processor (NFP) specific data structures. The P4 compiler automatically maps the different parts of the P4 program into the NFP internal resources. The P4 front-end compiler first compiles a P4 program to an intermediate representation (IR). The Netronome's P4 back-end compiler transpiles the IR into the Micro-C program, which can be compiled and linked to generate the NFP firmware using the network flow C compiler (NFCC) [35]. The firmware generated from the P4 code is loaded on multiple MEs, each of which can independently process packets according to the packet processing code written as a P4 program. Motivated by the portability of the implementation to a larger number of devices, i.e., P4 targets, the aim was to explore portable implementation purely in the P4 language. Netronome supports executing P4 programs written for the v1model architecture [36], a variation of a theoretical model defined by Portable Switch Architecture (PSA) [37,38]. Theoretically, the implementation would be portable to any P4 target with a v1model architecture, such as the Behavioral Model (BMv2). We used tools from Netronome SDK version 6.1-preview, the first version that supports the P4-16 language. The code of all LISPP implementations described in the following sections is publicly accessible [39].

4.2.1. Pure P4 Implementation

The biggest challenges in the P4-based FF3-1 implementation were the limitations directly imposed by the P4 language [40]. P4 language does not have the loop construct, which presents a serious challenge in implementing symmetric cryptographic algorithms consisting of many rounds to achieve data confusion and diffusion. Unrolling loops of the entire algorithm is not an option due to the size limit of the Code Store where the program code resides. The size limit of the Code Store comes to the fore due to the design limit specified in the Netronome SDK documentation that a maximum of 256 actions may be defined, meanwhile expecting an increased Code Store usage. For each action invocation, the Netronome P4 back-end compiler defines a new Micro-C function that provides the given action with the context (arguments) and invokes the action. Defining a new Micro-C function for each action invocation leads to a non-negligible increase in the program code size. Saving Code Store space becomes even more critical, considering that the Netronome P4 front-end compiler does not support P4 functions. That is why we had to overcome the non-existent loop construct limitation by resubmitting the packet for each round of the FF3-1 algorithm. Instead of replicating code for an entire FF3-1 round multiple times, invocation of the resubmit extern function from the v1model architecture returns the packet to the start of the ingress pipeline, representing a single FF3-1 round. The internal state of the FF3-1 algorithm is stored in the packet's metadata to save it across resubmission.

To implement AES encryption, we used a solution based on scrambled lookup tables [41]. The upside of this solution is that it performs all AES encryption in just one packet pass through the ingress pipeline and uses a pre-expanded AES key. The downside of this solution is the need for 160 match-action tables, a necessity arising from the P4 language's absence of array support. Each byte of the AES algorithm state requires an individual table because the same table cannot be applied more than once during a single packet pass through the ingress pipeline. All sixteen tables must be replicated for every round of the AES algorithm, resulting in the 160 tables mentioned above. Such a large number of tables harms latency [42], especially given that the P4 back-end compiler places all tables in External Memory, which has the longest access time.

The required number of tables exceeds Netronome SmartNIC's limit on the number of match-action tables used in the ingress pipeline of a P4 program [43]. That is why we had to reduce the number of tables to only five: four distinct for standard AES rounds and one for the last AES round. Table number reduction implies introducing additional packet resubmissions to implement AES encryption successfully. The most straightforward implementation would pass the packet through the ingress pipeline once for each of the 16 bytes in the AES state for every round of the AES algorithm. Such a naive implementation requires even $8 \times (1 + 9{\times}16 + 1{\times}16 + 1) = 1296$ resubmissions for each incoming packet. With a slightly more complex implementation, applying all four distinct tables for standard AES rounds in a single ingress pipeline pass, it is possible to reduce the packet resubmissions down to $8 \times (1 + 9{\times}4 + 1{\times}16 + 1) = 432$.

4.2.2. Packet Control and FF3-1 in P4 and AES in Micro-C Implementation

A large number of packet resubmissions causes a throughput well below the link capacity, as shown in Section 5. We tried to improve the throughput by replacing the parts of the P4 code with Micro-C. The Micro-C programming language is the most efficient way of programming the Agilio SmartNIC as it can take advantage of NFP architecture-specific data structures [44]. The Micro-C programming on the NFP slightly differs from the host-based generic C programming, as the NFP data structures and memories are specific to the NFP architecture.

We decided to port the complete AES encryption to the Micro-C language because AES encryption represents the most complex part of the FF3-1 algorithm and, in some SmartNICs, can be implemented using hardware acceleration. In order to achieve the highest possible throughput, the implementation is still based on scrambled lookup tables. The algorithm is considerably sped up by pre-computing part of the internal operations performed by the AES algorithm and storing the results in lookup tables [45]. Since the content of lookup tables is immutable, they do not have to be thread-local. Sharing these tables between threads leads to better memory space utilisation. Therefore, tables are explicitly marked as shared and placed in memories with the smallest latency. Only one of four tables for standard AES rounds is allocated to the fastest Local Memory because there is no more free space in the Local Memory due to its usage for register spilling. The remaining tables for standard AES rounds and the table for the last AES round are allocated to the slightly slower Cluster Local Scratch. All functions are inlined directly at the place of their invocation to gain an execution speedup.

Great attention has been paid to the types of data used due to the specifics of SmartNIC's hardware. The compiler supports 8-bit and 16-bit data types and their appropriate pointers, although at some potential performance cost. Still, users should not use 8-bit and 16-bit data types because access to quantities less than 32 bits (64 bits in MEM) generally involves additional operations to extract the appropriate bytes from the longword or quadword. Access through pointers to 8-bit and 16-bit types may also require runtime alignment of data, which is even more inefficient. Therefore, we have neither used 8-bit and 16-bit data types nor pointers. The AES state is defined as a 128-bit structure with four fields of 32-bit data type, and the given structure is transmitted exclusively by value. Implementing AES in Micro-C resulted in approximately 45 times increased throughput compared to the pure P4 implementation, as described in Section 5.

4.2.3. P4 Packet Control and Entire FF3-1 in Micro-C implementation

Finally, we ported the entire FF3-1 algorithm implementation from the P4-16 to the Micro-C language, leaving only packet parsing and filtering to the P4 code. The same coding principles imposed by SmartNIC's architecture were used for the FF3 implementation in Micro-C. In addition, an FF3-1 specific bit reversal operation was realised using lookup tables instead of bit masking and shifting. This implementation resulted in line rate LISPP operation.

5. Experimental Evaluation

LISPP performance was rigorously assessed using actual physical network devices rather than within a simulated environment. Two bare metal servers with dual Intel® Xeon® CPU E5-2660 processors and 40 GB of RAM each were used as the source and sink of the test traffic. These servers were connected through a Netronome Agilio CX programmable network interface card with two 10 Gbit/s ports installed in another bare metal server with dual Intel® Xeon® CPU E5-2680 processors and 64 GB of RAM, as depicted in Figure 8. The connections between the servers were through 10 Gbit/s ports on a switch. These ports were used solely in the testbed, so there was no interference of other cross traffic with the test traffic. MTU on all interfaces remained at 1500 bytes in all the experiments.

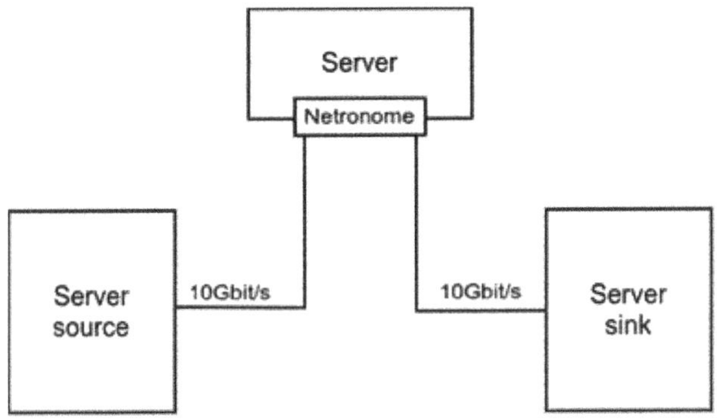

Figure 8. Testbed setup.

5.1. LISPP Performance Evaluation

In our performance evaluation, we employed three state-of-the-art tools for active network monitoring: iPerf2 [46] for the TCP throughput tests, PF_RING Zero Copy-based packet rate tests [47] and Sockperf [48] for latency tests. Each of these tools actively monitors by injecting test traffic to measure the performance characteristics of the underlying network. iPerf2 is designed to assess the maximum achievable TCP or UDP bandwidth. It establishes TCP sessions between the source and sink of the test traffic to gauge the highest achievable throughput. iPerf2 can achieve throughputs higher than 10 Gbit/s with default parameters and a single TCP stream [49]. PF_RING is a new type of network socket that improves the packet capture speed by avoiding any kernel intervention and can achieve up to 100 Gbit/s wire speed at any packet size. Sockperf is a network benchmarking utility designed for testing latency at a sub-nanosecond resolution. It is able to measure the latency of every single packet, even under a load of millions of packets per second.

Test packets were sent from the source to the sink server through the LISPP system on the Netronome SmartNIC. Table 2 gives iPerf2 TCP throughput for three different FF3-1 implementations described in Section 4.2. All measurements were made for IPv4 and IPv6 using /24 and /64 address masks, making the size of the encrypted and plaintext 24 and 80 bits, respectively. As a baseline measurement, we measured TCP throughput for a code which only passes the traffic between the two card interfaces without any modification (pass-through column).

Table 2. TCP throughput for three different FF3-1 implementations.

Implementation	Pure P4 (Described in Section 4.2.1)		P4 + AES in Micro-C (Described in Section 4.2.2)		P4 + FF3-1 in Micro-C (Described in Section 4.2.3)	
Protocol	IPv4	IPv6	IPv4	IPv6	IPv4	IPv6
Pass-through [Gbit/s]	9.38	9.26	9.38	9.26	9.38	9.26
Full FF3-1 [Gbit/s]	0.156	0.149	7.27	6.97	9.38	9.26

Measurement results reveal that the last implementation (entire FF3-1 in Micro-C) achieves maximum TCP throughput, equal to the wire speed. In contrast, the first two implementations—pure P4 and the one with only AES in Micro-C show significantly lower throughputs. Such a performance difference suggests that the compiling from P4 to the executable code in the domain of complex packet processing for the particular hardware configuration is not optimal. Therefore, while it is possible to achieve code portability to the other platforms by using only P4, the performance of the code is not assured. Results also show a difference in IPv4 and IPv6 TCP throughput. This difference can be attributed to the difference in header sizes, which is larger for IPv6, thus yielding less useful data bandwidth for the TCP stream, although the endpoint processing effects should not be neglected. In addition to the TCP throughput tests, one-way throughput tests using PF_RING Zero Copy-based packet streams showed that the system can achieve line rate throughput, i.e., 10 Gbit/s throughput, when the entire FF3-1 implementation is in Micro-C.

The per-packet processing overhead introduced by LISPP remains constant across all packet sizes, as it exclusively involves the encryption/decryption of specific packet header fields. These fields maintain the same size in every packet processed by the LISPP device. As a result, this overhead is not influenced by the size of the packet's payload, ensuring uniformity regardless of changes in payload dimensions. Therefore, the performance constraints of LISPP are more aptly gauged by the maximum achievable packet rate rather than by standard throughput measures. This aspect gains particular significance in the context of TCP, which often employs packets sized at the MTU for large-scale data transfers, a trend commonly observed in TCP throughput testing. It is also important to emphasise that the LISPP system is fully operational on the Netronome network interface card, without requiring any interaction with the hosting server apart from the initial code compilation and subsequent configuration upload, resulting in negligible CPU usage on the host computer to which the Netronome network interface card is connected.

We conducted tests to assess the impact of the test packet size on achievable packet rates and packet latency using the LISPP implementation with FF3-1 in Micro-C. Figure 9a displays the achieved throughput and packet rate using PF_RING Zero Copy-based packet streams. With packet sizes exceeding 600 bytes, the traffic fully saturates the links between the test servers at 10 Gbit/s, resulting in lower recorded packet rates (e.g., a 10 Gbit/s link is completely saturated by sending either 1.25 million 1000 byte-size packets or approximately 0.9 million 1400 byte-size packet per second). Tests with packets smaller than 600 bytes reveal the highest packet rates attainable with LISPP on Netronome cards, reaching 2.64 Mpps for IPv6 and 2.15 Mpps for IPv4. Figure 9b presents the one-way latency incurred by LISPP for packet sizes ranging from 100 to 1400 bytes. As a reference, we have measured one-way latency for packets that pass through the Netronome card without any LISPP processing (measurement denoted as "wire" latency in Figure 9b). Notably, the LISPP implementation with a complete FF3-1 in Micro-C consistently adds approximately 60 microseconds of latency, irrespective of packet size, thereby validating the hypothesis outlined in the previous paragraph.

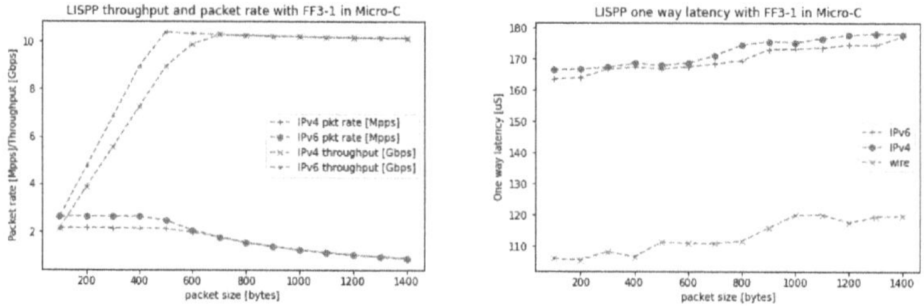

(a) (b)

Figure 9. (**a**) Throughput and achieved packet rate for various packet sizes; (**b**) Per-packet one-way latency for various packet sizes.

The data presented in Figure 9b also indicates that the latency induced by LISPP is independent of the ciphertext size. The added latency remained consistent for IPv4, with a 24-bit ciphertext, and IPv6, featuring an 80-bit ciphertext. This finding suggests that LISPP is scalable with the size (number of devices) of the protected network, which defines the size of the host part of the address to be encrypted. Ultimately, the key parameter that defines the performance limits of the system is the number of packets per second that the system can forward. For handling larger packet throughputs, additional testing with more powerful network accelerators and implementation of load balancing are necessary.

5.2. Lower Number of AES Rounds

FF3-1 is an 8-round algorithm that invokes the AES algorithm in each round, which has ten internal rounds. In our implementation, in which FF3-1 is entirely developed in Micro-C, AES takes 54% of the processor time. Decreasing the number of rounds in any of these algorithms would improve processing time and overall algorithm throughput. However, such an intervention comes with a potential decrease in algorithm security. Recent cryptanalytic papers showed that the complexity of the attack on FPE [29] depends on the number of Feistel rounds, which means that lowering the number of Feistel rounds will decrease attack complexity, which is not a viable option.

On the other hand, during the evaluation of the last stage candidates for the AES algorithm, it was discovered that the output of the Rijndael algorithm (which later became AES) appeared to be random after three rounds. Subsequent rounds produce randomness similar to that already obtained at round 3 [50]. Since the AES algorithm in FF3-1 is used as a random number generator rather than for data encryption, we argue that performance gains can be obtained by lowering the number of rounds in the AES algorithm without sacrificing the strength of the FPE algorithm. We conducted randomness tests on the series of FF3-1 encrypted host addresses. Address series were obtained by encrypting a 9-bit host address and 16-bit port. In each run, we picked 20 random host addresses for which we iterated all 65,536 different ports and analysed the series of obtained encrypted address values. We did the same tests for FF3-1 implementations with AES with 3 to 10 rounds. Runs and Discrete Fourier Transform (DFT) spectral analysis tests from the common randomness batteries of tests [51] were performed. In each case, the hypothesis that the output series is random was confirmed. Figure 10a shows the standard deviations of empirical frequencies of appearance of all possible IP addresses in the 9-bit address range for varying numbers of AES rounds. As can be seen from the image, standard deviations have approximately the same value regardless of the number of rounds, meaning that the variability of empirical frequencies does not change with the number of rounds. Figure 10b shows a DFT magnitude of an encrypted address series obtained using FF3-1 with 3-round AES. Visual inspection shows that the spectrum seems flat for the whole range of frequency values without a single value exceeding the peak threshold value, confirming that the encrypted address series behaves like a random series. This suggests that with FF3-1, performance gains can be obtained by lowering the number of AES rounds without sacrificing the algorithm's security. However, a more detailed analysis of this hypothesis in the field of cryptanalysis is needed.

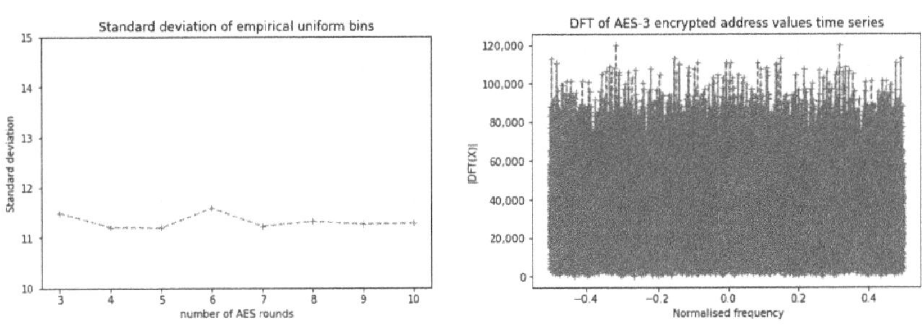

(a) (b)

Figure 10. (**a**) The standard deviation of empirical frequencies of appearance of all possible encrypted IP addresses in the 9-bit address range for varying numbers of AES rounds; (**b**) DFT magnitude of an encrypted address series obtained using FF3-1 with 3-round AES.

In the third batch of performance tests, we analysed the potential packet rate increase by reducing the number of FF3-1 rounds. We pushed 10,000,000 100-byte packets per second through the LISPP system and measured the number of packets that arrived on the sink side. Figure 11a shows the number of packets the system can process per second. The entire FF3-1 implementation in Micro-C can process more than 2 million packets per second, while the packet rate can be almost doubled using 3-round AES as a round function. It is interesting to notice that the network interface card achieved higher packet rates for IPv6 packets, which is probably a consequence of the simpler IPv6 header processing in the card (no header checksum).

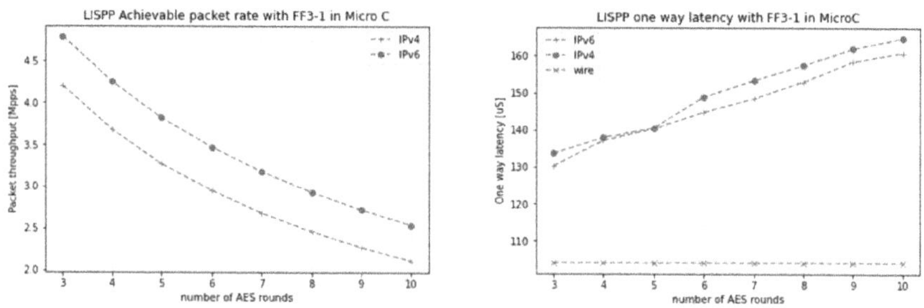

(a) **(b)**

Figure 11. (**a**) The number of packets per second the LISPP system can process for IPv4 and IPv6; (**b**) Per-packet one-way latency.

Finally, we measured the additional latency introduced due to the LISPP packet header field encryption with fewer FF3-1 rounds. Figure 11b gives the per-packet one-way latency measured by the Sockperf tool. The latency added by LISPP packet processing is between 30 and 60 microseconds for the FF3-1 implementations with 3-round and 10-round AES, respectively. Such an additional latency is negligible compared to the latencies on international links and corresponds to the signal propagation latency between the nodes, which are only 10 to 20 km away.

6. Conclusions

In this paper, we report on the research results of using FPE algorithms for privacy protection on the network layer. Designed and implemented system, LISPP, based on the FF3-1 FPE algorithm, can obfuscate source IP addresses and ports fully transparently with minimal additional one-way latency for both IPv4 and IPv6. Its performance on SmartNICs in real network environments is adequate for use in production networks. Therefore, the key conclusion is that the FPE algorithms are a viable option for packet header obfuscation and privacy protection.

Other important conclusions from this research are related to the experiences from the system implementation. Although P4 language is advertised as target-independent, its performance for processor-intensive applications on the particular target device is still highly dependent on the underlying hardware architecture. While code functionality is the same on different targets, its performance is far from optimal, and there are no automated optimisation options. This suggests that the compilation process from P4 to the specific hardware architectures, especially for processor-intensive applications like novel cryptographic algorithms without hardware acceleration, can be significantly improved.

Our further research activities will be in two key directions: integrating LISPP with the onion routing control plane to achieve lower latency than traditional onion routing systems and further optimising the FPE performance on multiprocessor targets using lightweight cryptographic algorithms.

Author Contributions: M.M.: investigation, methodology, software, validation, visualisation, and writing the original draft; U.R.: investigation, validation, and writing-review and editing; P.V.: conceptualisation, methodology, visualisation, supervision and writing the original draft. All authors have read and agreed to the published version of the manuscript.

Funding: This research was partially financially supported by the Ministry of Science, Technological Development, and Innovation of the Republic of Serbia (contract number 451-03-68/2022-14/200103).

Data Availability Statement: The data presented in this study are available on request from the corresponding author.

Acknowledgments: The authors want to express their thanks to Marinos Dimolianis from the National Technical University of Athens, Greece, for his valuable advice and help with the code deployment on Netronome cards.

Conflicts of Interest: The authors declare no conflict of interest.

References

1. Hoang, N.P.; Niaki, A.A.; Gill, P.; Polychronakis, M. Domain Name Encryption Is Not Enough: Privacy Leakage via IP-Based Website Fingerprinting. In Proceedings of the Privacy Enhancing Technologies (PETS), Virtual Event, 12–16 July 2021; pp. 420–440. [CrossRef]
2. Yan, Z.; Lee, J.H. The Road to DNS Privacy. *Future Gener. Comput. Syst.* **2020**, *112*, 604–611. [CrossRef]
3. Khormali, A.; Park, J.; Alasmary, H.; Anwar, A.; Saad, M.; Mohaisen, D. Domain Name System Security and Privacy: A Contemporary Survey. *Comput. Netw.* **2021**, *185*, 107699. [CrossRef]
4. Saidi, S.J.; Gasser, O.; Smaragdakis, G. One Bad Apple Can Spoil Your IPv6 Privacy. *ACM SIGCOMM Comput. Commun. Rev.* **2022**, *52*, 10–19. [CrossRef]
5. GDPR.Eu. What Is Considered Personal Data under the EU GDPR? Available online: https://gdpr.eu/eu-gdpr-personal-data/ (accessed on 17 October 2023).
6. Tor Project | Anonymity Online. Available online: https://www.torproject.org/ (accessed on 17 October 2023).
7. HTTPS Encryption on the Web–Google Transparency Report. Available online: https://transparencyreport.google.com/https/overview?hl=en (accessed on 17 October 2023).
8. Lee, J.K.; Koo, B.; Roh, D.; Kim, W.H.; Kwon, D. *Format-Preserving Encryption Algorithms Using Families of Tweakable Blockciphers*; Lecture Notes in Computer Science (including subseries Lecture Notes in Artificial Intelligence and Lecture Notes in Bioinformatics); Springer: Berlin/Heidelberg, Germany, 2014; Volume 8949, pp. 132–159. [CrossRef]
9. Dworkin, M. *Recommendation for Block Cipher Modes of Operation: Methods for Format-Preserving Encryption*; NIST: Gaithersburg, MD, USA, 2016. [CrossRef]
10. AlSabeh, A.; Khoury, J.; Kfoury, E.; Crichigno, J.; Bou-Harb, E. A Survey on Security Applications of P4 Programmable Switches and a STRIDE-Based Vulnerability Assessment. *Comput. Netw.* **2022**, *207*, 108800. [CrossRef]
11. Hauser, F.; Häberle, M.; Merling, D.; Lindner, S.; Gurevich, V.; Zeiger, F.; Frank, R.; Menth, M. A Survey on Data Plane Programming with P4: Fundamentals, Advances, and Applied Research. *J. Netw. Comput. Appl.* **2023**, *212*, 103561. [CrossRef]
12. Chen, C.; Asoni, D.E.; Barrera, D.; Danezis, G.; Perrig, A. HORNET: High-Speed Onion Routing at the Network Layer. In Proceedings of the ACM Conference on Computer and Communications Security 2015, Denver, CO, USA, 12–16 October 2015; pp. 1441–1454. [CrossRef]
13. Chen, C.; Asoni, D.E.; Perrig, A.; Barrera, D.; Danezis, G.; Troncoso, C. TARANET: Traffic-Analysis Resistant Anonymity at the Network Layer. In Proceedings of the 3rd IEEE European Symposium on Security and Privacy, London, UK, 24–26 April 2018; pp. 137–152. [CrossRef]
14. Chen, C.; Perrig, A. PHI: Path-Hidden Lightweight Anonymity Protocol at Network Layer. In Proceedings of the Privacy Enhancing Technologies, Minneapolis, MN, USA, 18–21 July 2017; pp. 100–117. [CrossRef]
15. Hsiao, H.C.; Kim, T.H.J.; Perrig, A.; Yamada, A.; Nelson, S.C.; Gruteser, M.; Meng, W. LAP: Lightweight Anonymity and Privacy. In Proceedings of the 2012 IEEE Symposium on Security and Privacy, San Francisco, CA, USA, 20–23 May 2012; pp. 506–520. [CrossRef]
16. Sankey, J.; Wright, M. *Dovetail: Stronger Anonymity in Next-Generation Internet Routing*; Lecture Notes in Computer Science (including subseries Lecture Notes in Artificial Intelligence and Lecture Notes in Bioinformatics); Springer: Berlin/Heidelberg, Germany, 2014; Volume 8555, pp. 283–303. [CrossRef]
17. Kuhn, C.; Beck, M.; Strufe, T. Breaking and (Partially) Fixing Provably Secure Onion Routing. In Proceedings of the 2020 IEEE Symposium on Security and Privacy (SP), San Francisco, CA, USA, 18–21 May 2020; pp. 168–185. [CrossRef]
18. Datta, T.; Feamster, N.; Rexford, J.; Wang, L. {SPINE}: Surveillance Protection in the Network Elements. In Proceedings of the 9th USENIX Workshop on Free and Open Communications on the Internet (FOCI 19), Santa Clara, CA, USA, 13 August 2019.
19. Wang, L.; Kim, H.; Mittal, P.; Rexford, J. Programmable In-Network Obfuscation of Traffic. *arXiv* **2020**, arXiv:2006.00097.
20. AS6447-IPv6 BGP Table Statistics. Available online: https://bgp.potaroo.net/v6/as6447/ (accessed on 17 October 2023).

21. Raghavan, B.; Kohno, T.; Snoeren, A.C.; Wetherall, D. *Enlisting ISPs to Improve Online Privacy: Ip Address Mixing by Default*; Lecture Notes in Computer Science (including subseries Lecture Notes in Artificial Intelligence and Lecture Notes in Bioinformatics); Springer: Berlin/Heidelberg, Germany, 2009; Volume 5672, pp. 143–163. [CrossRef]
22. Gont, F.; Krishnan, S.; Narten, T.; Draves, R. *Temporary Address Extensions for Stateless Address Autoconfiguration in IPv6*; Internet Engineering Task Force RFC: Fremont, CA, USA, 2021. [CrossRef]
23. Herrmann, D.; Fuchs, K.P.; Lindemann, J.; Federrath, H. *EncDNS: A Lightweight Privacy-Preserving Name Resolution Service*; Lecture Notes in Computer Science (including subseries Lecture Notes in Artificial Intelligence and Lecture Notes in Bioinformatics); Springer: Berlin/Heidelberg, Germany, 2014; Volume 8712, pp. 37–55. [CrossRef]
24. Non-Paper on the Way Forward on Data Retention. Council of the European Union on June 2021. Available online: https://www.statewatch.org/media/2592/eu-council-data-retention-com-non-paper-wk-7294-2021.pdf (accessed on 17 October 2023).
25. Nguyen, T.A.; Kim, M.; Lee, J.; Min, D.; Lee, J.W.; Kim, D. Performability Evaluation of Switch-over Moving Target Defence Mechanisms in a Software Defined Networking Using Stochastic Reward Nets. *J. Netw. Comput. Appl.* **2022**, *199*, 103267. [CrossRef]
26. Hasty Pudding Specification. Available online: https://web.archive.org/web/20111007174344/http://richard.schroeppel.name:8015/hpc/hpc-spec (accessed on 17 October 2023).
27. Durak, F.B.; Vaudenay, S. *Breaking the FF3 Format-Preserving Encryption Standard over Small Domains*; Lecture Notes in Computer Science (including subseries Lecture Notes in Artificial Intelligence and Lecture Notes in Bioinformatics); Springer: Berlin/Heidelberg, Germany, 2017; Volume 10402, pp. 679–707. [CrossRef]
28. Dunkelman, O.; Kumar, A.; Lambooij, E.; Sanadhya, S.K. *Cryptanalysis of Feistel-Based Format-Preserving Encryption*; Paper 2020/1311; IACR Cryptology ePrint Archive: Bellevue, WA, USA, 2020.
29. Beyne, T. *Linear Cryptanalysis of FF3-1 and FEA*; Lecture Notes in Computer Science (including subseries Lecture Notes in Artificial Intelligence and Lecture Notes in Bioinformatics); Springer: Berlin/Heidelberg, Germany, 2021; Volume 12825, pp. 41–69. [CrossRef]
30. Amon, O.; Dunkelman, O.; Keller, N.; Ronen, E.; Shamir, A. *Three Third Generation Attacks on the Format Preserving Encryption Scheme FF3*; Lecture Notes in Computer Science (including subseries Lecture Notes in Artificial Intelligence and Lecture Notes in Bioinformatics); Springer: Berlin/Heidelberg, Germany, 2021; Volume 12697, pp. 127–154. [CrossRef]
31. Jang, W.; Lee, S.Y. A Format-Preserving Encryption FF1, FF3-1 Using Lightweight Block Ciphers LEA and, SPECK. In Proceedings of the ACM Symposium on Applied Computing, Brno, Czech Republic, 30 March–3 April 2020; pp. 369–375. [CrossRef]
32. NFP-4000 Theory of Operation. Available online: https://www.netronome.com/static/app/img/products/silicon-solutions/WP_NFP4000_TOO.pdf (accessed on 17 October 2023).
33. The Joy of Micro-C, Netronome. Available online: https://cdn.open-nfp.org/media/documents/the-joy-of-micro-c_fcjSfra.pdf (accessed on 17 October 2023).
34. Bosshart, P.; Daly, D.; Gibb, G.; Izzard, M.; McKeown, N.; Rexford, J.; Schlesinger, C.; Talayco, D.; Vahdat, A.; Varghese, G.; et al. P4: Programming Protocol-Independent Packet Processors. *ACM SIGCOMM Comput. Commun. Rev.* **2014**, *44*, 87–95. [CrossRef]
35. Programming Netronome Agilio®SmartNICs. Available online: https://www.netronome.com/media/documents/WP_NFP_Programming_Model.pdf (accessed on 17 October 2023).
36. Architecture for Simple Switch-V1model.P4. Available online: https://github.com/p4lang/p4c/blob/main/p4include/v1model.p4 (accessed on 17 October 2023).
37. P416 Portable Switch Architecture (PSA) Version 1.2. The P4.Org Architecture Working Group 2022-12-22. Available online: https://p4.org/p4-spec/docs/PSA-v1.2.pdf (accessed on 17 October 2023).
38. Gomez, J.; Kfoury, E.F.; Crichigno, J.; Srivastava, G. A Survey on TCP Enhancements Using P4-Programmable Devices. *Comput. Netw.* **2022**, *212*, 109030. [CrossRef]
39. LISPP: A Lightweight Stateless Network Layer Privacy Protection System. Available online: https://github.com/marko-micovic/lispp (accessed on 17 October 2023).
40. Kaur, S.; Kumar, K.; Aggarwal, N. A Review on P4-Programmable Data Planes: Architecture, Research Efforts, and Future Directions. *Comput. Commun.* **2021**, *170*, 109–129. [CrossRef]
41. Chen, X. Implementing AES Encryption on Programmable Switches via Scrambled Lookup Tables. In Proceedings of the 2020 ACM SIGCOMM Workshop on Secure Programmable Network Infrastructure (SPIN), Virtual Event, 10–14 August 2020; pp. 8–14. [CrossRef]
42. Harkous, H.; Jarschel, M.; He, M.; Priest, R.; Kellerer, W. Towards Understanding the Performance of P4 Programmable Hardware. In Proceedings of the 2019 ACM/IEEE Symposium on Architectures for Networking and Communications Systems, ANCS 2019, Cambridge, UK, 24–25 September 2019. [CrossRef]
43. Viegas, P.B.; de Castro, A.G.; Lorenzon, A.F.; Rossi, F.D.; Luizelli, M.C. *The Actual Cost of Programmable SmartNICs: Diving into the Existing Limits*; Lecture Notes in Networks and Systems; Springer: Berlin/Heidelberg, Germany, 2021; Volume 225, pp. 181–194. [CrossRef]
44. Programming NFP with P4 and C. Available online: https://www.netronome.com/media/documents/WP_Programming_with_P4_and_C.pdf (accessed on 17 October 2023).

45. Bertoni, G.; Breveglieri, L.; Fragneto, P.; Macchetti, M.; Marchesin, S. *Efficient Software Implementation of AES on 32-Bit Platforms*; Lecture Notes in Computer Science (including subseries Lecture Notes in Artificial Intelligence and Lecture Notes in Bioinformatics); Springer: Berlin/Heidelberg, Germany, 2003; Volume 2523, pp. 159–171. [CrossRef]
46. A TCP, UDP, and SCTP Network Bandwidth Measurement Tool. Available online: https://github.com/esnet/iperf (accessed on 17 October 2023).
47. PF_RING ZC (Zero Copy), Multi-10 Gbit RX/TX Packet Processing from Hosts and Virtual Machines. Available online: https://www.ntop.org/products/packet-capture/pf_ring/ (accessed on 17 October 2023).
48. Mellanox Network Benchmarking Utility. Available online: https://github.com/Mellanox/sockperf (accessed on 17 October 2023).
49. Lopes, R.; Rand, D.; Chown, T.; Golub, I.; Vuletic, P. Network Performance Tests over the 100G BELLA Link between GÉANT and RNP. 2023. Available online: https://resources.geant.org/wp-content/uploads/2023/02/GN4-3_White-Paper_Network-Performance-Tests-Over-100G-BELLA-Link.pdf (accessed on 15 November 2023).
50. Randomness Testing of the Advanced Encryption Standard Finalist Candidates. Booz-Allen and Hamilton Inc Mclean Va. Available online: https://nvlpubs.nist.gov/nistpubs/Legacy/IR/nistir6483.pdf (accessed on 17 October 2023).
51. Random Number Generators: An Evaluation and Comparison of Random.Org and Some Commonly Used Generators. Management Science and Information Systems Studies Project Report, the Distributed Computing Group, Trinity College Dublin, Ireland. Available online: https://www.random.org/analysis/Analysis2005.pdf (accessed on 17 October 2023).

electronics

MDPI

Article

Deployment and Implementation Aspects of Radio Frequency Fingerprinting in Cybersecurity of Smart Grids

Maaz Ali Awan [1], Yaser Dalveren [1], Ferhat Ozgur Catak [2,*] and Ali Kara [3,*]

[1] Department of Electrical and Electronics Engineering, Atilim University, Ankara 06830, Turkey; awan.maaz@student.atilim.edu.tr (M.A.A.); yaser.dalveren@atilim.edu.tr (Y.D.)

[2] Electrical Engineering and Computer Science, University of Stavanger, 4021 Rogaland, Norway

[3] Department of Electrical and Electronics Engineering, Gazi University, Ankara 06570, Turkey

* Correspondence: f.ozgur.catak@uis.no (F.O.C.); akara@gazi.edu.tr (A.K.)

Abstract: Smart grids incorporate diverse power equipment used for energy optimization in intelligent cities. This equipment may use Internet of Things (IoT) devices and services in the future. To ensure stable operation of smart grids, cybersecurity of IoT is paramount. To this end, use of cryptographic security methods is prevalent in existing IoT. Non-cryptographic methods such as radio frequency fingerprinting (RFF) have been on the horizon for a few decades but are limited to academic research or military interest. RFF is a physical layer security feature that leverages hardware impairments in radios of IoT devices for classification and rogue device detection. The article discusses the potential of RFF in wireless communication of IoT devices to augment the cybersecurity of smart grids. The characteristics of a deep learning (DL)-aided RFF system are presented. Subsequently, a deployment framework of RFF for smart grids is presented with implementation and regulatory aspects. The article culminates with a discussion of existing challenges and potential research directions for maturation of RFF.

Keywords: radio frequency fingerprinting; machine learning; deep learning; software-defined radio; Internet of Things; cybersecurity; smart city; smart grid

Citation: Awan, M.A.; Dalveren, Y.; Catak, F.O.; Kara, A. Deployment and Implementation Aspects of Radio Frequency Fingerprinting in Cybersecurity of Smart Grids. *Electronics* **2023**, *12*, 4914. https://doi.org/10.3390/electronics12244914

Academic Editors: Dariusz Rzońca and Tomasz Rak

Received: 23 October 2023
Revised: 30 November 2023
Accepted: 4 December 2023
Published: 6 December 2023

1. Introduction

Over the past few years, smart cities have experienced substantial growth and expanded their horizon considerably. Notably, recent breakthroughs in IoT have opened exciting avenues, serving as pivotal technological foundations for smart cities [1]. These advancements facilitate the creation and automation of cutting-edge services and sophisticated applications tailored to the diverse needs of urban communities, thus benefiting a wide range of city stakeholders. Figure 1 illustrates key components of a modern smart city.

Complementary to these advancements, smart grids are transformative for smart cities, optimizing energy usage in real time and pre-empting potential problems [2]. Smart grids are an essential national asset for any country, playing a crucial role in modernizing energy infrastructure. Traditional power grids comprise power generation, transformation, transmission, and distribution. Smart grids incorporate diverse power equipment and may incorporate IoT devices that sense humidity, temperature, immersion, vibration, current leakage, and record video data. Pointed IoT equipment may enable implementation of intelligent power systems [3–5]. These IoT devices establish wireless device-to-device (D2D) communication at the physical layer [6]. Each network has a gateway for data concentration which constitutes the network layer, and the control station serves the application layer of IoT in smart grids. While smart grids offer immense benefits to smart cities, they also present significant security challenges. A substantial review was conducted by Alsuwian et al. [7] concerning cybersecurity threats in IoT of smart grids. These networks in smart grids operate at the intersection of the physical layer, network layer, and application layer, making them susceptible to cyber threats at multiple levels. Therefore, their security at all levels is paramount. The interconnected nature of smart grids entails that vulnerability at one layer may cascade across the entire system, potentially leading to widespread disruptions.

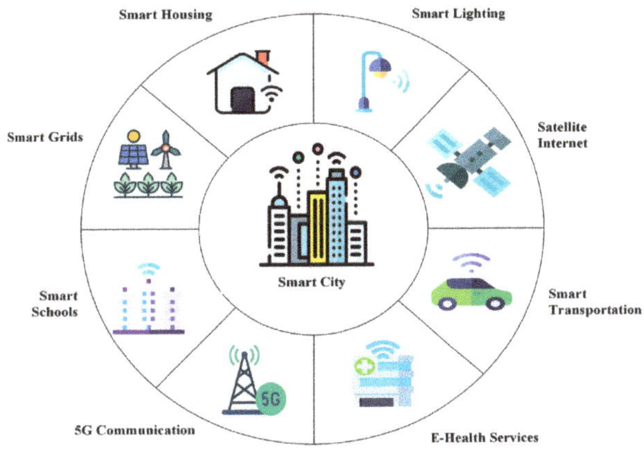

Figure 1. Concept of smart city [1].

Security requirements in a wireless network encompass several critical aspects to safeguard the integrity and privacy of data transmission [8]. Authenticity is paramount because unauthorized access is a major security concern, especially in life-critical IoT applications [9]. In existing IoT devices, coprocessors are employed for symmetric key-based encryption such as AES-128 and AES-256 [10]. However, maintenance of these keys is an administrative overhead and must be mitigated through the use of public/private key pairs [11]. Such pairs are mathematically correlated and allow for enhanced data security compared to symmetric cryptography. Nevertheless, generating mathematically complex key pairs using true random number generators is not a possibility on most low resource IoT devices for the time being. It is envisioned that IoT of the future may benefit from the massive potential of quantum cryptography in the post-quantum computing era. High-efficiency quantum digital signature (QDS) protocols are being developed using asymmetric quantum keys [12]. Another novel concept, Internet of Predictable Things (IoPT), could be employed in mitigation of cyberattacks using energy forecasting in smart grids with machine learning (ML) aids to detect anomalous data patterns [13]. These directions possess substance in the improvement of cybersecurity for future IoT.

The authentication challenge extends to confidentiality and integrity compelling drastic measures to limit access to sensitive data, allowing only intended users to view or modify it. Lastly, availability is mandatory for allowing authorized users to reliably access network resources whenever and wherever needed. Physical layer security measures such as the long-range frequency hopping spread spectrum (LR-FHSS) [14] are gaining popularity due to integration in contemporary long-range (LoRa)-based IoT devices. In the event of jamming or interference, frequency hopping at multiple channels can ensure better link availability. These security requirements collectively form the foundation of a robust and reliable wireless network. Wired networks rely on physical cables for node connections, while wireless networks are more vulnerable due to their broadcast nature making them susceptible to eavesdropping, denial-of-service (DoS), spoofing, man-in-the-middle (MITM) attacks, and message falsification. Cryptographic techniques are commonly used to prevent eavesdropping, ensuring identity verification. In IoT, security gaps exist due to reverse engineering threats and challenges in rapidly installing cryptographic protocols on insecure devices. On the other hand, noncryptographic methods, such as device-specific signal pattern analysis, complement traditional cryptography by identifying known devices and detecting rogue ones, offering essential security without modifying the IoT devices.

Radio frequency fingerprinting (RFF), being a noncryptographic method, has been in use for a few decades now. However, its potential as a physical layer security feature in wireless sensor networks of IoT has been gaining popularity recently [15,16]. RFF uses hardware impairments in the radio section of an IoT device for classification. These impairments are unique to each IoT device due to inherent nonlinearities in the manufacturing process of these inexpensive devices. The components of an IoT device are shown in Figure 2. It is imperative to point out that the aim of most studies on RFF has been device authentication at the physical layer—threat elimination at the first line of defense.

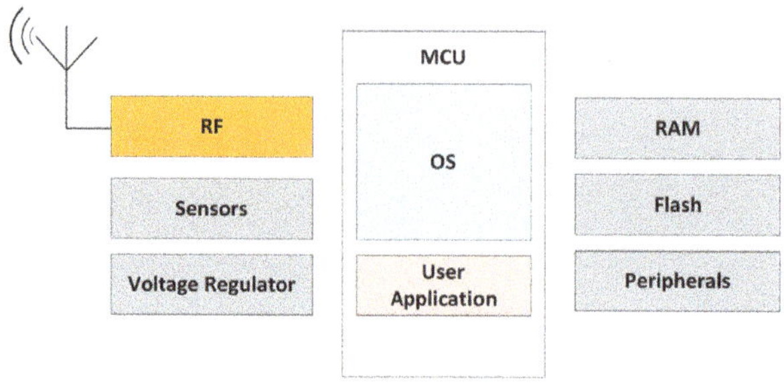

Figure 2. Components of an IoT device.

To this end, a gap exists in studying RFF deployment in a smart grid use case. This article primarily explores the feasibility of integrating RFF into the existing infrastructure of wireless sensor networks in smart grids. The body of this work investigates the potential of RFF as a physical layer security feature for the said application and presents a deployment framework.

The remainder of the article is structured as follows: Section 2 provides an overview of related research. Section 3 illustrates ingredients of a typical DL-aided RFF system. Security challenges and associated discussions are detailed in Section 4. Smart grids as a use case for RFF deployment is covered in Section 5. The ensuing Section 6 argues the potential challenges and directions for future research. Finally, Section 7 concludes the article.

2. Related Work

The core concept behind RFF involves the extraction of distinct patterns or features from devices and utilizing them as signatures for device classification. Previously, a wide range of features including but not limited to physical (PHY) layer and medium access control (MAC) layer have been employed in RFF. However, some straightforward identifiers like Internet Protocol (IP) addresses, MAC addresses, and international mobile station equipment identity (IMEI) numbers are susceptible to spoofing. Similarly, received signal strength indication (RSSI) and channel state information (CSI) could be affected by mobility and environmental changes. A recent research focus has been the investigation of features that are intrinsic to a specific device, possess stability over time, and are challenging for malicious actors to replicate. The authors' prior work has contributed to the advancement of RFF while it was still a topic of academic research. However, the focus of this work is the discussion of practical deployment aspects of RFF in real-world applications. The following subsections cover the related work from all the domains associated with this body of work.

2.1. Previous Work

The primary focus of previous work has been the development of cost-effective techniques for extracting RF fingerprints. In this context, a method for modifying transient signals was presented [17]. Emphasizing cost-effectiveness, a modular RF front end for RFF analysis of Bluetooth signals was offered [18]. Given the passive nature of RFF, modular solutions are particularly relevant, enabling a single RFF system to classify multiple IoT devices without any modification. A common use case of classifying Bluetooth radios of cellular devices was addressed in [19]. Firstly, the signal was preprocessed through transient signal decomposition using variational mode decomposition (VMD). Subsequently, a linear support vector machine (LSVM) was employed as a classifier. A comparative study on classifiers, considering varying signal-to-noise ratio (SNR) levels and dataset sizes was organized in [20]. The experimental results yielded excellent classification accuracy even at low SNR values, implying tremendous relevance in real-world scenarios. Adhering to open science principles, a rich dataset to aid the research community in advancing RFF technology was submitted in [21]. The goal was to aid prospective researchers with the inclusion of an acquisition method for gathering Bluetooth signals. Another potential avenue for academic exploration is the application of RFF for localization. Addressing this, [22] presented a discussion on recent advances and challenges surrounding RFF localization in outdoor environments.

2.2. Cybersecurity in Smart Grids

As smart grids emerge to play an integral role in the evolution of smart cities, a rising need to overhaul their cybersecurity is imminent. The threat spectrum of cyberattacks being faced by smart girds is tremendous [23]. Various organizations, such as the National Institute of Standards and Technology (NIST) and the Smart Grid Interoperability Panel (SGiP), are shaping security requirements for smart grids. Authentication and authorization are central to the overall security of smart grids. Per the guidelines for smart grid cybersecurity published by NIST [24], the focus of security has been limited to cryptographic techniques only. A detailed framework for key management and associated operational issues was provided in the referred document. Key management can be improved using physically unclonable functions (PUF). Generation of PUF hinges on the intrinsic uniqueness within the integrated circuit of a device. A key generated by a device employing PUF can only be regenerated by the same device. This characteristic is leveraged by the utility to authenticate data generated by smart meters [25]. With developed countries increasingly embracing smart grids, the security concerns and potential remedies have become a focal point for researchers and industry experts [26]. Ongoing endeavors are directed towards securing the network and application layers of IoT in smart grids. Remarkably, the non-cryptographic security techniques in IoT for smart grids have not been extensively studied. This represents a novel area where RFF may emerge as a promising candidate.

2.3. Historical and Contemporary Use of RFF

The classification of signals using passive radio frequency (RF) receivers enhanced by artificial intelligence has a historical precedent dating back three decades. Initial use of RFF involved the classification of signals from multiple radar sources leveraging their distinct attributes [27]. More recently, RFF has gained popularity in IoT with experiments on wireless devices using frequency, magnitude, phase offsets, and in-phase and quadrate (I/Q) imbalance as differentiating features [28]. Utilizing RFF for device authentication finds its most straightforward application in RFID systems [29]. The cited studies exhibit the relevance of RFF in various legacy and contemporary applications.

2.4. Physical Layer Security in Wireless Communication

There have been substantial studies concerning physical layer security in wireless communication of IoT devices. In the era of ML and DL, physical, network, and application layers of IoT are susceptible to security threats [30]. As adversarial attacks grow more and more complex, security measures on all layers of communication networks are emerging on the horizon. In this regard, classification efforts on cellular phones using their integrated physical components have been conducted [31]. Non-cryptographic methods for user authentication and device identification in static and mobile wireless networks have seen academic interest [32]. Nevertheless, there are advantages, limitations, and implementation challenges associated with these novel methods. A literature review of relevant studies underscore the potential of physical layer security in wireless communication between IoT devices for authentication.

2.5. Machine Learning in RFF

Emitter-specific hardware attributes can be leveraged without the use of machine learning employing expert features in RFF extraction algorithms such as signal phase [33]. However, this approach is over-reliant on the quality of the received signal, which is not practical in actual scenarios as wireless signals undergo drastic changes in amplitude and phase due to channel effects. Conversely, DL-aided RFF has gained popularity due to its ability to detect unique features in datasets. This approach has made the identification and classification problem scalable to cater unseen devices. More precisely, convolutional neural networks (CNN) have exhibited even more accurate results [34]. Automated feature extraction in DL has proven to be a potent solution, surpassing traditional methods employing only the handcrafted features. However, hybrid models have exhibited even better results when a handcrafted feature such as carrier frequency offset (CFO) is used in unison with DL [35]. An examination of reference studies reveals a multitude of prevalent ML, DL, and hybrid methods. The choice of a specific model hinges on the adopted representation of the RF signal, whether it be I/Q, spectrogram, or fast Fourier transform (FFT).

3. Typical DL-Aided RFF System

The two major domains in an RFF system comprise RF and DL. The choice of an SDR architecture for the RF domain is governed by its flexible nature to process raw waveforms and a wide range of operating frequencies. For the DL part, the host processor serves as a platform for training a neural network (NN) on a given dataset followed by classification in the inference stage. The following subsections provide some explanation for the process of RF signal acquisition followed by the rationale for pre-processing before the signal is subject to the training and inference stage.

3.1. RF Signal Acquisition

The first step in RFF comprises the RF signal acquisition. To make the signal fitting for the classification stage, there is a need to pre-process the signal. The collection of signals followed by pre-processing collectively constitutes the signal acquisition process. The requirement for pre-processing stems from the problem statement inherent in the RFF-based device classification. The classical wireless communication model serves a simple mathematical explanation. For the sake of simplicity, the high-frequency carrier component is omitted. Baseband signal at the input of the RFF system, $y(t)$, can then be given:

$$y(t) = G(h(\tau, t)) * F^K(x(t)) + n(t), \tag{1}$$

where $x(t)$ is the theoretical modulated signal. $G(\cdot)$ denotes the hardware effects of the receiver and $h(\tau, t)$ is the impulse response of time dispersive wireless channel with delay τ. $F^K(\cdot)$ signifies the transmitter specific effect of device under test (DUT), K, $n(t)$ is the additive white Gaussian noise (AWGN), and $*$ is the convolution operation. The goal of RFF is to extract $F^K(\cdot)$, unique to each hardware and difficult to clone or tamper. There are, however, some common hurdles in the development of a robust RFF. Firstly, the transmitter specific $F^K(\cdot)$ is miniscule and overly reliant on the signal quality [33]. One approach could be to artificially create artifacts in the transmitter, but this could hamper communication performance. Moreover, in practical wireless channels, the received signal $y(t)$ undergoes amplitude and phase dispersion due to channel impulse response $h(\tau, t)$. Therefore, the NN shown in Figure 3 has the tendency to make inaccurate predictions, since $h(\tau, t)$ is not predictable and may vary significantly between training and inference. As already highlighted, DL-aided RFF systems have shown performance improvement; however, DL relies on the assumption that the data points follow an independent and identical distribution (i.i.d). In other words, statistical parameters, such as mean and variance, must remain consistent across the entire dataset. The varying impulse response of a time-dispersive channel could therefore be a cause for a DL model to generalize poorly on unseen data. In dynamic scenarios, the variance of $h(\tau, t)$ is an even bigger challenge. In addition to the channel variance, the relative motion between the communicating nodes induces Doppler shift given by the following expression:

$$\Delta f = f_c \frac{c}{v} cos(\theta), \tag{2}$$

where Δf is the Doppler shift in the carrier frequency f_c due to relative motion between the communicating nodes having a relative velocity v at an angle θ, and c is the speed of light. The effect of Doppler shift causes signal degradation, which in turn affects the classification performance. CFO is another challenge that is prevalent in inexpensive radios; by virtue, low-cost crystal oscillators have accuracies in the excess of multiple tens of parts per million (PPM). Amidst these challenges, there is a burgeoning requirement to pre-process the signal before it is stacked in a dataset to train the NN. Pre-processing comprises signal conditioning, as employed in any legacy radio receiver, for accurate symbol detection. The most important aspect of pre-processing is to mitigate the channel effects, since it is the most unpredictable variable in the entire process. Various mitigation methods are prevalent in the literature, such as the channel-independent spectrogram for narrowband communication channels that experience very little change in a short time interval [36]. Another direction is data augmentation where a channel simulator may aid in training the NN on simulated channel conditions. This approach can minimize the channel effects since a NN trained on a dataset containing diverse channel conditions shall generalize much better in the inference stage. Nonetheless, rationale for the requirement of channel equalization is clear and justified. Detailed discussion of the implementation of channel equalizers is beyond the scope of this work. More importantly, it must be realized that synchronization is a mandatory step for channel equalization. Among other issues, the effect of the receiver $G(\cdot)$ must not alter the classification performance. A practically deployable RFF system must be agnostic to the effects of the receiver. A NN trained on one RFF system must be able to perform equally well on the other if there is a need to replace it in the event of failure. Lastly, normalization of the received signal is performed to bar the NN from using signal strength as a feature for training.

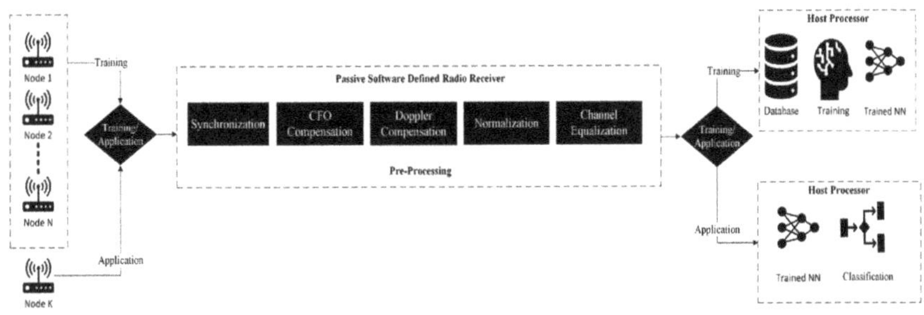

Figure 3. Typical DL-based RFF system.

3.2. Deep Learning

Figure 3 illustrates the working of a modern RFF system as a two-stage process, training and inference. As evident, training comprises receiving samples from N unique devices and an RF signal is received from device K in the inference stage to differentiate between a legitimate and rogue transmitter on a per packet basis. The mathematical model for the classification problem ensues. Let D^{train}, a training dataset from N devices be given by

$$D^{train} = \{(y_m, P_m)\}_{m=1}^{M_{train}}, \tag{3}$$

where y_m is the mth training sample and P_m is the respective output of the one-hot encoding function $O(\cdot)$ for the mth DUT label, given by

$$P_m = O(l_m), \tag{4}$$

where l_m is the ground truth DUT label of the mth training sample. If M_{train} is the total number of training samples in a neural network $f(y;\Theta)$, parameters Θ can be optimized using D^{train} by the following expression:

$$\Theta = \underset{\Theta}{argmin} \frac{1}{M_{train}} \sum\nolimits_{(y,p)\in\ D^{train}} L_{ce}\left(f(y;\Theta), p\right), \tag{5}$$

where $L_{ce}(\cdot)$ is the cross-entropy loss. In the inference stage, the receiver captures a signal y' and feeds it into the well-trained neural network $f(y;\Theta)$ for prediction. A probability vector $\hat{p}$ is obtained in the inference stage as

$$\hat{p} = f(y';\Theta), \tag{6}$$

where $\hat{p} = \{\hat{p}_1, \ldots, \hat{p}_k, \ldots, \hat{p}_N\}$ is a probability vector over all the N DUTs, and $\hat{p}_K$ is the estimated probability for the Kth DUT. The predicted device label $\hat{l}$ is derived by simply selecting the index of the element with the highest probability as defined below.

$$\hat{l} = \underset{k}{\text{argmax}}(\hat{p}). \tag{7}$$

The model outlined above serves as the foundation for device classification, utilizing labels derived from a predefined dataset. To declare an unknown device as rogue, each element from the set $\hat{p}$ must exhibit a probability value below a predetermined threshold. This criterion designates a device as absent from the roster of legitimate devices, thereby classifying it as rogue. This ability of the NN to identify unseen devices adds scalability to the system and makes the classification step an open-set problem.

To summarize, a typical DL-aided RFF system must have a common set of attributes. The scope of this article is to present a practically deployable RFF system. Therefore, based on state-of-the-art and literature reviews of relevant dissertations [37,38] and an elaborate survey [39], essential features of a practical DL-aided RFF system are listed:

1. Synchronization.
2. CFO Compensation.
3. Doppler Compensation.
4. Normalization.
5. Channel Equalization.
6. Receiver Agnostic.
7. Scalability.

4. IoT in Smart Grids

The US Department of Energy defines smart grids as modernized electrical grids that leverage advanced technology to enhance the efficiency, reliability, and sustainability of electricity generation, distribution, and consumption [40]. They incorporate various power generation sources, including customer-generated energy, solar, wind, and more. Understanding the role of IoT in smart grids and the security challenges it presents is crucial before delving into discussions about the necessity to bolster cybersecurity.

4.1. D2D Wireless Communication in Smart Grids

The effectiveness of smart grids is rooted in their ability to anticipate fluctuations in energy supply, optimize grid operations, and promptly respond to changes in demand and power failures. This capability not only strengthens grid stability but also contributes to the reduction in energy wastage, enhancing overall sustainability [41]. Central to the realization of this concept is D2D wireless communication between IoT devices at the control center, the power station, and consumers. Figure 4 shows the evolution of power grids. The dotted lines mark the communication network, which is crucial in achieving the functionality of smart grids.

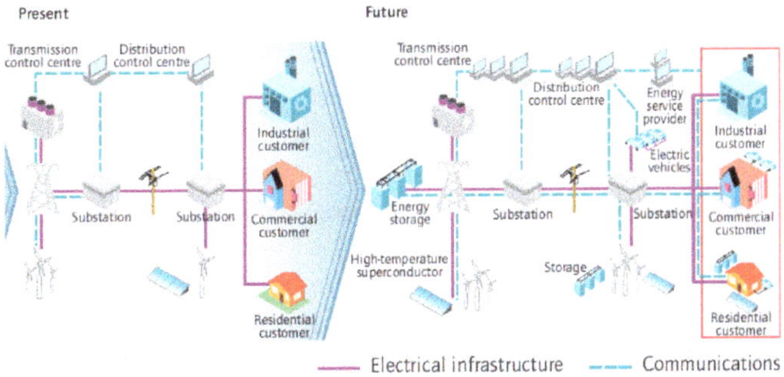

Figure 4. Evolution from conventional to smart grids [41].

4.2. Security Challenges in Wireless Communication

In wired networks, nodes are physically linked by cables. Conversely, wireless networks face heightened vulnerability due to their broadcast nature. They are susceptible to various malicious attacks, such as eavesdropping [42], denial-of-service (DoS) [43], spoofing [44], man-in-the-middle (MITM) [45], message falsification/injection [46], etc. To ensure confidentiality and authentication, existing systems commonly use cryptographic techniques to prevent eavesdropping and unauthorized access to networks [47,48].

Conventional cryptography ensures identity verification using techniques like message authentication codes, digital signatures, and challenge-response sessions [49]. However, in widely distributed IoT, security gaps persist due to reverse engineering threats [50], impracticality of rapid cryptographic protocol installation in insecure devices [51], and inefficacy against hijacked devices.

In a post-quantum computing era, the above cited challenges could be overcome using quantum cryptography. For instance, quantum light could be used to generate inherently unforgeable quantum cryptograms [52]. These cryptograms have exhibited the potential to be used in practical applications with near-term technology. Future IoT may benefit tremendously at the application layer as a solution to vulnerabilities present in symmetric cryptographic schemes. Non-cryptographic methods, such as device-specific signal pattern analysis, supplement traditional cryptography by identifying known devices and detecting rogue ones [53]. These approaches are crucial for enhancement of cybersecurity in IoT, without requiring major system modifications [54].

4.3. Cybersecurity in Smart Grids

The layered architecture in IoT of smart grids is illustrated in Figure 5 [55]. At the physical layer, data from sensors, actuators, and smart meters are collected at the gateways. At the network layer, data from multiple gateways are concentrated and relayed to the application layer operating on servers in the control center using legacy communication methods. The goal of cybersecurity in IoT is to ensure protection at every layer; the same is applicable in smart grids as well. A closer look at the threat spectrum being faced by smart grids underscores the importance of device authentication [56,57], although physical layer intrusion detection systems have the capacity to perform device authentication at the first stage of defense in wireless networks [58]. But, to this end, there has not been a study on the implementation of physical layer security measures in wireless communication between IoT devices of smart grids for authentication. To fill this gap, RFF emerges as a potential solution and this article builds the case for discussion on the associated deployment aspects in smart grids.

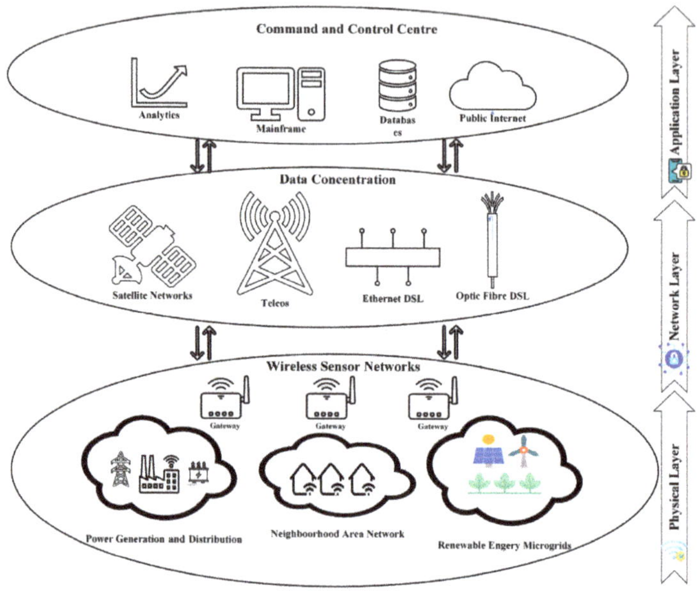

Figure 5. Cyber security in smart grids [55].

5. Deployment of RFF in Smart Grids

The aim of this article is to conduct a feasibility study and discuss practical deployment consideration apropos of the use of RFF in smart grids. Existing IoT frameworks have been considered for seamless integration of RFF with minimal changes. The core idea is to present RFF as an addition to existing IoT infrastructure instead of reinventing the wheel. The following sub-sections provide considerations and requirements for deployment of RFF in smart grids.

5.1. Network Considerations

In line with the aim of this article, performance metrics of existing IoT serve as a good starting point. Coverage and energy efficiency are important metrics for choosing a network topology [59]. Furthermore, data rate, range, application layer security, and localization are important factors for selecting a particular low-power wide-area network (LPWAN) [60]. From a practical standpoint, cost and scalability hold particular significance [61]. In the UK, smart meters communicate via cellular networks, utilizing 2G or 3G waveforms [62]. However, the use of a long-range wide-area network (LoRaWAN), a star-of-star network topology, in advanced metering infrastructure has been reported as well [63,64]. Given the novelty of RFF and the consideration of performance metrics including cost, energy efficiency, network topology, and communication range, LPWAN is a suitable candidate for the deployment of RFF.

5.2. Security Considerations

Cybersecurity experts have expressed concerns, revealing that 70% of IoT devices are vulnerable to cyberattacks [65]. The wireless sensor network of IoT exhibits vulnerabilities across various layers, and cyberattacks can manifest at different stages [66]. Likewise, LPWAN is not exempt from cyber threats [67]. Wireless sensor networks in smart grids comprise IoT devices equipped with temperature, humidity, light, and wind sensors. The threat from rogue IoT devices to generate falsified data is a significant concern. For instance, exaggerated sensor readings from a smart meter could lead to an unwarranted stimulus from the control station. The limitations of existing security schemes have been discussed in the introduction section of this article. Considering the vulnerability of higher layers to attacks, a novel approach is to secure the physical layer of D2D wireless communication across the network. It is proposed that this extra layer of security should always be in the loop for all end-to-end data transactions between IoT devices in the network.

5.3. Proposed RFF Framework

A key facet of smart grid infrastructure is the real-time estimation of household loads [40,41]. This requirement can be effectively addressed by smart energy meters transmitting data wirelessly at regular intervals. However, this simple task becomes challenging from a cybersecurity perspective in the presence of rogue devices. This scenario is accurately addressed in the physical layer security framework of RFF, as depicted in Figure 6. The proposed configuration ensures that all data transmission from the sensors must pass through the physical security barrier of the RFF system before reaching the control station. The star-of-stars network topology ensures that all the sensors first concentrate their data at their respective gateways. Hosted on the IoT gateways, RFF serves as a filter to allow readings from only legitimate sources while filtering the rogue ones on a per packet basis. Since these gateways can send and receive wireless data, they can filter data from rogue gateways as well.

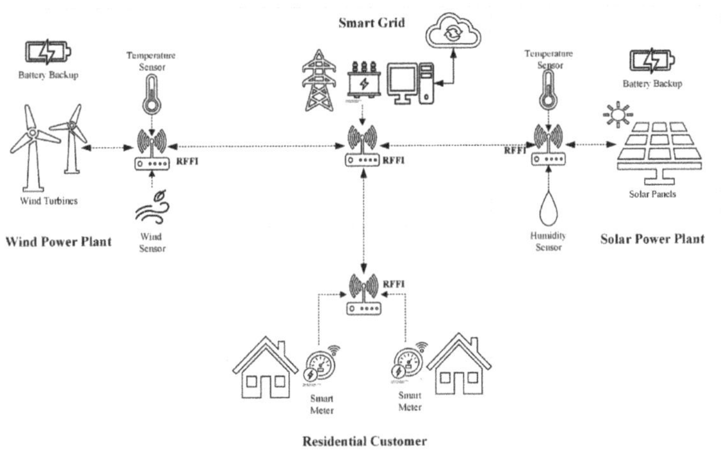

Figure 6. Proposed RFF framework for smart grids.

It is worth mentioning that within a mini star network, multiple gateways could be employed for time-based direction of arrival estimation. This can be extremely helpful in the localization of a rogue device followed by necessary remediation. The IoT devices equipped with sensors communicate unidirectionally with their respective IoT gateways. However, to cater for dynamic load requirements, the control station may issue commands to renewable energy plants, directing them to release stored energy into the system or increase power generation. This requires bidirectional communication in line with the fundamental characteristics of a smart grid [40,41]. This bidirectional communication offers a significant challenge for deployment of RFF in existing low-resource IoT devices, which is discussed in Section 6.

5.4. Performance Considerations

Before a technology is deemed suitable for practical deployment, it is important to estimate its performance considering real-world conditions. The aim of presenting a typical DL-aided RFF system in Section 3 was to highlight the hurdles in achieving the desired outcome. The key performance indicator (KPI) of an RFF system is its classification accuracy.

There has not been a study on the estimation of this KPI in a smart grid use case. However, the authors' previous work in [19] covered the performance comparison of various ML-aided classifiers with different SNR values of the received signal. Table 1 summarizes the experimental results from that study. The results show decent performance even in low SNR conditions. Given that the IoT devices in wireless sensor networks of smart grids are deployed in a static setting, empirical propagation measurements in urban environments may serve as a good reference for RSSI estimation [68]. Figure 7 provides a path loss curve in decibels (dB) against the distance between communicating nodes. Using the locations of smart meters, sensors, and IoT gateways, the expected RSSI could be estimated at the RFF receiver.

Table 1. Comparison of classifiers with various levels of SNR [19].

Classifier	SNR (dB)		
	(8–10)	**(12–15)**	**(18–23)**
L-SVM	79.3%	82.1%	90.5%
Complex Tree	66.8%	68.8%	85.4%
LDA	76.6%	77.8%	83.6%

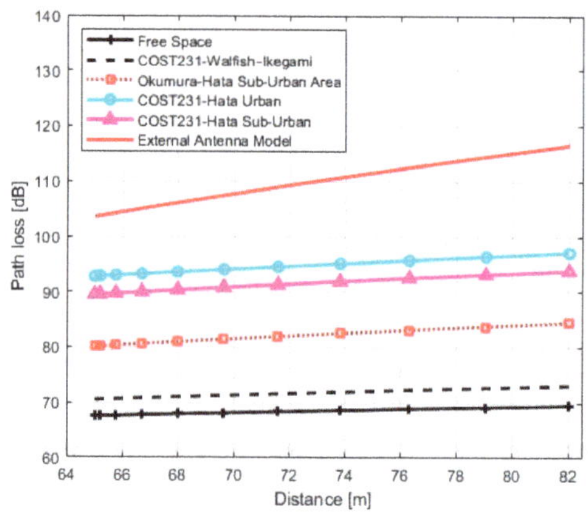

Figure 7. Empirical propagation model in urban environment [69].

Subsequently, the resultant SNR could be used to estimate the classification accuracy using Table 1. It is pointed out that the scope of this study is not limited to a specific smart grid. It is expected that through careful decision making in the selection of appropriate classifier and signal attributes, decent classification accuracy can be achieved, even with low SNR. The classification accuracies of various signal representations for as many as 60 unique LoRa devices are given in Table 2. It may be noted that there is another important aspect in gauging the performance of an RFF system: the time required for training. It is only reasonable to assume that installation, repair, and maintenance of IoT devices in smart grids is likely to be conducted by electric supply companies. Hence, this one-time training activity, even in a practical deployment scenario, may be tolerable given the extraordinary classification performance achieved as a trade-off. Therefore, for a smart grid use case, training time may not be treated as a KPI. Referring to the star network topology outlined in Section 5.3, each wireless sensor network incorporates an IoT gateway. These gateways have been proposed as an optimal site for RFF, ensuring comprehensive access to all IoT devices within the network for accurate classification. Considering the performance metrics across a large set of devices, the findings from referred studies can be reasonably extrapolated as a valuable reference for the smart grid.

Table 2. Classification accuracy of different models with required training time [37].

Signal Representation	DL Model	Accuracy			Number of Parameters	Training Time (minutes)
		w/o CFO Comp.	w/o CFO Comp.	Hybrid		
I/Q samples	MLP	54.08%	55.73%	78.26%	19,018,009	25
	CNN	64.10%	92.26%	98.11%	4,361,545	75
	LSTM	61.16%	89.54%	95.14%	4,267,289	70
FFT results	MLP	55.44%	94.48%	96.17%	19,018,009	25
	CNN	61.14%	82.10%	85.58%	4,361,545	75
	LSTM	49.20%	58.26%	82.81%	4,267,289	69
Spectrogram	MLP	88.60%	91.82%	95.95%	8,821,017	22
	CNN	83.53%	95.35%	96.40%	1,545,193	20
	LSTM	68.16%	89.50%	98.04%	3,427,609	80

5.5. Implementation Aspects

The RFF for smart grids emerges as a highly feasible solution for deployment, primarily owing to its cost-effectiveness and seamless integration capabilities within existing systems. Positioned at the intersection of two prominent domains, RF and ML, RFF may seem intricate from a technical perspective, but from the user's perspective, it can be offered as a plug-and-play solution, hence, simplifying its adoption into existing IoT. Smart grids, being a critical infrastructure from an operation standpoint, can benefit from the passive nature of RFF systems during training as well as inference stages. This can be helpful in ensuring uninterrupted functionality of the smart grids during the deployment process. RFF systems do not necessitate integration into every IoT device. Instead, they can be intelligently deployed only into IoT gateways and leverage the available processing prowess. Moreover, power efficiency poses no significant challenge since RFF systems operate in passive mode, necessitating no significant power requirement. Considering RFF is deployed as a technology, the hardware infrastructure overhead is minimal. In the features of a typical DL-aided RFF system, the ability to be receiver agnostic was discussed as a desirable feature. It would be a highly recommended feature in the event of a device failure, allowing hot replacement but not necessitating training the NN again. Lastly, an RFF system for smart grids was proposed as an open-set solution. This signifies that once the NN is trained on all legitimate IoT devices, any number of rogue devices could be detected [38]. This scalability further adds to the practicality of RFF. Overall, cost effectiveness, power efficiency, low deployment overhead, and scalability make RFF an appropriate practical choice. It is noteworthy that mobility-induced challenges such as antenna cross-polarization loss and Doppler shift may not pose significant hurdles within the context. This assertion is based on the observation that RFF gateways and IoT sensors predominantly exhibit static characteristics in the said application. These elements further simplify the implementation process.

5.6. Regulatory Requirements

The adherence to regulatory standards for RF-based systems stands as a crucial concern. Every country delineates unique requirements governing the utilization of frequency bands. Moreover, there is a limit on maximum permissible power levels for RF transmission. However, RFF, being a passive technology, poses no challenges in this regard. Since the addition of RFF has been proposed for existing LPWAN, the use of industrial, scientific, and medical (ISM) bands for operation is possible. The use of LPWAN in unlicensed bands is a viable direction for smart cities [53]. Having no additional regulatory compliance contributes to the overall feasibility and cost-effectiveness [54] of implementing RFF technology in wireless sensor networks of smart grids. However, the SDR of an RFF system may require EMC certification [69] subject to user needs.

6. Challenges and Future Directions

Being a novel technology and an unprecedented use case in smart grids, RFF entails challenges as well as significant potential for growth in the future. The aim of this section is to underscore the existing challenges and their potential solutions that can significantly advance the deployment of RFF in real-world applications. Additionally, prospective research directions aimed at the maturation of RFF as a technology are deliberated.

6.1. Challenges

RFF for wireless sensor networks of smart grids faces a multifaceted set of challenges. To start, long-term deviation in hardware impairments remains a largely uncharted territory. There has not been a study on long-term operational performance of RFF in IoT. Additionally, bidirectional communication security remains a notable challenge, particularly in scenarios where IoT devices are deployed as receivers. Due to limited resources available on these devices, identification of rogue gateways using RFF is not possible at the present. Addressing these multifarious challenges constitutes a burgeoning area of academic research. The longevity and robustness of RFF technology in the evolving landscape of wireless sensor networks of smart grids needs to be closely monitored in the years to come. Moreover, the emergence of deep generative attackers employing generative adversarial networks is a growing apprehension. These attackers pose a significant threat to device identification even at the physical layer. By leveraging these models, malicious entities can effectively train highly realistic signal or data packet generators capable of mimicking the signal characteristics of legitimate devices. This threat can overcome the ability of RFF systems to identify rogue devices as the success rate of spoofing attacks may increase from less than 10% to approximately 80% [70]. Another significant challenge lies in the availability of abundant datasets for conducting research and experimentation. Addressing these challenges can further add to the potential of RFF as a practical solution for the enhancement of cybersecurity in smart grids.

6.2. Future Research

Research efforts in the realm of RFF are required for channel estimation and equalization. This area holds immense potential for enhancing the reliability and performance of RFF systems in practical scenarios. Specifically, researchers can focus on developing advanced channel estimation techniques that effectively counteract signal distortion caused by time-dispersive channels. However, long training sequences (LTS) can be used to achieve high classification accuracy in 802.11 devices even if the training samples are collected from diverse locations [71]. This research direction has massive potential to benefit LPWAN as well. Simultaneously, the design of a receiver chain that minimizes the combined impact of the channel and receiver components is of paramount importance. Such research efforts can aid in the collection of I/Q datasets that closely resemble the originally transmitted signals, thereby bolstering the overall resilience and classification accuracy of RFF in real-world deployment scenarios. There is another issue in scenarios where IoT devices may be spoofed from a rogue RFF gateway, mimicking its hardware attributes. Such threats may be mitigated using multiple input multiple output (MIMO) receivers. Such localization methods can aid in estimating the difference between the expected and actual position of an IoT device. This additional check can be very useful, especially in smart grids, since the devices in the network are static. But these research directions remain unexplored to this end. Moreover, as already cited in the previous section, there is a pressing need for the collection and publication of open-source datasets. The creation of such datasets will not only facilitate a deeper understanding of RFF as a technology but also empower researchers to develop and validate new algorithms and models effectively. A few datasets have been published in [21,72], but this trend is limited. By fostering an environment of open data sharing and collaboration, the research community can collaborate in improving RFF as a technology for practical deployment in real-word scenarios.

7. Conclusions

The article argues for the potential of RFF as a physical layer security feature for wireless communication between IoT devices of smart grids. It underscores the importance of smart grids and identifies associated cybersecurity threats. It offers RFF as a complementary addition to contemporary cryptographic methods in existing IoT. Characteristics of a typical DL-aided RFF system were presented and the rationale behind design choices was highlighted. Previous work and the reference literature were reviewed as a substantial starting point. Cybersecurity aspects, network architecture, regulatory considerations, and implementation aspects of RFF for smart grids were deliberated. The article culminates with a discussion on the existing limitations and future research directions to improve RFF as a technology and its utilization as a long-term solution for smart grids.

Author Contributions: Investigation, resources, visualization, writing—original draft preparation, M.A.A.; conceptualization, M.A.A., Y.D. and A.K.; validation, supervision, writing—review and editing, Y.D., F.O.C. and A.K.; project administration, A.K. All authors have read and agreed to the published version of the manuscript.

Funding: This research received no external funding.

Data Availability Statement: Data is contained within the article.

Acknowledgments: This work was supported in part by Gazi University under grant FGA-2022-8043.

Conflicts of Interest: The authors declare no conflict of interest.

References

1. Bellini, P.; Nesi, P.; Pantaleo, G. IoT-Enabled Smart Cities: A Review of Concepts, Frameworks and Key Technologies. *Appl. Sci.* **2022**, *12*, 1607. [CrossRef]
2. Mehmood, Y.; Ahmad, F.; Yaqoob, I.; Adnane, A.; Imran, M.; Guizani, S. Internet-of-Things-Based Smart Cities: Recent Advances and Challenges. *IEEE Commun. Mag.* **2017**, *55*, 16–24. [CrossRef]
3. Chen, S.; Wen, H.; Wu, J.; Lei, W.; Hou, W.; Liu, W.; Xu, A.; Jiang, Y. Internet of Things Based Smart Grids Supported by Intelligent Edge Computing. *IEEE Access* **2019**, *7*, 74089–74102. [CrossRef]
4. Rana, M.M.; Xiang, W.; Wang, E. IoT-Based State Estimation for Microgrids. *IEEE Internet Things J.* **2018**, *5*, 1345–1346. [CrossRef]
5. Babar, M.; Tariq, M.U.; Jan, M.A. Secure and Resilient Demand Side Management Engine Using Machine Learning for IoT-Enabled Smart Grid. *Sustain. Cities Soc.* **2020**, *62*, 102370. [CrossRef]
6. Abujubbeh, M.; Al-Turjman, F.; Fahrioglu, M. Software-Defined Wireless Sensor Networks in Smart Grids: An Overview. *Sustain. Cities Soc.* **2019**, *51*, 101754. [CrossRef]
7. Alsuwian, T.; Shahid Butt, A.; Amin, A.A. Smart Grid Cyber Security Enhancement: Challenges and Solutions—A Review. *Sustainability* **2022**, *14*, 14226. [CrossRef]
8. Zou, Y.; Zhu, J.; Wang, X.; Hanzo, L. A Survey on Wireless Security: Technical Challenges, Recent Advances, and Future Trends. *Proc. IEEE* **2016**, *104*, 1727–1765. [CrossRef]
9. Halperin, D.; Heydt-Benjamin, T.S.; Ransford, B.; Clark, S.S.; Defend, B.; Morgan, W.; Fu, K.; Kohno, T.; Maisel, W.H. Pacemakers and Implantable Cardiac Defibrillators: Software Radio Attacks and Zero-Power Defenses. In Proceedings of the 2008 IEEE Symposium on Security and Privacy, Oakland, CA, USA, 18–22 May 2008; pp. 129–142. [CrossRef]
10. Kumar, S.; Deora, S.S. Comparative Analysis of Security Techniques in Internet of Things. In Proceedings of the 2022 Seventh International Conference on Parallel, Distributed and Grid Computing (PDGC), Solan, India, 25 November 2022; pp. 407–412. [CrossRef]
11. Word to the Wise. Cryptography with Alice and Bob. 2014. Available online: https://wordtothewise.com/2014/09/cryptography-alice-bob/ (accessed on 18 September 2023).
12. Yin, H.-L.; Fu, Y.; Li, C.-L.; Weng, C.-X.; Li, B.-H.; Gu, J.; Lu, Y.-S.; Huang, S.; Chen, Z.-B. Experimental Quantum Secure Network with Digital Signatures and Encryption. *Natl. Sci. Rev.* **2023**, *10*, nwac228. [CrossRef]
13. Semtech. Application Note: LR-FHSS System Performance, AN1200.64 Rev 1.2. 2022. Available online: https://semtech.my.salesforce.com/sfc/p/#E0000000JelG/a/3n000000v6Za/sHlDztpPfXWzJd7mr01Yj7CaMR0Uxbqy71YmSVpxxIw (accessed on 19 September 2023).
14. Tian, Q.; Lin, Y.; Guo, X.; Wen, J.; Fang, Y.; Rodriguez, J.; Mumtaz, S. New Security Mechanisms of High-Reliability IoT Communication Based on Radio Frequency Fingerprint. *IEEE Internet Things J.* **2019**, *6*, 7980–7987. [CrossRef]
15. Cali, U.; Kuzlu, M.; Sharma, V.; Pipattanasomporn, M.; Catak, F.O. Internet of Predictable Things (IoPT) Framework to Increase Cyber-Physical System Resiliency. *arXiv* **2021**, arXiv:2101.07816.

16. Nouichi, D.; Abdelsalam, M.; Nasir, Q.; Abbas, S. IoT Devices Security Using RF Fingerprinting. In Proceedings of the 2019 Advances in Science and Engineering Technology International Conferences (ASET), Dubai, United Arab Emirates, 26 March–10 April 2019; pp. 1–7. [CrossRef]
17. Ali, A.M.; Uzundurukan, E.; Kara, A. Improvements on Transient Signal Detection for RF Fingerprinting. In Proceedings of the 2017 25th Signal Processing and Communications Applications Conference (SIU), Antalya, Turkey, 15–18 May 2017; pp. 1–4. [CrossRef]
18. Uzundurukan, E.; Ali, A.M.; Kara, A. Design of Low-Cost Modular RF Front End for RF Fingerprinting of Bluetooth Signals. In Proceedings of the 2017 25th Signal Processing and Communications Applications Conference (SIU), Antalya, Turkey, 15–18 May 2017; pp. 1–4. [CrossRef]
19. Ali, A.M.; Uzundurukan, E.; Kara, A. Assessment of Features and Classifiers for Bluetooth RF Fingerprinting. *IEEE Access* **2019**, *7*, 50524–50535. [CrossRef]
20. Aghnaiya, A.; Ali, A.M.; Kara, A. Variational Mode Decomposition-Based Radio Frequency Fingerprinting of Bluetooth Devices. *IEEE Access* **2019**, *7*, 144054–144058. [CrossRef]
21. Uzundurukan, E.; Dalveren, Y.; Kara, A. A Database for the Radio Frequency Fingerprinting of Bluetooth Devices. *Data* **2020**, *5*, 55. [CrossRef]
22. Dogan, D.; Dalveren, Y.; Kara, A. A Mini-Review on Radio Frequency Fingerprinting Localization in Outdoor Environments: Recent Advances and Challenges. In Proceedings of the 2022 14th International Conference on Communications (COMM), Bucharest, Romania, 16 June 2022; pp. 1–5. [CrossRef]
23. Yan, Y.; Qian, Y.; Sharif, H.; Tipper, D. A Survey on Smart Grid Communication Infrastructures: Motivations, Requirements and Challenges. *IEEE Commun. Surv. Tutor.* **2013**, *15*, 5–20. [CrossRef]
24. *Guidelines for Smart Grid Cybersecurity*; National Institute of Standards and Technology: Gaithersburg, MD, USA, 2014. Available online: https://nvlpubs.nist.gov/nistpubs/ir/2014/NIST.IR.7628r1.pdf (accessed on 25 September 2023). [CrossRef]
25. Nabeel, M.; Kerr, S.; Ding, X.; Bertino, E. Authentication and Key Management for Advanced Metering Infrastructures Utilizing Physically Unclonable Functions. In Proceedings of the 2012 IEEE Third International Conference on Smart Grid Communications (SmartGridComm), Tainan, Taiwan, 5–8 November 2012; pp. 324–329. [CrossRef]
26. Komninos, N.; Philippou, E.; Pitsillides, A. Survey in Smart Grid and Smart Home Security: Issues, Challenges and Countermeasures. *IEEE Commun. Surv. Tutor.* **2014**, *16*, 1933–1954. [CrossRef]
27. Willson, G.B. Radar classification using a neural network. In *Applications of Artificial Neural Networks*; SPIE: Bellingham, WA, USA, 1990; Volume 1294, pp. 200–210. [CrossRef]
28. Candore, A.; Kocabas, O.; Koushanfar, F. Robust Stable Radiometric Fingerprinting for Wireless Devices. In Proceedings of the 2009 IEEE International Workshop on Hardware-Oriented Security and Trust, San Francisco, CA, USA, 27 July 2009; pp. 43–49. [CrossRef]
29. Danev, B.; Capkun, S.; Jayaram Masti, R.; Benjamin, T.S. Towards Practical Identification of HF RFID Devices. *ACM Trans. Inf. Syst. Secur.* **2012**, *15*, 1–24. [CrossRef]
30. Al-Garadi, M.A.; Mohamed, A.; Al-Ali, A.K.; Du, X.; Ali, I.; Guizani, M. A Survey of Machine and Deep Learning Methods for Internet of Things (IoT) Security. *IEEE Commun. Surv. Tutor.* **2020**, *22*, 1646–1685. [CrossRef]
31. Baldini, G.; Steri, G. A Survey of Techniques for the Identification of Mobile Phones Using the Physical Fingerprints of the Built-In Components. *IEEE Commun. Surv. Tutor.* **2017**, *19*, 1761–1789. [CrossRef]
32. Zeng, K.; Govindan, K.; Mohapatra, P. Non-Cryptographic Authentication and Identification in Wireless Networks. *IEEE Wirel. Commun.* **2010**, *17*, 56–62. [CrossRef]
33. Hall, J.; Barbeau, M.; Kranakis, E. Detection of Transients in Radio Frequency Fingerprinting Using Signal Phase. *Wirel. Opt. Commun.* **2003**, *9*, 13–18.
34. Riyaz, S.; Sankhe, K.; Ioannidis, S.; Chowdhury, K. Deep Learning Convolutional Neural Networks for Radio Identification. *IEEE Commun. Mag.* **2018**, *56*, 146–152. [CrossRef]
35. Shen, G.; Zhang, J.; Marshall, A.; Peng, L.; Wang, X. Radio Frequency Fingerprint Identification for LoRa Using Deep Learning. *IEEE J. Sel. Areas Commun.* **2021**, *39*, 2604–2616. [CrossRef]
36. Shen, G.; Zhang, J.; Marshall, A.; Cavallaro, J.R. Towards Scalable and Channel-Robust Radio Frequency Fingerprint Identification for LoRa. *IEEE Trans. Inf. Forensics Secur.* **2022**, *17*, 774–787. [CrossRef]
37. Shen, G. Deep Learning Enhanced Radio Frequency Fingerprint Identification for LoRa. Ph.D. Thesis, University of Liverpool, Liverpool, UK, June 2023.
38. Andrews, S.D. Extensions to Radio Frequency Fingerprinting. Ph.D. Thesis, Virginia Polytechnic Institute and State University, Blacksburg, VA, USA, 2019.
39. Youssef, K.; Bouchard, L.; Haigh, K.; Silovsky, J.; Thapa, B.; Valk, C. Vander Machine Learning Approach to RF Transmitter Identification. *IEEE J. Radio Freq. Identif.* **2018**, *2*, 197–205. [CrossRef]
40. U.S. Department of Energy. Smart Grid. Available online: https://www.energy.gov/oe/services/technology-development/smart-grid (accessed on 16 September 2023).
41. Txone Networks. Achieving Energy Transformation: Building a Cyber Resilient Smart Grid. Available online: https://media.txone.com/prod/uploads/2023/04/Achieving-Energy-Transformation-Building-a-Cyber-Resilient-Smart-Grid-TXOne-WP-202303.pdf (accessed on 16 September 2023).

42. Lakshmanan, S.; Tsao, C.-L.; Sivakumar, R.; Sundaresan, K. Securing Wireless Data Networks against Eavesdropping Using Smart Antennas. In Proceedings of the 2008 the 28th International Conference on Distributed Computing Systems, Beijing, China, 17–20 June 2008; pp. 19–27. [CrossRef]
43. Raymond, D.R.; Midkiff, S.F. Denial-of-Service in Wireless Sensor Networks: Attacks and Defenses. *IEEE Pervasive Comput.* **2008**, *7*, 74–81. [CrossRef]
44. Kannhavong, B.; Nakayama, H.; Nemoto, Y.; Kato, N.; Jamalipour, A. A Survey of Routing Attacks in Mobile Ad Hoc Networks. *IEEE Wirel. Commun.* **2007**, *14*, 85–91. [CrossRef]
45. Meyer, U.; Wetzel, S. A Man-in-the-Middle Attack on UMTS. In Proceedings of the 3rd ACM Workshop on Wireless Security, Philadelphia, PA, USA, 1 October 2004; ACM: New York, NY, USA, 2004; pp. 90–97. [CrossRef]
46. Ohigashi, T.; Morii, M. A practical message falsification attack on WPA. In Proceedings of the 2009 Joint Workshop on Information Security, Kaohsiung, Taiwan, 6–7 August 2009.
47. Paar, C.; Pelzl, J. *Understanding Cryptography: A Textbook for Students and Practitioners*; Springer: Berlin/Heidelberg, Germany, 2011; ISBN 9783642041006.
48. Elliott, C. Quantum Cryptography. *IEEE Secur. Priv.* **2004**, *2*, 57–61. [CrossRef]
49. Wang, J.; Liu, Y.; Niu, S.; Song, H.; Jing, W.; Yuan, J. Blockchain Enabled Verification for Cellular-Connected Unmanned Aircraft System Networking. *Future Gener. Comput. Syst.* **2021**, *123*, 233–244. [CrossRef]
50. Shwartz, O.; Mathov, Y.; Bohadana, M.; Elovici, Y.; Oren, Y. Opening Pandora's Box: Effective Techniques for Reverse Engineering IoT Devices. In Proceedings of the Smart Card Research and Advanced Applications: 16th International Conference, CARDIS 2017, Lugano, Switzerland, 13–15 November 2017; pp. 1–21. [CrossRef]
51. Lakew, Y.F.; Singh, A.K.; Bhatia, S. Assessing and Exploiting Security Vulnerabilities of Unmanned Aerial Vehicles. In *Smart Systems and IoT: Innovations in Computing*; Springer: Berlin/Heidelberg, Germany, 2020; pp. 701–710. [CrossRef]
52. Schiansky, P.; Kalb, J.; Sztatecsny, E.; Roehsner, M.-C.; Guggemos, T.; Trenti, A.; Bozzio, M.; Walther, P. Demonstration of Quantum-Digital Payments. *Nat. Commun.* **2023**, *14*, 3849. [CrossRef]
53. Wang, W.; Sun, Z.; Piao, S.; Zhu, B.; Ren, K. Wireless Physical-Layer Identification: Modeling and Validation. *IEEE Trans. Inf. Forensics Secur.* **2016**, *11*, 2091–2106. [CrossRef]
54. Brik, V.; Banerjee, S.; Gruteser, M.; Oh, S. Wireless Device Identification with Radiometric Signatures. In Proceedings of the 14th ACM International Conference on Mobile Computing and Networking, San Francisco, CA, USA, 14 September 2008; ACM: New York, NY, USA, 2008; pp. 116–127. [CrossRef]
55. Communication Network Interdependencies in Smart Grids: Methodology for the Identification of Critical Communication Network Links and Components. Available online: https://www.enisa.europa.eu/publications/communication-network-interdependencies-in-smart-grids (accessed on 1 October 2023).
56. Kimani, K.; Oduol, V.; Langat, K. Cyber Security Challenges for IoT-Based Smart Grid Networks. *Int. J. Crit. Infrastruct. Prot.* **2019**, *25*, 36–49. [CrossRef]
57. Wang, W.; Lu, Z. Cyber Security in the Smart Grid: Survey and Challenges. *Comput. Netw.* **2013**, *57*, 1344–1371. [CrossRef]
58. Yousaf, A.; Loan, A.; Babiceanu, R.F.; Yousaf, O. Physical-layer Intrusion Detection System for Smart Jamming Attacks. *Trans. Emerg. Telecommun. Technol.* **2017**, *28*, e3189. [CrossRef]
59. Mekki, K.; Bajic, E.; Chaxel, F.; Meyer, F. A Comparative Study of LPWAN Technologies for Large-Scale IoT Deployment. *ICT Express* **2019**, *5*, 1–7. [CrossRef]
60. Perez, M.; Sierra-Sanchez, F.E.; Chaparro, F.; Chaves, D.M.; Paez-Rueda, C.-I.; Galindo, G.P.; Fajardo, A. Coverage and Energy-Efficiency Experimental Test Performance for a Comparative Evaluation of Unlicensed LPWAN: LoRaWAN and SigFox. *IEEE Access* **2022**, *10*, 97183–97196. [CrossRef]
61. Hossain, M.I.; Markendahl, J.I. Comparison of LPWAN Technologies: Cost Structure and Scalability. *Wirel. Pers. Commun.* **2021**, *121*, 887–903. [CrossRef]
62. The Full Story on UK Smart Meters. Available online: https://www.smartme.co.uk/smets-2.html (accessed on 20 September 2023).
63. Agung Enriko, I.K.; Zaenal Abidin, A.; Noor, A.S. Design and Implementation of LoRaWAN-Based Smart Meter System for Rural Electrification. In Proceedings of the 2021 International Conference on Green Energy, Computing and Sustainable Technology (GECOST), Miri, Malaysia, 7 July 2021; pp. 1–5. [CrossRef]
64. Gallardo, J.L.; Ahmed, M.A.; Jara, N. LoRa IoT-Based Architecture for Advanced Metering Infrastructure in Residential Smart Grid. *IEEE Access* **2021**, *9*, 124295–124312. [CrossRef]
65. Rawlinson, K. HP Study Reveals 70 Percent of Internet of Things Devices Vulnerable to Attack. Available online: https://www.proquest.com/docview/1549571608 (accessed on 3 December 2023).
66. Brar, H.S.; Kumar, G. Cybercrimes: A Proposed Taxonomy and Challenges. *J. Comput. Netw. Commun.* **2018**, *2018*, 1798659. [CrossRef]
67. Bouzidi, M.; Amro, A.; Dalveren, Y.; Alaya Cheikh, F.; Derawi, M. LPWAN Cyber Security Risk Analysis: Building a Secure IQRF Solution. *Sensors* **2023**, *23*, 2078. [CrossRef]
68. Bouzidi, M.; Mohamed, M.; Dalveren, Y.; Moldsvor, A.; Cheikh, F.A.; Derawi, M. Propagation Measurements for IQRF Network in an Urban Environment. *Sensors* **2022**, *22*, 7012. [CrossRef] [PubMed]
69. *EN 300 220-1 V2.4.1*; Technical Characteristics and Test Methods. European Committee for Electrotechnical Standardization: Brussels, Belgium, January 2012.

70. Shi, Y.; Davaslioglu, K.; Sagduyu, Y.E. Generative Adversarial Network for Wireless Signal Spoofing. In Proceedings of the ACM Workshop on Wireless Security and Machine Learning, Miami, FL, USA, 15 May 2019; ACM: New York, NY, USA, 2019; pp. 55–60. [CrossRef]
71. Li, G.; Yu, J.; Xing, Y.; Hu, A. Location-Invariant Physical Layer Identification Approach for Wi-Fi Devices. *IEEE Access* **2019**, *7*, 106974–106986. [CrossRef]
72. Sankhe, K.; Belgiovine, M.; Zhou, F.; Riyaz, S.; Ioannidis, S.; Chowdhury, K. ORACLE: Optimized Radio Classification through Convolutional Neural Networks. In Proceedings of the IEEE INFOCOM 2019—IEEE Conference on Computer Communications, Paris, France, 29 April–2 May 2019.

electronics MDPI

Article

Securing the Future: A Resourceful Jamming Detection Method Utilizing the EVM Metric for Next-Generation Communication Systems

Cem Örnek [1,2,*] and Mesut Kartal [2]

1 Radar and Electronic Warfare Systems Business Sector, Aselsan Inc., Ankara 06830, Turkey
2 Electronics and Communication Engineering, Istanbul Technical University, Istanbul 34469, Turkey; kartalme@itu.edu.tr
* Correspondence: ornek19@itu.edu.tr; Tel.: +90-536-976-5667

Abstract: This paper addresses the escalating threat of malicious jamming in next-generation communication systems, propelled by their continuous advancement in speed, latency, and connectivity. Recognizing the imperative for communication security, we propose an efficient jamming detection method with distinct innovations and contributions. Motivated by the growing sophistication of jamming techniques, we advocate the adoption of the error vector magnitude (EVM) metric, measured in IQ symbols, deviating from traditional received signal strength and bit error rate-based measurements. Our method achieves enhanced jamming detection sensitivity, surpassing existing approaches. Furthermore, it introduces low complexity, ensuring resource-effective detection. Crucially, our approach provides vital jammer frequency information, enhancing counteraction capabilities against jamming attacks. It demonstrates stable results against varying system parameters, such as modulation type and code rate, thereby contributing to adaptability. Emphasizing practicality, the method seamlessly integrates into 5G and LTE systems without imposing additional overhead. Versatility is demonstrated through successful operations in diverse scenarios that are run by extended simulation conditions. Theoretical analysis substantiates these advantages, reinforcing the validity of our methodology. The study's success is further validated through laboratory experiments, providing empirical evidence of its effectiveness. The proposed method represents a significant step toward fortifying next-generation communication systems against evolving jamming threats.

Keywords: jamming detection; EVM; 5G; resource block

Citation: Örnek, C.; Kartal, M. Securing the Future: A Resourceful Jamming Detection Method Utilizing the EVM Metric for Next-Generation Communication Systems. *Electronics* **2023**, *12*, 4948. https://doi.org/10.3390/electronics12244948

Academic Editors: Dariusz Rzońca and Tomasz Rak

Received: 1 November 2023
Revised: 30 November 2023
Accepted: 4 December 2023
Published: 9 December 2023

1. Introduction

5G and beyond communication systems are revolutionizing communication in today's rapidly evolving technological landscape. These systems provide a significant increase in access to Internet-based services with high speeds, low latency, and wide bandwidth. They offer users a seamless experience across multiple devices, facilitating integration between mobile devices, desktops, and other platforms. They also support innovative features and applications, enabling technologies such as augmented reality, remote interventions, and the Internet of things. Owing to their flexibility and future-proof adaptability, these systems play a key role in digital transformation, bringing a more efficient, secure, and rich experience to the world of communications. However, all of these features also open up the possibility for malicious jammers to attack more targets and corrupt more data. Therefore, fast, accurate, and effective detection of jamming attacks is vital for increasing the defense capabilities of systems.

Several jamming detection methods are proposed for wireless networks [1,2]. A significant number of these utilize received signal strength (RSS) measurements. The authors of [3–7] obtain the RSS by estimating the spectrum of the received signal and observe the effect of jamming signals on the RSS. In other studies, the optimal RSS thresholds for jamming detection are determined with likelihood tests performed by considering the jamming presence and absence hypotheses. This method is widely proposed for massive SIMO [8], massive MIMO [9], LTE [10], direct-sequence spread-spectrum (DSSS) [11], wireless sensor networks [12], ad hoc networks [13], cognitive radio networks [14] and satellite communication [15] systems.

In addition, many studies propose the use of RSS-based metrics in combination with bit error rate (BER)-based metrics such as throughput, packet error rate, packet delivery ratio, packet sent ratio, packet loss rate, and bad packet ratio. Accordingly, the effects of jamming are observed jointly on the RSS- and BER-based metrics, and jamming detection threshold levels are set on these metrics. The jamming detection performance of such methods is demonstrated with simulations in [16–20], and with experimental studies as well as simulations in [21–23]. On the other hand, in [24–27], these metrics are used to train machine learning algorithms such as support vector machines, neural networks, and random forests for jamming detection. In addition to the aforementioned metrics, the chip error rate [28] and inter-arrival time [29] are other metrics examined for jamming detection.

Machine learning algorithms are also trained using spectrogram images [30], IQ samples [31], time-domain signal samples [32], and FFT samples [33,34] for jamming detection. Although machine learning algorithms are becoming increasingly popular, the issues of training these algorithms, collecting sufficient data for training, adapting to varying jamming strategies, and integrating them into the system architecture with minimal overhead must be considered.

Subspace analysis methods are the other methods used in jamming detection. Such methods use eigenvalue [35] or singular-value [36] analyses to identify the subspaces formed by the signal and jamming. However, the jamming detection success of such methods requires the jamming level to be sufficiently higher than the legitimate signal level.

In our previous study [37], the EVM vs. RB metric was proposed to detect jamming attacks in 5G networks. The error vector magnitude (EVM) is measured for each resource block (RB) in the received signal and jamming signals are then detected at RBs where the EVM upper threshold is not met. The success of EVM vs. RB in terms of sensitivity compared to classical BER-based methods was verified with simulations containing only a limited number of scenarios. The EVM metric is also used in studies [38,39] to study jamming effects in OFDM systems. However, these studies have aimed to identify the jamming strategies that cause the greatest damage to the system.

In this paper, we extend the work for the EVM vs. RB measurement and list below all the innovations and contributions achieved:

1. EVM metric utilization: The paper advocates for the utilization of the EVM metric measured in IQ symbols, a departure from the commonly used classical RSS and BER based metrics in the literature.
2. Enhanced sensitivity: The proposed method demonstrates a significant improvement in jamming detection sensitivity compared to existing approaches. Although low-power hidden jamming signals that cannot be detected using conventional metrics do not cause denial of service, they can limit the data transmission rate. Due to the EVM's ability to detect small variations in jamming level, jamming signals hidden in an extreme form 20 dB below the legal signal are also successfully detected.
3. Low complexity: For next-generation networks with low latency requirements, it is advantageous that the proposed method has a low complexity of $O(N)$. This advantage also contributes to the fast response of the system for anti-jamming measures.
4. Jammer frequency information: The proposed method calculates the EVM metric for each RB in the received signal. Since RBs represent the frequency domain, the frequency bands in which jamming attacks occur are also revealed. This important information, which is not provided by most methods, offers an important background for countermeasure steps such as jammer localization [40] and antijamming frequency planning. In addition, the concepts of ambient backscattering and RF energy harvesting [41,42] are recently proposed as solutions to the battery problems of IoT devices.

By using the jamming frequency information provided by our method, these devices can be tuned to the correct jamming frequencies and, as a result, jamming energy, which is usually emitted at high RF powers, can be utilized.

5. Reliability: The EVM vs. RB measurement provides a stable jamming detection performance against varying system parameters such as modulation degree and code rate. However, BER-based methods are affected by the variations of these parameters and provide unreliable results.
6. Usability and compatibility: In LTE and 5G systems, it is known that reference IQ symbols are also sent in the transmitted data packet to enable the UE to estimate the channel. The EVM metric used by the proposed method is calculated using these reference symbols that are already in the system architecture. Thus, the proposed method can be easily integrated into the system without the need for changes in system operation or hardware. Moreover, since jamming detection can be performed using a single threshold level for the EVM metric, there is no need for any pre-operational training and validation phases. As a result, the proposed method is suitable for LTE, 5G, and beyond communication systems, which include IQ modulation and resource block (RB) architectures.
7. Theoretical analysis support: All presented advantages are substantiated with thorough theoretical analysis, reinforcing the validity and efficacy of the proposed jamming detection methodology.
8. Versatility in system scenarios: The proposed method's successful operation in different system scenarios is underscored by extending the simulation conditions to cover the sub-6 GHz frequency region usage, different numerology (OFDM subcarrier spacing) usage, line-of-sight (LOS) and non-line-of-sight (NLOS) channel cases, MIMO structures, and millimeter-wave (mmWave) band usage scenarios.
9. Laboratory experiment validation: The study's success is conclusively demonstrated through experiments conducted in a laboratory environment, providing empirical evidence of the method's effectiveness.

2. System Model

The effectiveness of the proposed method is demonstrated on a 5G downlink data-transmission infrastructure. For this purpose, the process steps shown in Figure 1 are implemented in MATLAB [43] by considering the 3GPP standards [44–48].

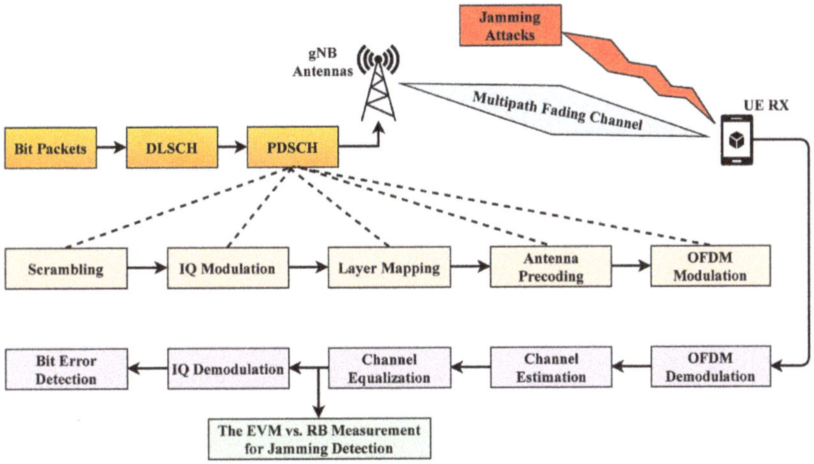

Figure 1. 5G Downlink Data Transmission.

First, the data bits are generated at the gNB (base station); they are then subjected to "cyclic redundancy check" insertion and "low-density parity-check" coding [44] at the downlink shared channel (DLSCH) step [45]. This permits the UE to detect and correct bit errors. The obtained code words are then transferred to the physical downlink shared channel (PDSCH) stage.

In the PDSCH stage [46], the code words are first scrambled so that the broadcast cannot be decoded by unautorized devices. IQ modulation is then performed, providing one of the QPSK, 16-QAM, 64-QAM or 256-QAM options [47].

The obtained IQ symbols are mapped to the MIMO transmitter antennas in the layer-mapping phase. In addition, demodulation reference signals (DM-RS) [47], which are reference IQ symbols required for channel estimation in the UE side, are also included in the data symbols.

The IQ symbols are modulated into the RF band using OFDM. The smallest frequency grid required for downlink transmission is called a resource element, which corresponds to one OFDM subcarrier frequency. A group of 12 consecutive subcarriers (resource elements) in the frequency domain form a resource block (RB). The total bandwidth allocated to a UE is expressed in the number of RBs, and the concept of RB is used throughout the rest of the paper.

Finally, the obtained RF signal is transmitted via MIMO antennas. The mentioned MIMO-OFDM system is detailed in Figure 2. There are N_T transmitter and N_R receiver antennas in the system.

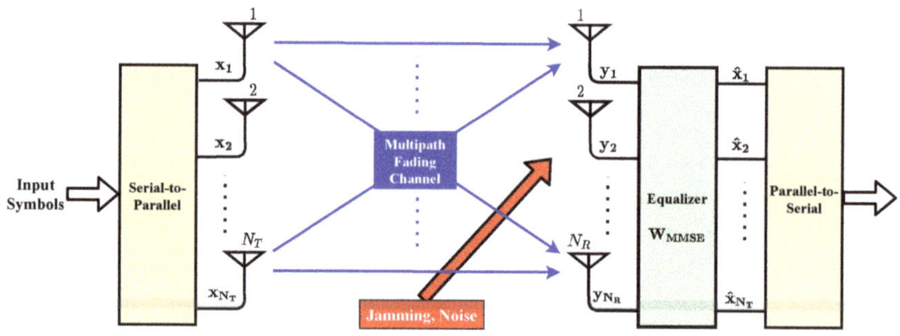

Figure 2. MIMO-OFDM Transmit–Receive Model.

The CDL (Clustered Delay Line) channel model, specified by 3GPP [48] for 5G and beyond communication systems, represents a realistic channel structure with clustered multipath components, each exhibiting Rayleigh fading characteristics. This model aligns with industry standards and is well-suited for the simulation of wireless communication systems, allowing us realistic capture of the effects of multipath propagation and fading in our study. Hence, the overall multipath channel can be expressed by an **H** matrix with each element following a Rayleigh distribution.

$$\mathbf{H} = \begin{bmatrix} \mathbf{h}_{1,1} & \dots & \mathbf{h}_{1,N_T} \\ \vdots & \ddots & \vdots \\ \mathbf{h}_{N_R,1} & \dots & \mathbf{h}_{N_R,N_T} \end{bmatrix}, \tag{1}$$

where $\mathbf{h}_{i,j} = \left[h_{i,j}[L-1] \dots h_{i,j}[0]\right]$ is the channel between the ith receiver and jth transmitter antennas, and L is the maximum channel length of all $N_R \times N_T$ links. The statistical properties of $h_{i,j}[l] (l = 0, \dots, L-1)$ and $\mathbf{h}_{i,j}$ can be summarized as follows:

$$\mathbb{E}\{h_{i,j}[l]\} = 0, \tag{2}$$

$$\mathbb{E}\{|h_{i,j}[l]|^2\} = 1, \tag{3}$$

$$\mathbb{E}\{h_{i,j}[l]h^*_{m,n}[l]\} = 0 \text{ if } i \neq m \text{ or } j \neq n, \text{ and so} \tag{4}$$

$$\mathbb{E}\{\mathbf{h}_{i,j}\mathbf{h}^H_{i,j}\} = L. \tag{5}$$

The received signal samples at time instant k are expressed as follows:

$$\mathbf{y}[k] = \sqrt{\frac{P_T}{N_T}} \begin{bmatrix} \mathbf{h}_{1,1} & \dots & \mathbf{h}_{1,N_T} \\ \vdots & \vdots & \vdots \\ \mathbf{h}_{N_R,1} & \dots & \mathbf{h}_{N_R,N_T} \end{bmatrix} \begin{bmatrix} \mathbf{x}_1[k] \\ \vdots \\ \mathbf{x}_{N_T}[k] \end{bmatrix} + \mathbf{v}[k], \tag{6}$$

where P_T represents the average transmitted symbol power, $\mathbf{v}[k] = \mathbf{j}[k] + \mathbf{n}[k]$ represents the sum of the received jamming and noise vectors, and

$$\mathbf{x}_j[k] = \begin{bmatrix} x_j[k-L+1] \\ \vdots \\ x_j[k] \end{bmatrix} \tag{7}$$

is the vector of the transmitted symbols, each with an average power of one unit, that is, $\sigma_x^2 = 1$.

T received vector samples can be combined into a single matrix as

$$\begin{aligned} \mathbf{Y} = [\mathbf{y}[k] \dots \mathbf{y}[k+T-1]] &= \sqrt{\frac{P_T}{N_T}} \begin{bmatrix} \mathbf{h}_{1,1} & \dots & \mathbf{h}_{1,N_T} \\ \vdots & \vdots & \vdots \\ \mathbf{h}_{N_R,1} & \dots & \mathbf{h}_{N_R,N_T} \end{bmatrix} \begin{bmatrix} \mathbf{X}_1 \\ \vdots \\ \mathbf{X}_{N_T} \end{bmatrix} + \mathbf{V} \\ &= \sqrt{\frac{P_T}{N_T}} \mathbf{H}\mathbf{X} + \mathbf{V}, \end{aligned} \tag{8}$$

where $\mathbf{X}_j$ and $\mathbf{V}$ are written as

$$\mathbf{X}_j = \begin{bmatrix} x_j[k-L+1] & x_j[k-L+2] & \dots & x_j[k-L+T] \\ \vdots & \vdots & \dots & \vdots \\ x_j[k-1] & x_j[k] & \dots & x_j[k+T-2] \\ x_j[k] & x_j[k+1] & \dots & x_j[k+T-1] \end{bmatrix} \text{ and} \tag{9}$$

$$\mathbf{V} = \begin{bmatrix} v_1[k] & v_1[k+1] & \dots & v_1[k+T-1] \\ v_2[k] & v_2[k+1] & \dots & v_2[k+T-1] \\ \vdots & \vdots & \dots & \vdots \\ v_{N_R}[k] & v_{N_R}[k+1] & \dots & v_{N_R}[k+T-1] \end{bmatrix}. \tag{10}$$

After receiving $\mathbf{Y}$, the Minimum Mean Squared Error (MMSE) equalizer is used to mitigate the negative effects caused by the channel, such as fading. The MMSE equalization matrix, $\mathbf{W}_{MMSE}$ [49], is calculated as

$$\begin{aligned} \mathbf{W}_{MMSE} &= \sqrt{\frac{N_T}{P_T}} \left(\mathbf{H}^H\mathbf{H} + \frac{P_V N_T}{P_T} \mathbf{I}_{N_T} \right)^{-1} \mathbf{H}^H \\ &= \sqrt{\frac{N_T}{P_T}} \mathbf{B}\mathbf{H}^H, \end{aligned} \tag{11}$$

where $P_V = P_J + P_N$ is the sum of jamming (P_J) and noise (P_N) powers and

$$\mathbf{B} = \left(\mathbf{H}^H\mathbf{H} + \frac{P_V N_T}{P_T}\mathbf{I}_{N_T}\right)^{-1}. \tag{12}$$

To estimate the transmitted IQ symbols, the equalizer is applied as follows:

$$\begin{aligned}\hat{\mathbf{X}} = \mathbf{W}_{MMSE}\mathbf{Y} &= \sqrt{\frac{N_T}{P_T}}\mathbf{B}\mathbf{H}^H\mathbf{Y} \\ &= \mathbf{B}\mathbf{H}^H\mathbf{H}\mathbf{X} + \sqrt{\frac{N_T}{P_T}}\mathbf{B}\mathbf{H}^H\mathbf{V} \\ &= \mathbf{B}\mathbf{C}\mathbf{X} + \sqrt{\frac{N_T}{P_T}}\mathbf{B}\mathbf{H}^H\mathbf{V},\end{aligned} \tag{13}$$

where $\mathbf{C} = \mathbf{H}^H\mathbf{H}$.

At this point, the proposed error vector magnitude (EVM) metric for jamming detection is calculated using the reference and estimated IQ symbols as follows:

$$EVM_n = \sqrt{\frac{e_n^2}{\frac{1}{N}\sum_{n=1}^{N}(i_n^2 + q_n^2)}}, \tag{14}$$

where

- n denotes the index of the IQ symbol,
- N is the total number of symbols used for calculation,
- $e_n^2 = (i_n - \hat{i_n})^2 + (q_n - \hat{q_n})^2$ is the power of the error caused by the jamming and noise,
- i_n and q_n are the reference in-phase and quadrature values of the nth symbol ($x_n = i_n + jq_n$),
- $\hat{i_n}$ and $\hat{q_n}$ are the estimated in-phase and quadrature values of the nth symbol ($\hat{x_n} = \hat{i_n} + j\hat{q_n}$),
- $\frac{1}{N}\sum_{n=1}^{N}(i_n^2 + q_n^2)$ represents the average power of the reference symbols.

As shown in Equation (14), each of the N symbols is used once for vectoral difference calculation. Therefore, the computational complexity of the EVM is in terms of the first power of N, that is, $O(N)$. Consequently, the computational complexity of our jamming detection method using the EVM metric has a low value of $O(N)$.

For EVM calculation, both reference and estimated symbols are required. The natural flow of next-generation communication systems, such as LTE and 5G, includes the transmission of reference symbols. In this manner, without any pre-training and without changing the system architecture, we calculate the EVM metric and detect the presence of a jamming signal by checking whether the EVM exceeds a single threshold level. This makes the proposed method very advantageous in terms of integrability into real-world scenarios.

In addition, Equation (14) indicates that EVM is proportional to the square root of the jamming plus noise-to-signal ratio ($JNSR_{\hat{x}}$). Therefore, to perform EVM analysis, it is necessary to extract the signal, jamming, and noise power components from $\hat{\mathbf{X}}$. For this purpose, it is convenient to calculate the covariance matrix of $\hat{\mathbf{X}}$. Using the statistical independence property [49] and Equation (5), the covariance matrix is calculated as follows:

$$\begin{aligned}\mathbf{\Sigma}_{\hat{X}\hat{X}} &= \mathbf{\Sigma}_{BB}\mathbf{\Sigma}_{CC}\mathbf{\Sigma}_{XX} + \frac{N_T}{P_T}\mathbf{\Sigma}_{BB}\mathbf{\Sigma}_{HH}\mathbf{\Sigma}_{VV} \\ &= \left(\sigma_b^2 L^2 \sigma_x^2 + \frac{N_T}{P_T}\sigma_b^2 L \sigma_v^2\right)\mathbf{I}_{\mathbf{N_T}},\end{aligned} \tag{15}$$

where $\sigma_x^2 = 1$ as expressed in Equation (7), and σ_v^2 is $P_V = P_J + P_N$. The diagonals of $\mathbf{\Sigma}_{\hat{X}\hat{X}}$ indicate the total power of each $\hat{x_n}$. Thus, $JNSR_{\hat{x}}$ can be obtained as follows:

$$JNSR_{\hat{x}} = \frac{N_T}{P_T}\frac{\sigma_b^2 L \sigma_v^2}{\sigma_b^2 L^2 \sigma_x^2} = \frac{P_V N_T}{P_T L}. \tag{16}$$

Consequently, it is revealed that the EVM is related to the parameters given in Equation (17).

$$EVM_n \propto \sqrt{JNSR_{\hat{x}}} = \sqrt{\frac{P_V N_T}{P_T L}} = \sqrt{\frac{P_J + P_N}{(P_T/N_T)L}}. \tag{17}$$

As shown in Equation (17), EVM depends directly on the jamming power represented by P_J. Thus, the EVM metric can sense even small changes in the jamming level. However, for BER-based metrics, such as throughput and packet delivery ratio, to detect jamming signals, the jamming power must be strong enough to divert the received IQ symbols to the wrong regions in the constellation diagram, that is, to create a bit error. Because jamming signals below this jamming power do not create any bit errors, jamming is not sensed by BER-based metrics. Although such weak jamming signals do not cause denial of service, they may limit the data rate performance. Owing to the aforementioned ability of the EVM, these jamming signals can also be successfully detected.

EVM also depends on the transmitted symbol power, which is denoted as P_T. After equalization, P_T is normalized by N_T. Parameter L, on the other hand, is the expected improvement brought by the equalizer. This improvement is also mentioned in simulation results in Section 3.1.

Another conclusion is that the EVM metric is not affected by varying system parameters, such as modulation type and code rate, and as a result, jamming signals are stably detected. However, as shown in the results in Sections 3.2 and 3.3, BER-based metrics are affected by these system parameters and exhibit unreliable results.

The final EVM is expressed in both the RMS (18) and MAX (19). Because the maximum EVM can sense instantaneous distortions in the received signal, it can also detect more sophisticated jamming attacks that target a short-timed fragment of the legitimate signal. Such jammers are also called reactive or responsive jammers [50] and may adopt such short-time operating styles to minimize both their detectability and battery usage. Therefore, the maximum EVM is used in this study.

$$EVM_{RMS} = \sqrt{\frac{\sum_{n=1}^{N} EVM_n^2}{N}}, \tag{18}$$

$$EVM_{MAX} = \max_{n\in[1,\ldots,N]} EVM_n. \tag{19}$$

The maximum EVM is measured for each RB in the received signal. Thus, the EVM vs. RB data are obtained. Because the RBs represent the frequency domain, the EVM vs. RB data reveal the frequency bands attacked by the jammer. After this stage, the operations in the receiver side are completed with IQ demodulation and decoding, and the data bits are obtained. The BER and throughput are measured using the data bits, and these measurements are also observed for jamming detection, whereas the EVM vs. RB detects the jamming attack at an earlier stage. This capability brings extra speed along with low computational complexity.

3. Simulation Results

3.1. Base Scenario

The processing steps required for 5G downlink data transmission are explained in Section 2. The system parameters used in the processing steps are listed in Table 1.

Jammers may concentrate their RF energy into certain frequency bands using tone-type signals, or occupy a broader spectrum using chirp-type signals. Therefore, it is considered sufficient to examine the tone and chirp jammers in this study.

Table 1. Selected Data Transmission Parameters for the Base Scenario.

Parameter Name	Value	Explanation
Carrier Frequency	2.65 GHz	Frequency Range-1 for 5G
MIMO Structure	8×2	
MIMO Transmission Layers	2	
Fading Channel Model	CDL-C	Urban Macrocell Model, NLOS
OFDM Subcarrier Spacing (SCS)	30 kHz	$\mu = 1$ (numerology)
Assigned RBs	51	Transmission bandwidth close to 20 MHz with the 30 kHz SCS
IQ Modulation	16QAM	
Code Rate	490/1024	

First, in the no-jammer case, the RF power spectrum and EVM vs. RB are measured for the received signal, and the measurement results are shown in Figure 3. The observed fluctuations in the spectrum is caused by multipath fading. As explained in Section 2, an equalizer is used to minimize the fading effect on the received IQ symbols. To observe the effect of equalizer on the EVM data, EVM vs. RB is measured for both the unequalized and equalized IQ symbols, as shown in Figure 3b. The EVM vs. RB measurement obtained using unequalized IQ symbols directly reflects the fluctuation characteristics of the RF spectrum (red line in Figure 3b). On the other hand, the improvement brought about by the equalizer shows a decrease in the EVM data (blue line in Figure 3b). As shown in Equation (17), this improvement is expected.

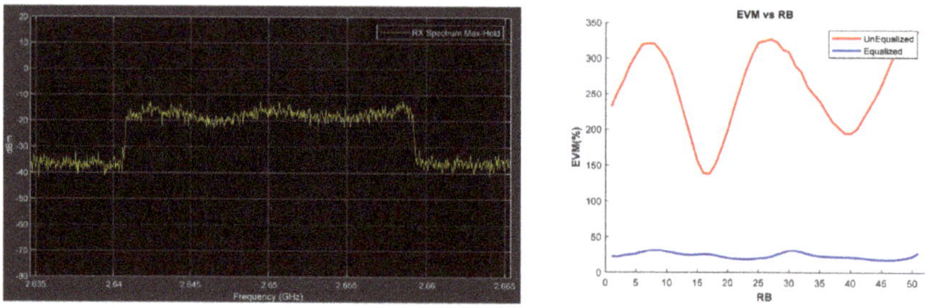

(**a**) (**b**)

Figure 3. No-Jammer Case, Obtained Throughput = %100 (BER = 0). (**a**) The Rx signal spectrum and (**b**) EVM vs. RB.

The same measurements are performed for different jamming cases, that is, tone and chirp jammers. Reviews related to tone jamming are given below, whereas repeated reviews for chirp jamming are provided in the Appendix A. The SJR parameter is selected as −5 dB for both jamming conditions. The observed changes in the RF power spectrum and EVM after the application of these jamming signals are shown in Figures 4 and A1, respectively. In the EVM vs. RB data obtained using unequalized symbols, jamming effects are observed in addition to fluctuations owing to the fading (red lines in Figures 4b and A1b). On the other hand, in the EVM vs. RB measurement taken with equalized symbols, the fluctuation is minimized owing to the equalizer, but jamming effects are still clearly observed (blue lines in Figures 4b and A1b)). Thus, in the EVM vs. RB data obtained with equalized symbols, jamming signals can be easily detected using a single threshold level, without considering any fluctuation effect in the data. Therefore, to avoid dealing with the fluctuation effect due to fading, EVM vs. RB measurement using equalized IQ symbols is proposed for jamming detection.

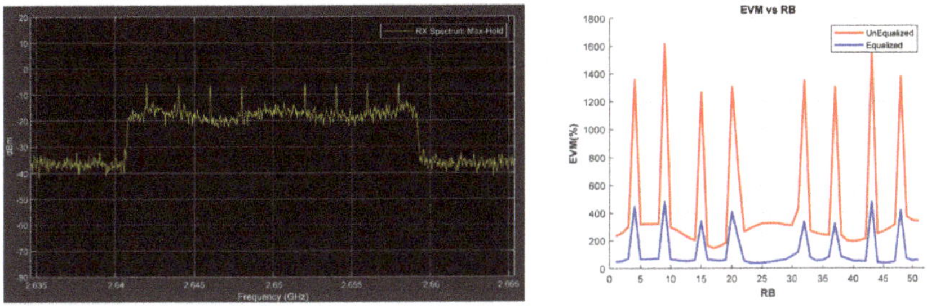

(**a**) (**b**)

Figure 4. Tone Jammer Case, SJR = −5 dB, Obtained Throughput = %0 (BER = 0.343). (**a**) shows the Rx signal spectrum and (**b**) shows the EVM vs. RB.

In the next simulation, where the SJR is increased to 10 dB, the received signal is contaminated with jammers of the same tone and chirp type. The RF spectrum and EVM vs. RB measurements are shown in Figures 5 and A2, respectively. In this SJR case, the jamming signals cannot be detected using the RF power spectrum, which provides RSS information, as shown in Figures 5a and A2a. In addition, the throughput measurement for both the jamming cases is 100%, which means that the jamming effect cannot be sensed using this BER-based metric. However, the EVM vs. RB measurements successfully detect these small jamming signals, as shown in Figures 5b and A2b. This reveals the success of the proposed method in terms of sensitivity compared with RSS- and BER-based methods.

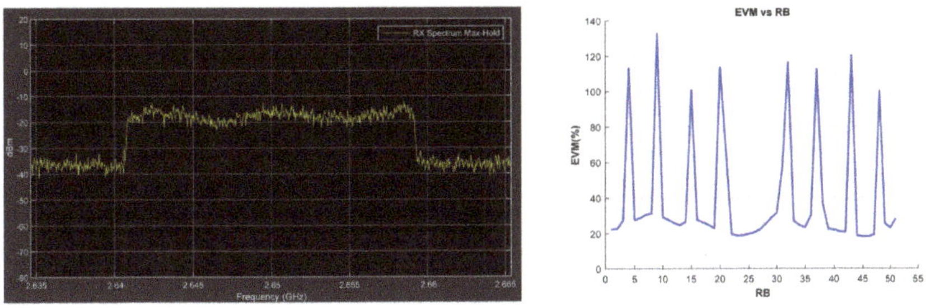

(**a**) (**b**)

Figure 5. Tone Jammer Case, SJR = 10 dB, Obtained Throughput = 100% (BER = 0), (**a**) the Rx signal spectrum and (**b**) EVM vs. RB.

EVM vs. RB, throughput, and BER are measured for various SJR values to examine the dependencies of the jamming detection metrics on SJR. According to the results shown in Figure 6d, jamming cannot be sensed using the throughput and BER observations when SJR exceeds 10 dB. However, using the EVM vs. RB measurement, jamming signals are successfully detected, even under extreme SJR conditions, such as 20 dB (Figure 6b). To demonstrate the performance of EVM vs. RB under other SJR conditions, the peak value of EVM vs. RB for each SJR is calculated and the results are shown in Figure 6c. It is concluded that jamming signals with an SJR of 25 dB can also be detected using EVM vs. RB. However, beyond 25 dB, EVM vs. RB also becomes unsuccessful in jamming detection.

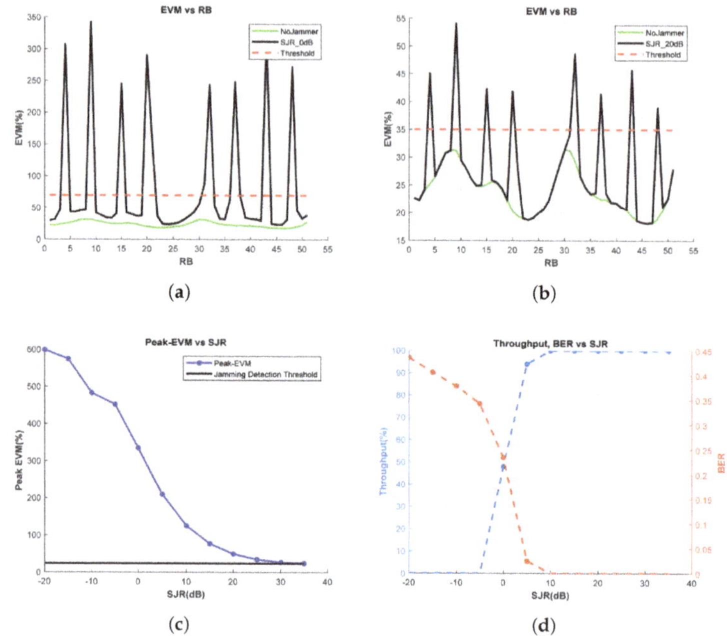

Figure 6. EVM vs. RB, BER and Throughput Measurements for Multitone Jammer. (**a**) EVM vs. RB for SJR = 0 dB, (**b**) EVM vs. RB for SJR = 20 dB, (**c**) Peak-EVM vs. SJR, and (**d**) throughput and BER vs. SJR.

The sensitivity performance of the proposed method for tone jamming is also valid for chirp jamming, as shown in Figure A3. In the following sections, the BER results are not presented alongside the throughput results, because, as shown in the figures, the BER is inversely proportional to the throughput and does not provide any additional information.

3.2. Reliability of the Proposed Method against Modulation Type Change

5G systems choose the appropriate M-PSK or M-QAM modulation types according to the data rates required by the UEs and channel availability. 16-QAM modulation is considered in the base scenario (Section 3.1). In this section, along with 16-QAM, QPSK and 64-QAM modulations are considered. Thus, jamming detection performances of EVM vs. RB and throughput metrics are examined against changes in the modulation type.

For the QPSK, 16-QAM, and 64-QAM modulation-type use cases, the peaks of EVM vs. RB are calculated for each SJR, and the results are presented in Figure 7a. Because the EVM measurement shows consistent results across modulation types, the proposed method can be safely used for jamming detection in system scenarios in which the modulation type changes.

On the other hand, Figure 7b shows the throghput results versus SJR for the use cases of the aforementioned modulation types. When SJR is 0 dB, the throughput for the QPSK case is 100%; therefore, no jamming signal is detected. If the system decides that there is no jamming threat by looking at this throughput result and then increases the modulation degree to 16-QAM or 64-QAM, it experiences a dramatic decrease in throughput. In other words, the jamming effect is sensed differently by using the throughput metric under different modulation-type usage conditions. However, the proposed measurement consistently detects jamming threats independently of the chosen modulation type, thereby possessing the capability to provide reliable guidance to the system.

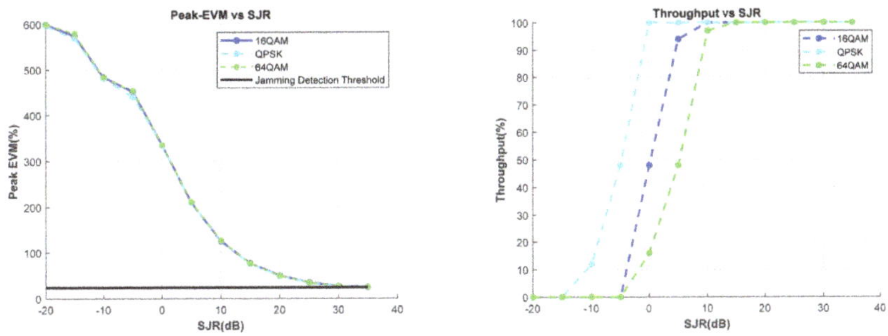

(a) (b)

Figure 7. Effect of the Modulation Type Change, Tone Jammer. (**a**) Peak-EVM vs. SJR, and (**b**) throughput vs. SJR.

The results in Figure A4 show that this reliability of the proposed method against changes in the modulation type is also achieved for the chirp jamming case.

3.3. Reliability of the Proposed Method against Code Rate Change

In 5G systems, the code rate parameter can also be changed depending on the requirements. A code rate of 490/1024 is considered for the base scenario. In this section, code rates of 245/1024 and 980/1024 are also considered.

For the aforementioned code rate use cases, the peaks of EVM vs. RB are calculated for each SJR, and the results are shown in Figure 8a. The proposed method provides stable results without being affected by the code rate parameter; therefore, it can be safely used for jamming detection in system scenarios in which the code rate changes.

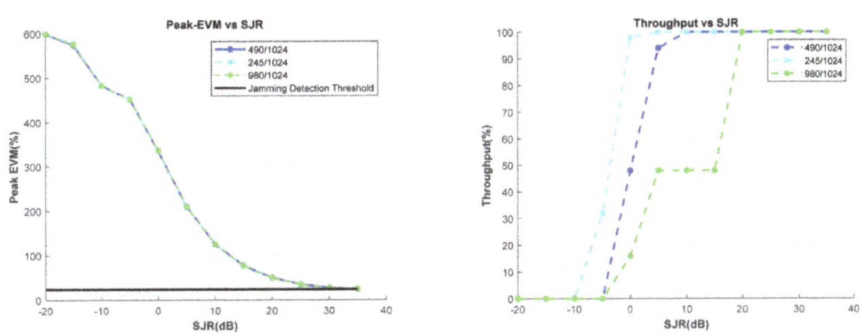

(a) (b)

Figure 8. Effect of the Code Rate Change, Tone Jammer. (**a**) Peak-EVM vs. SJR, and (**b**) Throughput vs. SJR.

However, Figure 8b shows that different throughput results are obtained for different code rate conditions for a fixed SJR case. For example, when the SJR is 0 dB and a code rate of 245/1024 is used, the throughput approaches 100%. Therefore, the jamming effect cannot be clearly observed. If the system relies on this and decides to increase the code rate to 490/1024 or 980/1024, the throughput decreases significantly. Meanwhile, the proposed measurement can prevent such incorrect decisions, because it detects jamming threats without being affected by code rate changes.

This achievement of the proposed method for the tone-jamming scenario is also valid under chirp-jamming, as shown in Figure A5.

3.4. Change in the OFDM Subcarrier Space (SCS)

In the previous sections, simulations are performed for 30 kHz OFDM SCS use; however, 5G networks can also use OFDM SCSs of 15, 60, 120, and 240 kHz to serve other applications with different bandwidth requirements. This flexible use of different OFDM SCS corresponds to the numerology term. However, only the 15 kHz SCS option is available for LTE networks. In this section, we demonstrate that the EVM vs. RB measurement successfully detects jamming attacks for different SCS use cases. For this purpose, simulations are performed for 15 and 60 kHz SCS selections.

The jamming signal types and jamming frequencies are the same as those described in the previous sections. When the SCS is reduced from 30 to 15 kHz, the transmission bandwidth is halved, resulting in half of the jamming frequencies occupying the spectrum (Figures 9a and A6a). Conversely, when the SCS is increased to 60 kHz, all jamming frequencies are observed in the transmission bandwidth (Figures 10a and A7a). Figures 9b, 10b, A6b and A7b show that the EVM vs. RB measurement successfully detects all jamming attacks included in the transmission bandwidth regardless of the OFDM SCS applied.

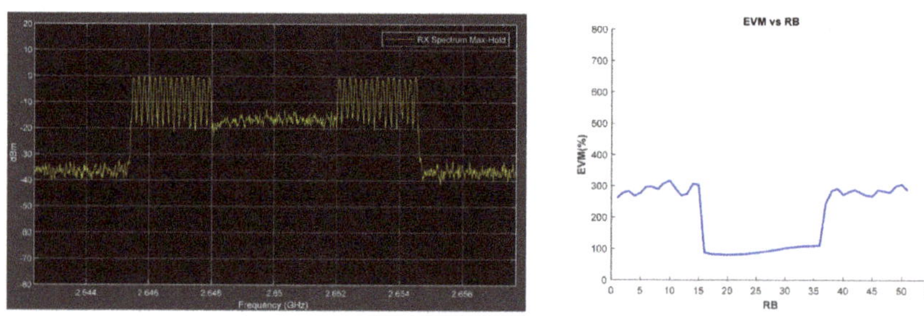

(a) (b)

Figure 9. Chirp Jammer Case, SJR = −10 dB, SCS = 15 kHz. (**a**) the Rx signal spectrum and (**b**) EVM vs. RB.

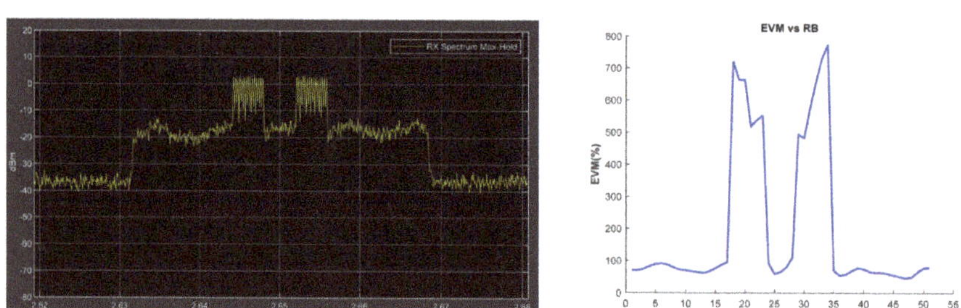

(a) (b)

Figure 10. Chirp Jammer Case, SJR = −10 dB, SCS = 60 kHz. (**a**) the Rx signal spectrum and (**b**) EVM vs. RB.

3.5. Jamming Detection for mmWave Conditions

Millimeter waves encompass frequencies of 24 GHz and above. Millimeter-wave (mmWave) bands offer increased bandwidth and data transfer rates, although they have a limited coverage range. Consequently, mmWave signals rely significantly on line-of-sight (LOS) propagation to ensure effective coverage.

In the previous sections, experiments are conducted on the utilization of 5G in the sub-6 GHz frequency range. In this section, on the other hand, the channel conditions are changed, taking into consideration the deployment of 5G in the mmWave frequency band, along with the corresponding channel conditions. In this context, the carrier frequency is adjusted to 28 GHz, the transmission channel type is set to CDL-D (LOS), and OFDM SCS is configured at 60 kHz. Figures 11 and 12 show that the EVM vs. RB metric can be successfully used to detect jamming attacks under mmWave data transmission conditions.

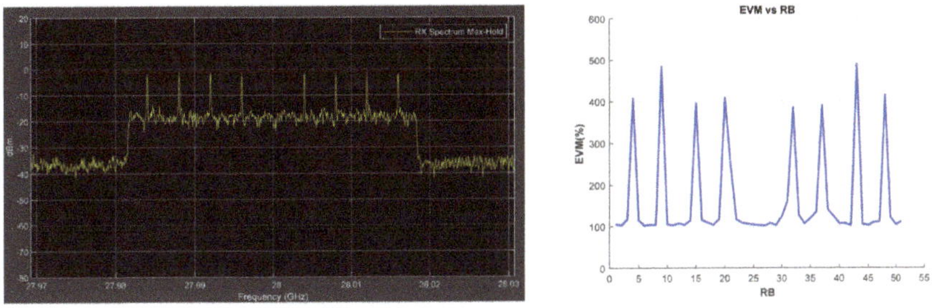

(**a**) (**b**)

Figure 11. Tone Jammer Case, MmWave Conditions, SJR = −10 dB. (**a**) the Rx signal spectrum and (**b**) EVM vs. RB.

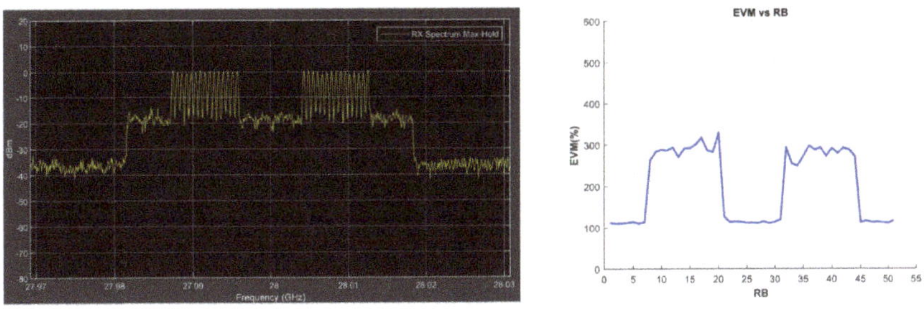

(**a**) (**b**)

Figure 12. Chirp Jammer Case, MmWave Conditions, SJR = −10 dB. (**a**) the Rx signal spectrum and (**b**) EVM vs. RB.

4. In-Lab Validation

In this section, the jamming detection performance of the EVM vs. RB measurement is demonstrated through experiments performed in a laboratory environment in addition to theoretical analysis and simulations. Because broadcasting interfering (jamming) signals alongside legitimate communication is illegal, experiments are performed in a closed-loop manner by adopting the following procedure to overcome this legal limitation:

First, the vector signal generator shown in Figure 13 generates a 5G signal by modulating the IQ symbols in the baseband to the RF band with OFDM. The IQ modulation, OFDM subcarrier spacing and number of OFDM subcarriers are set to 16-QAM, 30 kHz and 612, respectively, to make the generated signal similar to that in the base scenario (Section 3.1). The jamming signal, on the other hand, is generated in the RF band using the analog signal generator. The 5G signal is then contaminated with the jamming signal using the RF combiner module, and the resulting signal is transferred to the spectrum analyzer, which represents the receiver.

The spectrum analyzer calculates the RF power spectrum that provides the RSS information and performs RF demodulation to obtain IQ symbols. To calculate the EVM vs. RB data, the correct (reference) IQ symbols transmitted by the vector signal generator and jammed IQ symbols obtained by the spectrum analyzer are transferred to the test PC. EVM vs. RB data are then obtained by calculating the EVM metric using Equation (14) for each RB.

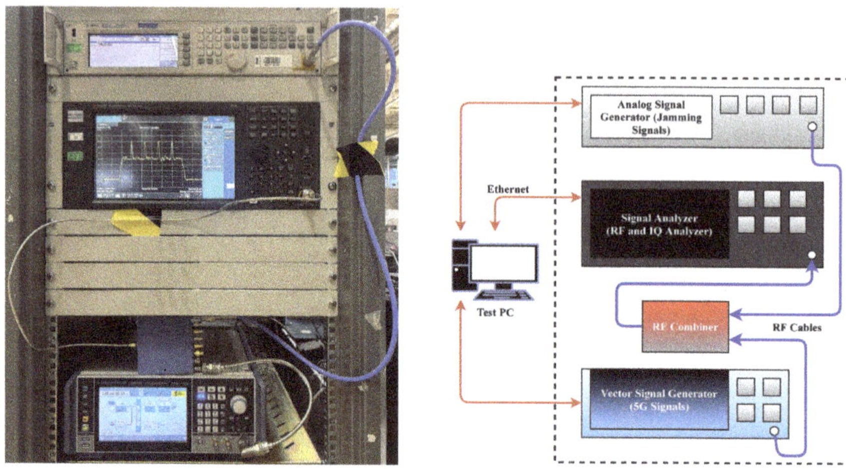

(**a**) (**b**)

Figure 13. (**a**) The hardware setup and (**b**) the block diagram of the setup.

The first experiment is conducted for a no-jammer scenario. Figure 14a shows the power spectrum of the received RF signal and Figure 14b shows EVM vs. RB results. It is observed that there is no jamming signal in the spectrum other than the 5G signal, and on the other hand, the EVM values are low as expected.

Figures 15 and 16 show the results for the tone and chirp jamming cases, respectively, where SJR is −10 dB. The effects of the jamming signals on the spectrum are clearly visible in Figures 15a and 16a. In parallel, the EVM vs. RB measurement successfully reveals jamming attacks for RBs corresponding to the frequency bands exposed to the jamming signals (Figures 15b and 16b).

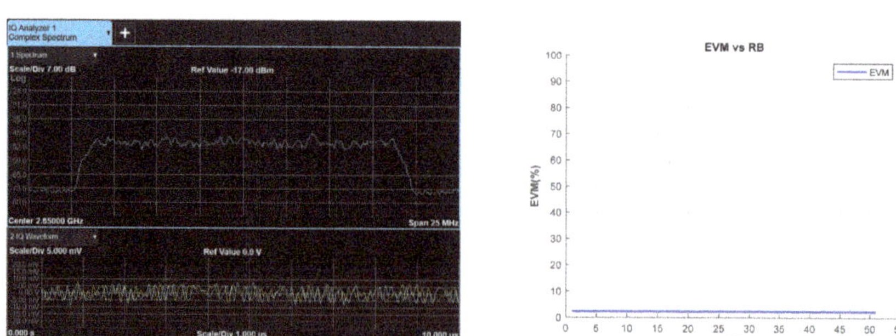

(**a**) (**b**)

Figure 14. No-Jammer Case, (**a**) the Rx signal spectrum and time-domain IQ waveform obtained after RF demodulation and (**b**) EVM vs. RB.

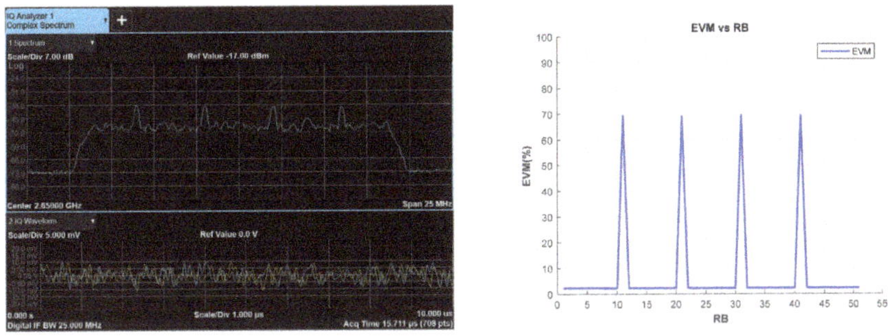

(a) (b)

Figure 15. Tone Jammer Case, SJR = −10 dB, (**a**) the Rx signal spectrum and time-domain IQ waveform obtained after RF demodulation and (**b**) EVM vs. RB.

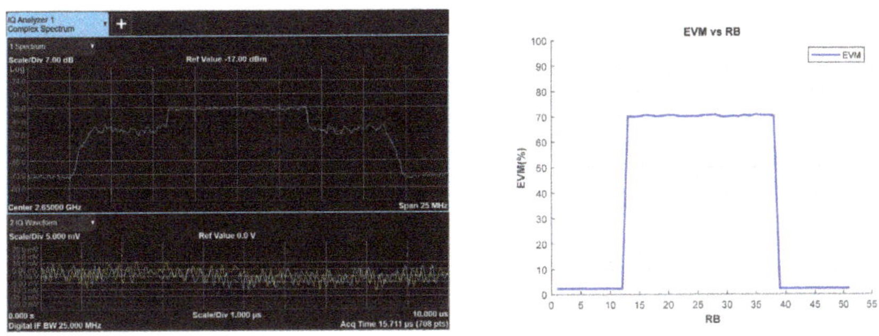

(a) (b)

Figure 16. Chirp Jammer Case, SJR = −10 dB, (**a**) the Rx signal spectrum and time-domain IQ waveform obtained after RF demodulation and (**b**) EVM vs. RB.

In the next experiment, to test the jamming detection sensitivity of both the RF power spectrum and the EVM-vs-RB metric, the SJR parameter is set to 0 dB by reducing the power of the jamming signals by 10 dB. The results obtained for the tone and chirp jamming cases are shown in Figures 17 and 18, respectively. As shown in Figures 17a and 18a, the jamming signals become no longer detectable in the RF power spectrum. However, the EVM-vs-RB metric (Figures 17b and 18b) can still clearly detect jamming threats hidden in the spectrum.

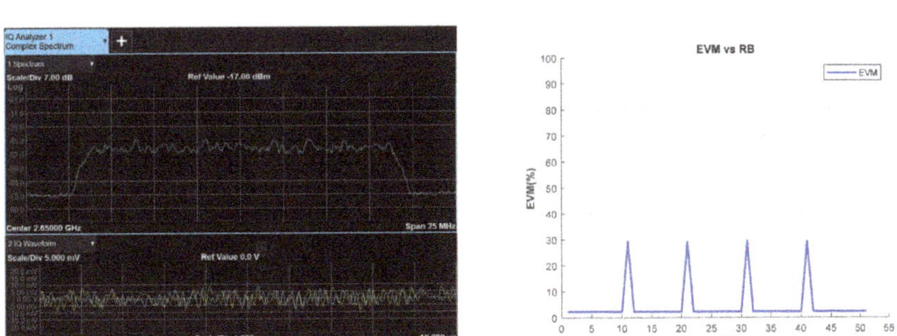

(a) (b)

Figure 17. Tone Jammer Case, SJR = 0 dB, (**a**) the Rx signal spectrum and time-domain IQ waveform obtained after RF demodulation and (**b**) EVM vs. RB.

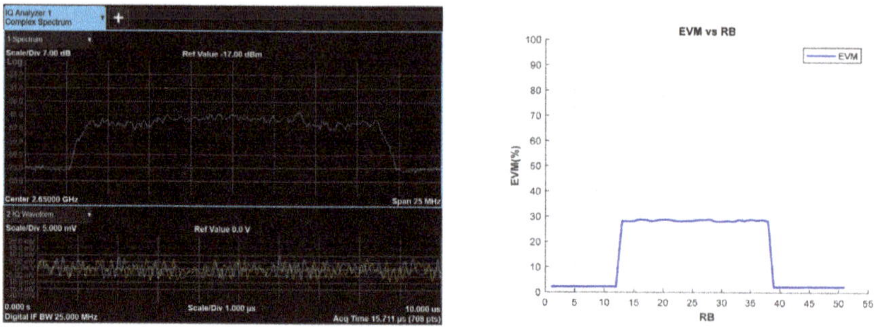

(a) (b)

Figure 18. Chirp Jammer Case, SJR = 0 dB, (**a**) the Rx signal spectrum and time-domain IQ waveform obtained after RF demodulation and (**b**) EVM vs. RB.

5. Discussion

The presented paper introduces a novel and efficient jamming detection method, EVM vs. RB, designed to enhance the security of next-generation communication systems against jamming attacks. The method is characterized by its ability to measure the EVM in the system, offering a direct perception of changes in jamming levels. The sensitivity success of the proposed method is a significant contribution to the field, as it enables robust detection even in the presence of small jamming signals that may remain unnoticed by other metrics.

A crucial aspect of the proposed method is its low complexity, operating at $O(N)$, and its independence from variable system parameters such as modulation degree and code rate. This independence ensures the method's adaptability to diverse communication scenarios, adding to its practicality and versatility in real-world applications.

The theoretical analysis of the proposed method begins with the construction of a 5G data transmission infrastructure based on international 3GPP standards. By incorporating a jamming attack into the system model, analytical expressions for received IQ symbols are calculated, leading to the derivation of the EVM expression. This analytical foundation establishes the groundwork for understanding the method's inner workings, particularly its capability to perceive changes in jamming levels directly.

Simulation results using MATLAB software [43] showcase the effectiveness of EVM vs. RB in providing the jammer's spectrum information. Comparative metrics, including power spectrum for Received Signal Strength (RSS), Bit Error Rate (BER), and BER-dependent throughput, are evaluated. The results demonstrate that EVM vs. RB outperforms these metrics in detecting jamming signals, even at an extreme Signal Jamming Ratio (SJR) of 25 dB. This robust performance underscores the method's resilience against varying jamming levels, reinforcing its potential as a reliable jamming detection solution.

Furthermore, the simulations reveal the stability of EVM vs. RB against changes in system parameters such as modulation degree and code rate. In contrast, metrics like throughput exhibit unreliability under such variations. This highlights the method's ability to maintain consistent performance across different communication scenarios, a critical factor for its widespread applicability.

The study extends its scope to various applications, including 5G's mmWave technology, demonstrating the versatility of EVM vs. RB across different communication technologies. The method's success is further validated through experimental studies conducted in a laboratory environment, providing empirical evidence of its effectiveness in real-world settings.

In conclusion, the proposed EVM vs. RB jamming detection method presents a compelling solution to enhance the security of next-generation communication systems. Its direct perception of jamming level changes, low complexity, and independence from variable system parameters contribute to its robustness and adaptability. The extensive theoretical analysis, simulations, and experimental studies collectively establish the method as a promising and practical tool in the ongoing efforts to safeguard communication systems against jamming attacks.

6. Conclusions

This paper introduces a capable jamming detection method to secure LTE, 5G, and next-generation communication systems. Through the utilization of the EVM metric measured in IQ symbols, the proposed approach diverges from traditional methods based on RSS- and BER-based measurements, thereby contributing to the advancement of jamming detection methodologies.

The achieved contributions of this research are multi-faceted. First, the utilization of the EVM metric demonstrates its effectiveness in enhancing jamming detection sensitivity, surpassing existing approaches and providing a more reliable solution. Moreover, the method introduces low computational complexity. On the other hand, the provision of jammer frequency information by measuring the EVM for each RB in the received signal, a critical aspect often lacking in other methods, further fortifies the system's capabilities in understanding and counteracting jamming attacks.

A notable strength of the proposed methodology is that it provides stable results against changes in system parameters such as modulation type and code rate. This stability contributes to the reliability of the results.

The verification methods employed in this study serve to reinforce the credibility of the proposed approach. The method's successful operation in diverse system scenarios, as highlighted through extended simulation conditions, underscores its versatility and applicability in real-world situations. Theoretical analyses provide a solid foundation for the presented advantages, establishing the validity and efficacy of the jamming detection methodology. Furthermore, the conclusive demonstration of the method's success in laboratory experiments offers empirical evidence, validating its effectiveness in practical settings.

Looking ahead, the future direction of this research aims to leverage the jammer frequency information provided by the proposed method. The intention is to develop an intelligent frequency assignment strategy for anti-jamming purposes. This forward-looking approach underscores the continuous evolution of the proposed methodology, with potential applications in optimizing communication systems against sophisticated jamming attacks.

In summary, this study not only introduces a novel jamming detection method, but also substantiates its effectiveness through theoretical analysis and empirical validation. The method's low computational complexity, adaptability to varying system parameters, and seamless integration into existing communication systems position it as a promising solution for securing LTE, 5G, and future communication networks against jamming attacks. The envisioned future direction further emphasizes the potential of this methodology to contribute to intelligent anti-jamming strategies.

Author Contributions: Conceptualization, C.Ö. and M.K.; Methodology, C.Ö. and M.K.; Software, C.Ö.; Validation, C.Ö. and M.K.; Formal analysis, C.Ö.; Investigation, M.K.; Resources, C.Ö.; Data curation, C.Ö.; Writing—original draft, C.Ö.; Writing—review & editing, C.Ö. and M.K.; Supervision, M.K. All authors have read and agreed to the published version of the manuscript.

Funding: This research received no external funding.

Data Availability Statement: Most data are included in the article. All data are available upon request from the corresponding author.

Conflicts of Interest: This work was supported by Aselsan Inc. The authors declare that the research was conducted in the absence of any commercial or financial relationships that could be construed as a potential conflict of interest.

Abbreviations

The following abbreviations are used in this manuscript:

3GPP	3rd Generation Partnership Project
5G	5th Generation of Cellular Networks
BER	Bit Error Rate
CDL	Clustered Delay Line
DLSCH	Downlink Shared Channel
DM-RS	Demodulation Reference Signals
DSSS	Direct-Sequence Spread-Spectrum
EVM	Error Vector Magnitude
FFT	Fast Fourrier Transform
IQ	In-phase and Quadrature
JNSR	Jamming plus Noise-to-Signal Ratio
LOS	Line of Sight
LTE	Long-Term Evolution
MIMO	Multiple Input, Multiple Output
MMSE	Minimum Mean Squared Error
NLOS	Non-Line of Sight
OFDM	Orthogonal Frequency Division Multiplexing
PDSCH	Physical Downlink Shared Channel
PSK	Phase Shift Keying
QAM	Quadrature Amplitude Modulation
QPSK	Quadrature Phase Shift Keying
RB	Resource Block
RF	Radio Frequency
RSS	Received Signal Strength
Rx	Receive
SCS	Subcarrier Spacing
SIMO	Single Input, Multiple Output
SJR	Signal-to-Jamming Ratio
Tx	Transmit
UE	User Equipment

Appendix A

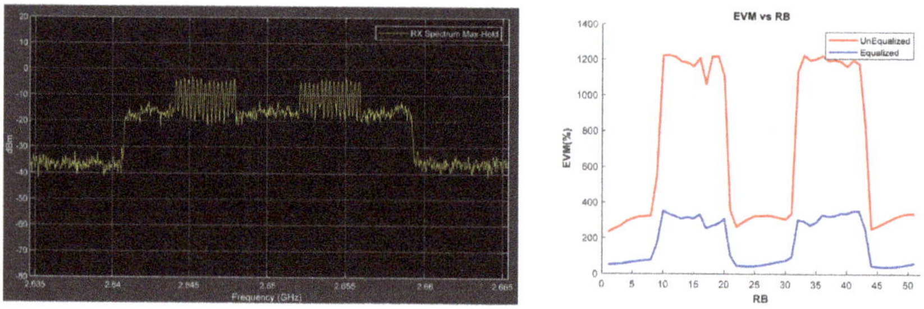

(a) (b)

Figure A1. Chirp Jammer Case, SJR = −5 dB, Obtained Throughput = %1 (BER = 0.344). (**a**) the Rx signal spectrum and (**b**) EVM vs. RB.

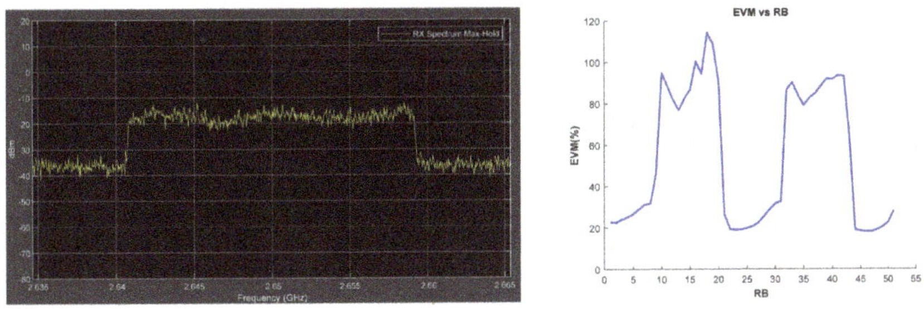

(a) (b)

Figure A2. Chirp Jammer Case, SJR = 10 dB, Obtained Throughput = %100 (BER = 0), (**a**) the Rx signal spectrum and (**b**) EVM vs. RB.

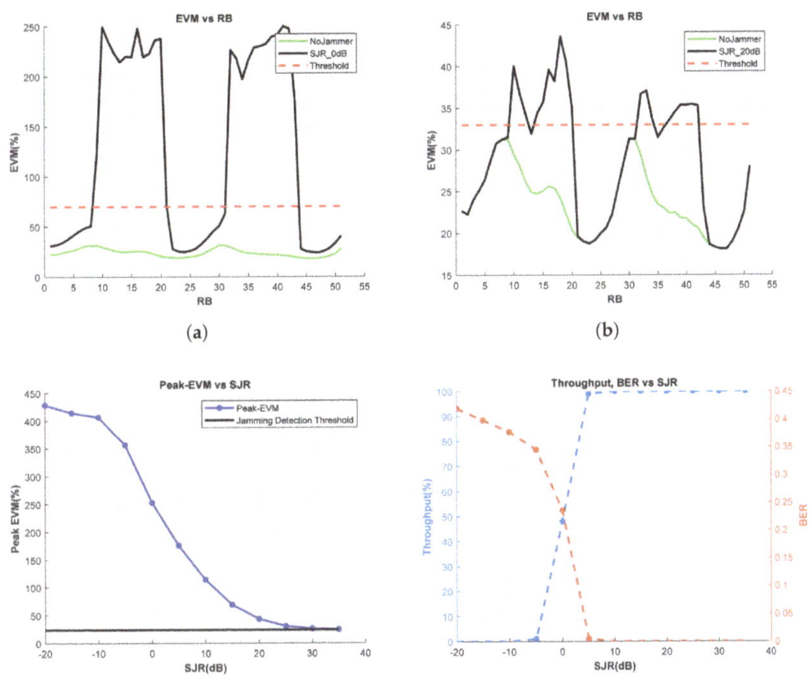

(a) (b) (c) (d)

Figure A3. EVM vs. RB , BER and Throughput Measurements for Chirp Jammer, (**a**) EVM vs. RB for SJR = 0 dB, (**b**) EVM vs. RB for SJR = 20 dB, (**c**) Peak-EVM vs. SJR, and (**d**) Throughput and BER vs. SJR.

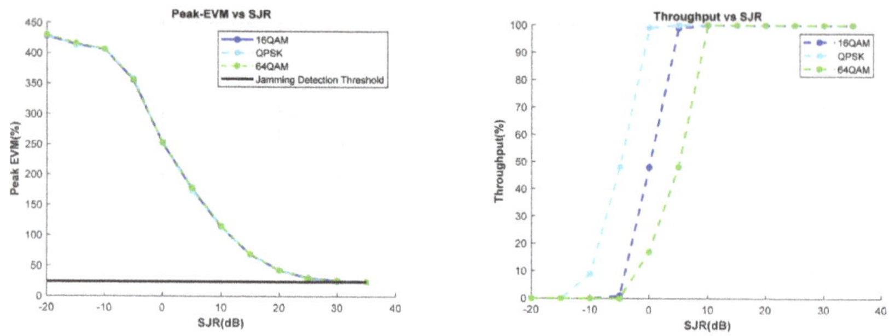

(**a**) (**b**)

Figure A4. Effect of the Modulation Type Change, Chirp Jammer. (**a**) Peak-EVM vs. SJR, and (**b**) Throughput vs. SJR.

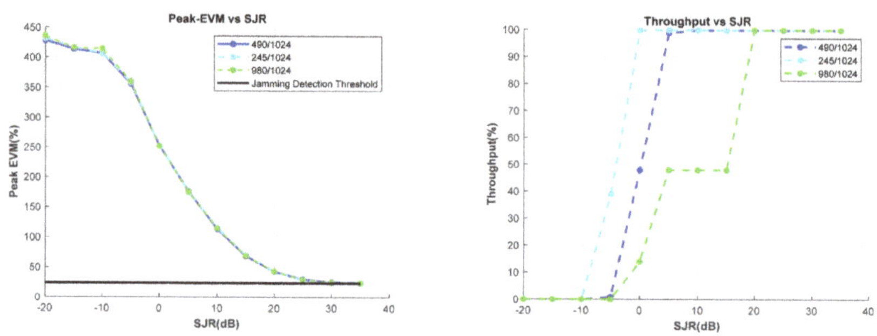

(**a**) (**b**)

Figure A5. Effect of the Code Rate Change, Chirp Jammer. (**a**) Peak-EVM vs. SJR, and (**b**) Throughput vs. SJR.

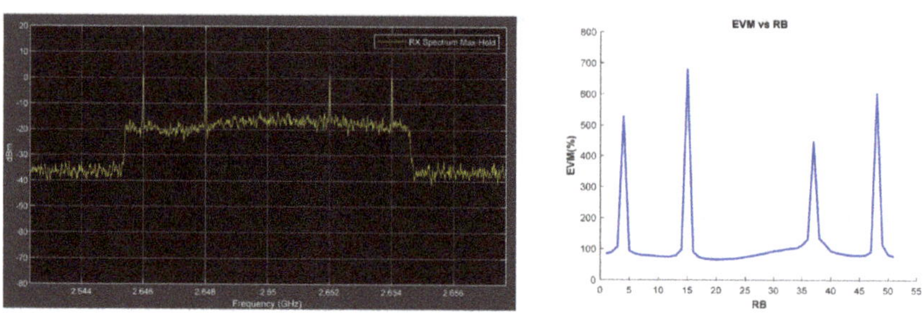

(**a**) (**b**)

Figure A6. Tone Jammer Case, SJR = −10 dB, SCS = 15 kHz. (**a**) the Rx signal spectrum and (**b**) EVM vs. RB.

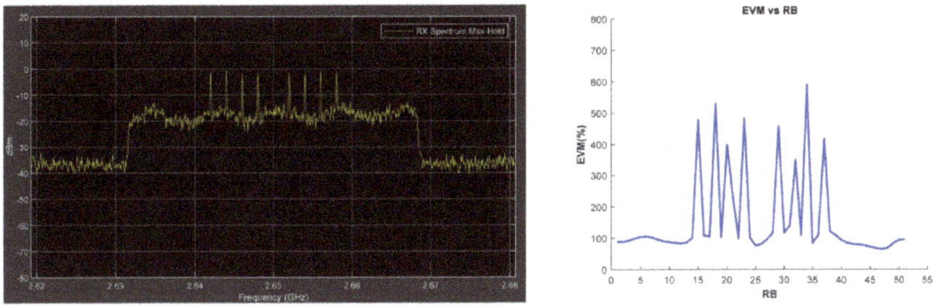

(a) (b)

Figure A7. Tone Jammer Case, SJR = −10 dB, SCS = 60 kHz. (**a**) the Rx signal spectrum and (**b**) EVM vs. RB.

References

1. Pirayesh, H.; Zeng, H. Jamming Attacks and Anti-Jamming Strategies in Wireless Networks: A Comprehensive Survey. *IEEE Commun. Surv. Tutor.* **2022**, *24*, 767–809. [CrossRef]
2. Grover, K.; Lim, A.; Yang, Q. Jamming and anti-jamming techniques in wireless networks: A survey. *Int. J. Ad Hoc Ubiquitous Comput.* **2014**, *17*, 197. [CrossRef]
3. Xu, H.; Cheng, Y.; Wang, P. Jamming Detection in Broadband Frequency Hopping Systems Based on Multi-Segment Signals Spectrum Clustering. *IEEE Access* **2021**, *9*, 29980–29992. [CrossRef]
4. Liu, Z.; Liu, H.; Xu, W.; Chen, Y. An Error-Minimizing Framework for Localizing Jammers in Wireless Networks. *IEEE Trans. Parallel Distrib. Syst.* **2014**, *25*, 508–517. [CrossRef]
5. Mughal, M.O.; Dabcevic, K.; Marcenaro, L.; Regazzoni, C.S. Compressed sensing based jammer detection algorithm for wide-band cognitive radio networks. In Proceedings of the 2015 3rd International Workshop on Compressed Sensing Theory and Its Applications to Radar, Sonar and Remote Sensing (CoSeRa), Pisa, Italy, 17–19 June 2015 ; pp. 119–123. [CrossRef]
6. Hamdy, A.; Digham, F.; Nasr, O.A.; Mourad, H.M. Automatic detection of jammer interference in GSM networks. In Proceedings of the 2018 International Conference on Innovative Trends in Computer Engineering (ITCE), Aswan, Egypt, 19–21 February 2018; pp. 248–252. [CrossRef]
7. Ferre, R.M.; Richter, P.; Fuente, A.D.L.; Lohan, E.S. In-lab validation of jammer detection and direction finding algorithms for GNSS. In Proceedings of the 2019 International Conference on Localization and GNSS (ICL-GNSS), Nuremberg, Germany, 4–6 June 2019; pp. 1–6. [CrossRef]
8. Xu, S.; Xu, W.; Pan, C.; Elkashlan, M. Detection of Jamming Attack in Non-Coherent Massive SIMO Systems. *IEEE Trans. Inf. Forensics Secur.* **2019**, *14*, 2387–2399. [CrossRef]
9. Akhlaghpasand, H.; Razavizadeh, S.M.; Björnson, E.; Do, T.T. Jamming Detection in Massive MIMO Systems. *IEEE Wirel. Commun. Lett.* **2018**, *7*, 242–245. [CrossRef]
10. Eygi, M.; Kurt, G.K. Jamming Detection: A Multicarrier Approach. In Proceedings of the 2018 26th Telecommunications Forum (TELFOR), Belgrade, Serbia, 20–21 November 2018; pp. 1–4. [CrossRef]
11. Chen, X.; Yang, W. Detection of Jamming in DSSS Systems Using FRESH Filters. In Proceedings of the 2020 IEEE 3rd International Conference of Safe Production and Informatization (IICSPI), Chongqing, China, 28-30 November 2020; pp. 320–325. [CrossRef]
12. Choi, J.; Mughal, M.O.; Choi, Y.; Kim, D.; Lopez-Salcedo, J.A.; Kim, S. CUSUM-based Joint Jammer Detection and Localization. In Proceedings of the 2018 IEEE International Symposium on Dynamic Spectrum Access Networks (DySPAN), Seoul, Republic of Korea, 22–25 October 2018; pp. 1–5. [CrossRef]
13. Eriksson, G.; Hansson, A. Derivation of detection times for a simple follower-jammer model used in mobile ad hoc-network simulations. In Proceedings of the 2019 International Conference on Military Communications and Information Systems (ICMCIS), Budva, Montenegro, 14–15 May 2019; pp. 1–5. [CrossRef]
14. Mohammadi, J.; Stańczak, S.; Zheng, M. Joint spectrum sensing and jamming detection with correlated channels in cognitive radio networks. In Proceedings of the 2015 IEEE International Conference on Communication Workshop (ICCW), London, UK, 8–12 June 2015; pp. 889–894. [CrossRef]
15. Liu, M.; Jin, L.; Shang, B. LSTM-Based Jamming Detection for Satellite Communication with Alpha-Stable Noise. In Proceedings of the 2021 IEEE Wireless Communications and Networking Conference Workshops (WCNCW), Nanjing, China, 29 March 2021; pp. 1–5. [CrossRef]
16. Malebary, S.; Xu, W.; Huang, C.-T. Jamming mobility in 802.11p networks: Modeling, evaluation, and detection. In Proceedings of the 2016 IEEE 35th International Performance Computing and Communications Conference (IPCCC), Las Vegas, NV, USA, 9–11 December 2016; pp. 1–7. [CrossRef]
17. Manju, V.C.; Kumar, M.S. Detection of jamming style DoS attack in Wireless Sensor Network. In Proceedings of the 2012 2nd IEEE International Conference on Parallel, Distributed and Grid Computing, Solan, India, 6–8 December 2012; pp. 563–567. [CrossRef]

18. Bodkhe, A.A.; Raut, A.R. Identifying Jammers in Wireless Sensor Network with an Approach to Defend Reactive Jammer. In Proceedings of the 2014 Fourth International Conference on Communication Systems and Network Technologies, Bhopal, India, 7–9 April 2014; pp. 89–92. [CrossRef]
19. Yu, B.; Zhang, L.-Y. An improved detection method for different types of jamming attacks in wireless networks. In Proceedings of the 2014 2nd International Conference on Systems and Informatics (ICSAI 2014), Shanghai, China, 15–17 November 2014; pp. 553–558. [CrossRef]
20. Marttinen, A.; Wyglinski, A.M.; Jäntti, R. Statistics-Based Jamming Detection Algorithm for Jamming Attacks against Tactical MANETs. In Proceedings of the 2014 IEEE Military Communications Conference, Baltimore, MD, USA, 6–8 October 2014; pp. 501–506. [CrossRef]
21. Sufyan, N.; Saqib, N.A.; Zia, M. Detection of jamming attacks in 802.11b wireless networks. *J. Wirel. Commun. Netw.* **2013**, *2013*, 208. [CrossRef]
22. Liu, G.; Liu, J.; Li, Y.; Xiao, L.; Tang, Y. Jamming Detection of Smartphones for WiFi Signals. In Proceedings of the 2015 IEEE 81st Vehicular Technology Conference (VTC Spring), Glasgow, UK, 11–14 May 2015; pp. 1–3. [CrossRef]
23. Duan, B.; Yin, D.; Cong, Y.; Zhou, H.; Xiang, X.; Shen, L. Anti-Jamming Path Planning for Unmanned Aerial Vehicles with Imperfect Jammer Information. In Proceedings of the 2018 IEEE International Conference on Robotics and Biomimetics (ROBIO), Kuala Lumpur, Malaysia, 12–15 December 2018; pp. 729–735. [CrossRef]
24. Arjoune, Y.; Salahdine, F.; Islam, M.S.; Ghribi, E.; Kaabouch, N. A Novel Jamming Attacks Detection Approach Based on Machine Learning for Wireless Communication. In Proceedings of the 2020 International Conference on Information Networking (ICOIN), Barcelona, Spain, 7–10 January 2020; pp. 459–464. [CrossRef]
25. Jahanshahi, J.A.; Ghorashi, S.A.; Eslami, M. A support vector machine based algorithm for jamming attacks detection in cellular networks. In Proceedings of the 2011 Wireless Advanced, London, UK, 20–22 June 2011; pp. 180–184. [CrossRef]
26. Puñal, O.; Aktaş, I.; Schnelke, C.-J.; Abidin, G.; Wehrle, K.; Gross, J. Machine learning-based jamming detection for IEEE 802.11: Design and experimental evaluation. In Proceedings of the IEEE International Symposium on a World of Wireless, Mobile and Multimedia Networks 2014, Sydney, NSW, Australia, 19 June 2014; pp. 1–10. [CrossRef]
27. Upadhyaya, B.; Sun, S.; Sikdar, B. Machine Learning-based Jamming Detection in Wireless IoT Networks. In Proceedings of the 2019 IEEE VTS Asia Pacific Wireless Communications Symposium (APWCS), Singapore, 28–30 August 2019; pp. 1–5. [CrossRef]
28. Spuhler, M.; Giustiniano, D.; Lenders, V.; Wilhelm, M.; Schmitt, J.B. Detection of Reactive Jamming in DSSS-based Wireless Communications. *IEEE Trans. Wirel. Commun.* **2014**, *13*, 1593–1603. [CrossRef]
29. Osanaiye, O.; Alfa, A.S.; Hancke, G.P. A Statistical Approach to Detect Jamming Attacks in Wireless Sensor Networks. *Sensors* **2018**, *18*, 1691. [CrossRef] [PubMed]
30. Morales Ferre, R.; de la Fuente, A.; Lohan, E.S. Jammer Classification in GNSS Bands Via Machine Learning Algorithms. *Sensors* **2019**, *19*, 4841. [CrossRef] [PubMed]
31. Shi, Y.; Davaslioglu, K.; Sagduyu, Y.E.; Headley, W.C.; Fowler, M.; Green, G. Deep Learning for RF Signal Classification in Unknown and Dynamic Spectrum Environments. In Proceedings of the 2019 IEEE International Symposium on Dynamic Spectrum Access Networks (DySPAN), Newark, NJ, USA, 11–14 November 2019; pp. 1–10. [CrossRef]
32. Li, T.; Wang, M.; Peng, D.; Yang, X. Identification of Jamming Factors in Electronic Information System Based on Deep Learning. In Proceedings of the 2018 IEEE 18th International Conference on Communication Technology (ICCT), Chongqing, China, 8–11 October 2018; pp. 1426–1430. [CrossRef]
33. Zhang, N.; Li, Y.; Shi, Y.; Shen, J. A CNN-Based Adaptive Federated Learning Approach for Communication Jamming Recognition. *Electronics* **2023**, *12*, 3425. [CrossRef]
34. Shen, J.; Li, Y.; Zhu, Y.; Wan, L. Cooperative Multi-Node Jamming Recognition Method Based on Deep Residual Network. *Electronics* **2022**, *11*, 3280. [CrossRef]
35. Vinogradova, J.; Björnson, E.; Larsson, E.G. Detection and mitigation of jamming attacks in massive MIMO systems using random matrix theory. In Proceedings of the 2016 IEEE 17th International Workshop on Signal Processing Advances in Wireless Communications (SPAWC), Edinburgh, UK, 3–6 July 2016; pp. 1–5. [CrossRef]
36. Yang, X.; Li, A.; Wei, M.; Zhang, X.; Lu, S.; Wang, W. Jamming Signal Detection Based on TSVD Method. In Proceedings of the 2020 IEEE International Conference on Advances in Electrical Engineering and Computer Applications(AEECA), Dalian, China, 25–27 August 2020; pp. 558–562. [CrossRef]
37. Örnek, C.; Kartal, M. An Efficient EVM Based Jamming Detection in 5G Networks. In Proceedings of the 2022 4th IEEE Middle East and North Africa COMMunications Conference (MENACOMM), Amman, Jordan, 6–8 December 2022; pp. 130–135. [CrossRef]
38. Alakoca, H.; Kurt, G.K.; Ayyıldız, C. PHY based Jamming attacks against OFDM systems: A measurement study. In Proceedings of the 2017 25th Telecommunication Forum (TELFOR), Belgrade, Serbia, 21–22 November 2017; pp. 1–4. [CrossRef]
39. Bilodeau-Robitaille, O.; Gagnon, F. Digital RF Memory Jamming on OFDM SISO. In Proceedings of the 2014 IEEE Military Communications Conference, Baltimore, MD, USA, 6–8 October 2014; pp. 1542–1548. [CrossRef]
40. Örnek, C.; Kartal, M. Work-in-Progress: An Efficient EVM Based Hybrid Jammer Localization Method for 5G Networks. In Proceedings of the 2023 IEEE International Black Sea Conference on Communications and Networking (BlackSeaCom), Istanbul, Turkiye, 4–7 July 2023; pp. 408–413. [CrossRef]
41. Zheng, K.; Jia, X.; Chi, K.; Liu, X. DDPG-Based Joint Time and Energy Management in Ambient Backscatter-Assisted Hybrid Underlay CRNs. *IEEE Trans. Commun.* **2023**, *71*, 441–456. [CrossRef]

42. Zheng, K.; Luo, R.; Wang, Z.; Liu, X.; Yao, Y. Short-Term and Long-Term Throughput Maximization in Mobile Wireless-Powered Internet of Things. *IEEE Internet Things J.* **2023** . [CrossRef]
43. *MATLAB R2021a*; The MathWorks, Inc.: Natick, MA, USA, 2021.
44. *3GPP TS 38.212. NR*; Multiplexing and Channel Coding. 3rd Generation Partnership Project; Technical Specification Group Radio Access Network: Sophia Antipolis Cedex, France, 2020.
45. *3GPP TS 38.202. NR*; Services Provided by the Physical Layer. 3rd Generation Partnership Project; Technical Specification Group Radio Access Network: Sophia Antipolis Cedex, France, 2020.
46. *3GPP TS 38.214. NR*; Physical Layer Procedures for Data. 3rd Generation Partnership Project; Technical Specification Group Radio Access Network: Sophia Antipolis Cedex, France, 2020.
47. *3GPP TS 38.211. NR*; Physical Channels and Modulation. 3rd Generation Partnership Project; Technical Specification Group Radio Access Network: Sophia Antipolis Cedex, France, 2020.
48. *3GPP TR 38.901. 5G*; Study on Channel Model for Frequencies from 0.5 to 100 GHz. Technical Specification Group Radio Access Network: Sophia Antipolis Cedex, France, 2020.
49. Paulraj, A.; Nabar, R.; Gore, D. *Introduction to Space-Time Wireless Communications*; Cambridge University Press: Cambridge, UK, 2003; pp. 84–186.
50. Arjoune, Y.; Faruque, S. Smart Jamming Attacks in 5G New Radio: A Review. In Proceedings of the 2020 10th Annual Computing and Communication Workshop and Conference (CCWC), Las Vegas, NV, USA, 6–8 January 2020; pp. 1010–1015. [CrossRef]

electronics

Article

Cybersecurity in Supply Chain Systems: The Farm-to-Fork Use Case

Helen C. Leligou [1], Alexandra Lakka [1], Panagiotis A. Karkazis [1,*], Joao Pita Costa [2], Eva Marin Tordera [3], Henrique Manuel Dinis Santos [4] and Antonio Alvarez Romero [5]

[1] Synelixis Solutions S.A., 34100 Chalkida, Greece; nleligou@synelixis.com (H.C.L.); lakka@synelixis.com (A.L.)
[2] XLAB, SI-1000 Ljubljana, Slovenia; joao.pitacosta@xlab.si
[3] X Lab, Universidad Politecnica de Catalunya, 08034 Barcelona, Spain; eva.marin@upc.edu
[4] ALGORITMI R&D Centre, University of Minho, 4710-057 Braga, Portugal; hsantos@dsi.uminho.pt
[5] Eviden, 28037 Madrid, Spain; antonio.alvarez@eviden.com
* Correspondence: pkarkazis@synelixis.com; Tel.: +30-6973249129

Abstract: Modern supply chains comprise an increasing number of actors which deploy different information technology systems that capture information of a diverse nature and diverse sources (from sensors to order information). While the benefits of the automatic exchange of information between these systems have been recognized and have led to their interconnection, protecting the whole supply chain from potential attacks is a challenging issue given the attack proliferation reported in the literature. In this paper, we present the FISHY platform, which anticipates protecting the whole supply chain from potential attacks by (a) adopting novel technologies and approaches including machine learning-based tools to detect security threats and recommend mitigation policies and (b) employing blockchain-based tools to provide evidence of the captured events and suggested policies. This platform is also easily expandable to protect against additional attacks in the future. We experiment with this platform in the farm-to-fork supply chain to prove its operation and capabilities. The results show that the FISHY platform can effectively be used to protect the supply chain and offers high flexibility to its users.

Keywords: cybersecurity; supply chain systems; blockchain; validation; security monitoring; attack mitigation

Citation: Leligou, H.C.; Lakka, A.; Karkazis, P.A.; Costa, J.P.; Tordera, E.M.; Santos, H.M.D.; Romero, A.A. Cybersecurity in Supply Chain Systems: The Farm-to-Fork Use Case. *Electronics* **2024**, *13*, 215. https://doi.org/10.3390/electronics13010215

Academic Editors: Dariusz Rzońca and Tomasz Rak

Received: 9 October 2023
Revised: 15 December 2023
Accepted: 20 December 2023
Published: 3 January 2024

1. Introduction

Supply chains today have become more and more complex, involving many different businesses and consumers that deploy and use diverse IT systems and applications. These IT systems usually involve IoT-based islands, robots or other smart devices next to sensors, servers and end-devices serving their users, as also happens in other sectors like health [1]. Examining cybersecurity in such a complex environment involving solutions of different types from different software companies is a very challenging problem. Today, platforms that target enhanced network security (like TERAFLOW, described in [2]) or Digital Single Market's E-Commerce Ecosystem (like ENSURESEC, described in [3]) or cloud level security are being developed. However, they target protection against a subset of the security threats applicable in the supply chains. Additionally, quantum computing has provided very promising results with respect (and not limited) to digital signatures (see [4–6]) this technology is not yet mature for being applied to the supply chain complex environment.

Should examining be challenging, ensuring protection is far more so, especially considering that the attacks targeting supply chains proliferate every day, as reported in [7]. Cybersecurity in supply chains has been recognized not only as a challenging task but as a very important task because it does not only affect a single entity (business or individual/consumer), but a series of actors in the chain. The intricacy of the supply chain attack is that it affects multiple actors at the same time, as clearly pointed out in [7]. For example, succeeding in inserting fake information in the information system of an actor in a supply chain may affect all its downstream counterparts. Such a security breach may put at risk food safety when the supply chain under consideration is the farm-to-fork supply chain.

The "farm-to-fork" (F2F) supply chain includes all the actors that contribute to the cultivation (farmer), to the transportation (transporter), to the storage (warehouse operator), to the wholesaler and to the retailer of the vegetables that the end consumer will purchase and consume with their forks. The security challenges and requirements of such a supply chain (as reported in [8]) primarily include (a) the need for end-to-end solutions for vulnerabilities and risks management, (b) the lack of evidence-based metrics for security assurance and trust guarantees, and (c) the cumbersome coordination in multi-actor and multi-vendor supply chains of ICT systems. These have been identified for the F2F supply chain, but they are common to other supply chains as well as in, e.g., smart factories. *The problem* (research question) in this environment is "how to ensure the security of the whole supply chain and not only of isolated IT systems when these systems can significantly differ in the types of security vulnerabilities they suffer from". Another research question is this: "could a platform that answers the above question be expandable to emerging threats?". *The challenge* is to design and deliver a platform/solution that can address multiple types of vulnerability while most security-oriented solutions today target specific vulnerabilities like IoT/edge or blockchain or network security aspects.

In this paper, we present a *platform that aims* at protecting the IT systems of supply chains from multiple types of attacks including blockchain-oriented, network-oriented and web application-oriented attacks by detecting them and then recommending and possibly enforcing mitigation policies in an automated way. We validate this approach in the farm-to-fork supply chain that uses state-of-the-art IT systems. The presented platform anticipates being (a) capable of detecting a variety of attacks, (b) flexible and configurable so as to protect diverse IT systems taking into consideration their internal organization, (c) able to recommend and capable of enforcing mitigation policies and (d) flexibly deployable on premise or on cloud.

For the evaluation of such a platform, it is imperative to perform the following:

(a) Carefully consider user interface aspects: for this reason, in the piloting round, we recruited people outside the FISHY teams for carrying out the evaluation of the UI and used the prepared user manual to do so.
(b) Examine and ensure that the functionality and value of all the FISHY components is validated.
(c) Check the extensibility of the FISHY platform to address additional attacks that may be considered in the future as important for the FISHY supply chains. To examine this possibility, we have used the MITRE ATT&CK framework [9]. This has also allowed us to ensure that FISHY employs techniques that are aligned with the state of the art (reflected in MITRE ATT&CK) and that the techniques we use in FISHY enable the detection of a wide set of additional attacks in the future.

2. A Cross-Solution Security Platform—The FISHY Platform

The FISHY platform is a coordinated framework for cyber-resilient supply chain systems. Its goal is to protect diverse IT systems towards enhancing the trust among the actors of the supply chain.

FISHY platform consists of multiple functional components which can either be deployed in the same premises as the IT systems under protection or can be deployed in a different cloud infrastructure. In the latter case, a minimal set of components needs to be deployed on the same premises as the IT system under protection to enable the flow of status information (e.g., logs) from the system under protection to the FISHY platform and vice versa.

The FISHY architecture (an initial version of which can be found in [10]) is shown in Figure 1. It consists of the following set of building modules: (1) Intent-based Resilience Orchestrator and Dashboard (IRO), (2) Security Assurance and Certification Manager

(SACM), (3) Trust and Incident Manager (TIM), (4) Enforcement and Dynamic Configuration (EDC), (5) Security and Privacy Data Space Infrastructure (SPI) and (6) Secure Infrastructure Abstraction (SIA).

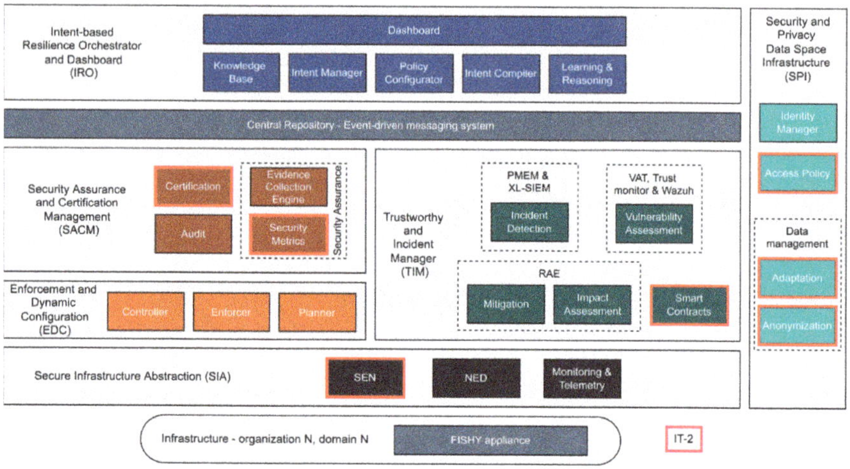

Figure 1. The architecture of the FISHY platform.

Next, we give an overview of each one of the major FISHY modules:

1. The IRO is in charge of interfacing with the security personnel/administrators of the IT systems to be protected (through the dashboard) to receive their security requirements and translate them within the FISHY platform into intents and, in turn, corresponding security workflows and policies. To be more specific, an intent is the set of data which describes the action a user can perform, for example, banning a malicious IP address [11]. It is through the IRO dashboard that the inspection of the detected security events and security control (e.g., enforcement of security policies as a response to a detected security attack) is made possible as well as the performance monitoring.
2. The SACM coordinates the monitoring process, the automated evidence-based security reporting and the certification towards ensuring that the required security policies are correctly implemented [12].
3. The TIM includes tools, such as incident detection, vulnerability and risk estimation, along with incident detection and management, with a goal of developing mechanisms, which ensure security assessment of the stakeholder's supply chains. These tools may also include machine learning-based mechanisms like those presented in with a comparison being presented in [13].
4. The EDC is in charge of security policies enforcement and configuring the specific infrastructure and network security functions (NSF) to ensure resilience. Automated remediation is thus made possible, as discussed in [12,14].
5. The SPI is in charge of identity management, access policy and data management procedures including several activities, such as access control, definition and enforcement of policies, and anonymization of the data and the tools for assessing the security of the stakeholder's devices [15].
6. The SIA module enables secure connectivity among different infrastructures (IoT, edge, cloud) and the FISHY platform, controlling connectivity and providing telemetry of the network, in order to adapt the received data to formats that the FISHY other modules can use [15].

Apart from the previously described modules, a central repository which also includes an event-driven messaging system is included, which is used to store and access information written by the FISHY components.

It is worth stressing that in this revised version of the architecture designed in the final year of the project, FISHY consortium realized that it would be beneficial of its exploitation and sustainability plans to adopt an architecture that would allow for easy integration of additional components (which we name "tools") detecting additional attacks or performing additional functionalities in the future [10]. The evolution of the architecture and further details of the workflow of the platform are provided in [10].

3. The F2F Systems under Consideration

Food security attracts continuously growing attention, as we all want to know the practices and conditions under which the food we consume has been cultivated in the farms, has been transported, has been stored and finally exposed to the shelves of the retailers. In the farm-to-fork (F2F) pilot, we distinguish the following five actors:

- The actor in the farm (user/administrator of the IoT island that is deployed in the farm);
- The actor of the transportation company which associates the products with the conditions under which the products are transported (captured by the IoT island deployed in the vehicle);
- The actor in the warehouse where the products are stored and associates the conditions under which the products are kept up to the point they are purchased by a consumer;
- The consumer who purchases the product and, based on the RFID tag attached to the product, can inspect the full history of the product;
- The administrator of the platform that gathers the information from all IoT islands and delivers it to the consumer.

In real life, there are additional actors of the same type (e.g., transportation and supermarket actors) who perform the same activities as the transporter and the warehouse manager. Each of the above represents a node in this supply chain and can be supplier and customer at the same time. For example, the actor from the transportation company represents a consumer for the farmer and a supplier for the actor of the warehouse.

We now briefly describe the F2F platform from a technical point of view. Such a system consists of multiple Internet of Things (IoT) islands registering data in different repositories and deploying different business logics. In the following figure, such an example system is presented on the left-hand side of the figure. For our study, we have selected a system that has already employed traditional authentication and authorization techniques along with state-of-the-art blockchain technology to offer a secure solution [16]. The IoT islands (shown at the bottom of Figure 2) inject traffic through the so-called federation adapters (FA) which are then responsible for storing the information in the consortium ledger. Once the product arrives at the supermarket shelves, the hashes of all relevant information are used to create a unique entry in the public distributed ledger technology (DLT) which is, in our implementation, the public Ethereum network with its hash stored in a third blockchain named KSI, which is a commercial blockchain solution. To provide an interface for the users to interact with the underlying platform, a supervisor web server has been implemented.

To protect any F2F platform, the security officers of/people responsible for the F2F platform must define the specific points they are interested in monitoring and protecting and facilitate the creation of "security probes". In our example, we have implemented the components that deliver to the FISHY platform information from four distinct points of the deployed F2F platform, as shown in the figure. The aforementioned F2F platform has been studied and from the specified distinct points we have identified four types of attacks of major interest. For each type of attack, we also specify the data that should be monitored in order to detect such an attack. The attack types and the relevant "metadata" follow:

- Type 1: Unauthorised device—wallet ID level. Metadata: {Attacker wallet ID, Expected Legitimate Wallet ID, Device name}.

- Type 2: Unauthorised device—Decentralised Identifier (DID) level (with DID characterizing the device). Metadata: {Attacker DID, Device name, Jwt}.
- Type 3: Unauthorised user. Metadata: {username, IP}.
- Type 4: Attack to Blockchain node. Metadata: {IP, port, incident type}.

The "security probes" in our example are points where logs are collected and passed to the FISHY platform so that it can analyse them to detect attacks and propose countermeasures and remediations. For example, entry points 1 and 2 are relevant to the registration of information in the farm, transportation and warehouse steps of the supply chain during which the information is stored in the ledger maintained per step. Entry point 3 is relevant to the consumer or administrator of platform and entry points 4a and 4b are relevant to the consortium level operations. The logs from these "security probes" are sent to the FISHY platform through the SIA module in the form of a JSON object which will include the following fields: Unique Universal ID (UUID), Timestamp (UTC timestamp), Type, Metadata.

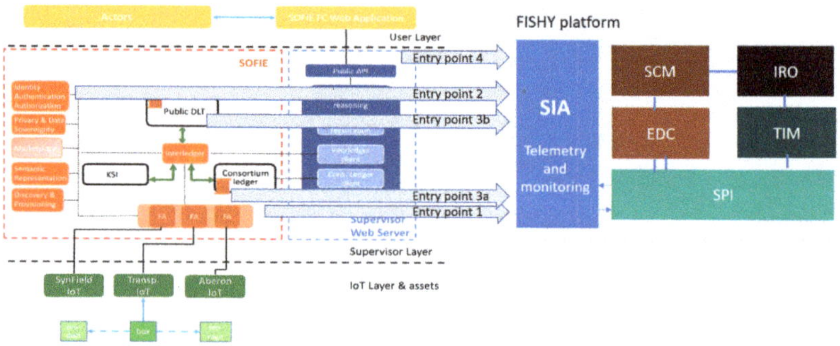

Figure 2. The F2F platform and its interconnection with the FISHY platform.

We have also discussed with other partners and decided to protect the F2F platform against additional attacks to extend the protection against additional attacks, if this is feasible and what extra actions are needed.

4. Evaluation and Discussion

Our aim is to evaluate FISHY platform from multiple perspectives ranging from technical to more commercial exploitation-oriented ones. For each of them, a different validation methodology has been adopted as will be explained in the next subsection. The aspects our evaluation has focus on include:

(a) Technical validation: We have validated that FISHY platform protects the considered platform in the defined attack scenarios (implementing or emulating the attack which is the typical methodology in attack detection, e.g., used in [17]). In this technical validation, the validation scenarios were selected based on the following criteria: (a) Attacks of interest to our customers. DDos attacks affect availability, and wallet or DID level attacks affect data integrity and privacy. These are the most important concerns in the farm-to-fork use cases. (b) Attacks of significant variety including "traditional" attacks (like DDoS attack and brute force attacks) and technology specific (blockchain specific) attacks.

(b) Additional attack detection capability with the existing tools (relevant to commercial exploitation): we studied whether the FISHY platform can protect against additional attacks outside those reported above using the currently deployed tools, which is closely related to the expandability of the platform;

(c) Commercial exploitation in diverse supply chain instances: we explored the value of offering multiple deployment options;

(d) Expandability with respect to the number and type of threat detection: we adopted the MITRE ATT@ACK framework to check how far such a platform could go in the number of attack types it can handle based on the "security probes" types we have adopted.

4.1. Evaluation of FISHY for Wallet ID Level-Oriented Attack

To carry out the technical validation for all attacks, i.e., to check whether FISHY platform efficiently detects the attacks under consideration in the farm-to-fork use case, the methodology we adopted was the following: we deployed the farm-to-fork platform in a dedicated infrastructure and developed code performing the considered attacks. The "reaction" of FISHY in the attempted attacks was monitored as well as the result in the farm-to-fork platform.

The aim is to confirm that the FISHY platform detects the attacks of type 1 titled "unauthorized device—wallet ID level". This is an attack that could occur in any of the IoT islands as, for example, the one deployed in the farm. For example, a malicious actor uses an unauthorized device and attempts to enter "fake" information in the F2F platform. In this platform, the IoT devices (through the so-called federation adapter—FA) register information about the fresh products, and in this registration, they use a wallet ID.

Assuming a malicious user intends to push to the platform fake information, they would use a device which has not been registered in the F2F platform. The information about data registration and the corresponding wallet IDs are passed to the FISHY TIM module through the security probes and the SIA deployed in the F2F platform premises. The F2F platform operator has appropriately configured the FISHY and more specifically the SACM module so that it recognizes which wallet IDs are legible or not. Thus, when the malicious user uses an unregistered wallet ID, SACM will detect this and will report a security event. In the following figure, the dashboard of the FISHY platform is shown in Figure 3. The instance presented here shows the events (one per row) detected by FISHY and their details as well as whether this event has been registered in the FISHY blockchain network (indicated by the green (check mark) symbol on the right-hand side of the event). This will trigger the IRO so that the relevant intent is identified, and a policy is suggested to the F2F platform operator. Once (s)he confirms (s)he agrees for the enforcement of the policy, the EDC undertakes its translation into a low-level policy, and it is passed to the F2F platform.

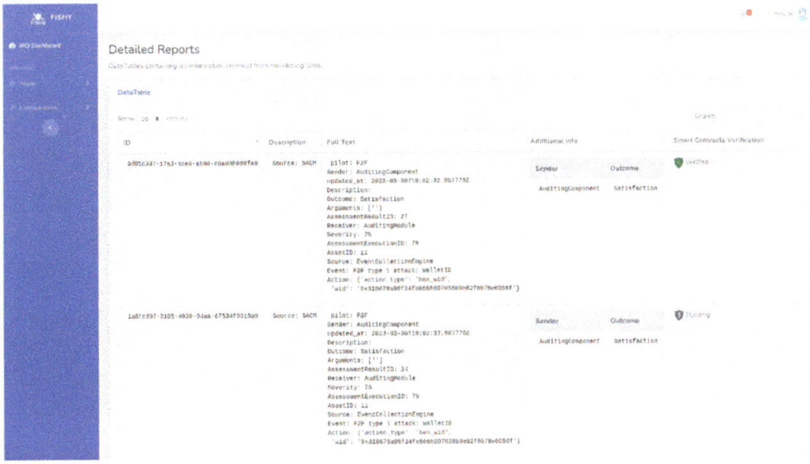

Figure 3. The FISHY dashboard presenting the detected event.

Now, the F2F platform will no longer communicate with the malicious federation adapter. Instead, the F2F platform displays a message to the attacker that the information (s)he tries to register is not accepted (as shown in the red box in the Figure 4).

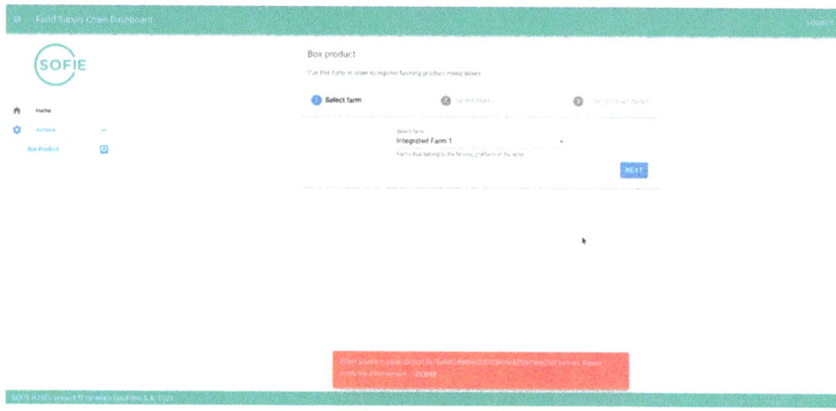

Figure 4. Screenshot from the F2F platform where the inability of the malicious user to enter information is shown.

4.2. Evaluation of FISHY for Attack to Blockchain Node

The aim in this section is to demonstrate that the FISHY platform detects the attacks of type 4, titled "Attack to blockchain node". This is an attack more likely to occur from a knowledgeable person to insert fake information in the blockchain used by the F2F platform. Let us assume the attacker tries to compromise the blockchain node, trying to connect to the blockchain node from a device with an IP address that is not whitelisted for the F2F premises.

The malicious actor could try to construct a request to the F2F blockchain, to try to insert fake information, for example, an unauthorized "Farm" platform, as depicted in Figure 5. The adversary, in order to prepare for the attack, can attempt to gain information on the nodes of the F2F blockchain network by exploiting the Tessera transaction manager of the nodes. Figure 6 shows the results of the exploitation of the Tessera endpoints. The user gains knowledge of the public keys of the nodes, which (s)he can use to sign their transaction and send to the blockchain.

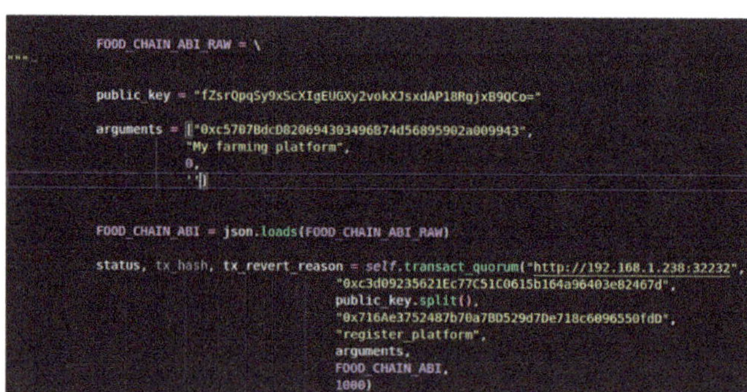

Figure 5. The adversary attempts to register fake information to the blockchain.

Should an external connection from an unknown IP occur, then the FISHY platform and more specifically SACM tool is notified as shown in Figure 7. In this validation, SIA, SPI, TIM and IRO were involved.

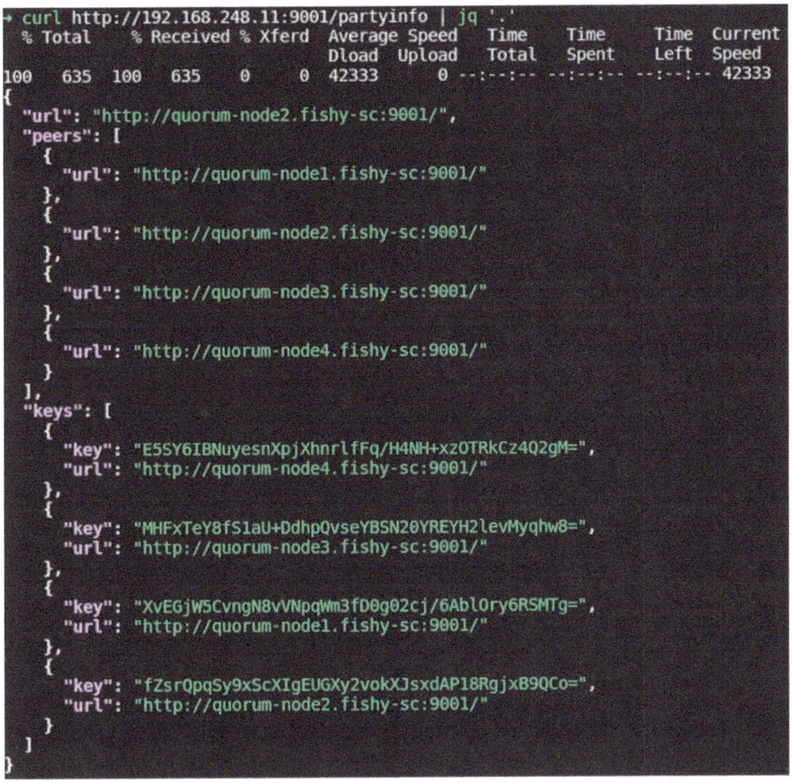

Figure 6. The adversary exploits the endpoints of the Tessera transaction manager of the nodes to find their public keys.

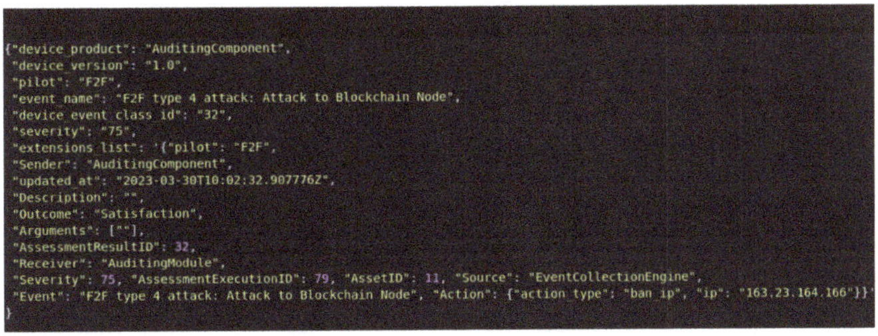

Figure 7. SACM monitors the IPs being connected to the blockchain node and checks whether these are whitelisted IP addresses.

Next, FISHY platform proposes a policy to be enforced. This policy is a ban-IP policy and is generated in IRO and turned to a low-level policy by EDC which then enforces it in the F2F use case, as shown in Figure 8. The end result is that the connection of the adversary node is terminated.

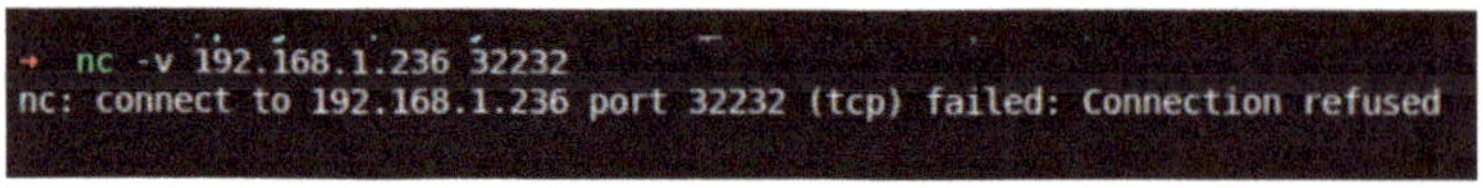

Figure 8. The malicious user can no longer connect to the blockchain node.

4.3. Evaluation of FISHY for DDoS Attack

The aim in this section is to demonstrate that the FISHY platform detects distributed denial of service (DDoS) attacks. For the F2F platform, the availability of the services is extremely important, due to the economic loss to the actors that rely on the F2F platform (e.g., retailers that use the platform to guarantee the safety of the supply chain) that can be caused by downtimes. Therefore, it is important for FISHY to be able to protect the platform against this type of attack.

To do this, the real-time network traffic is captured from the platform and then it is sent continuously to the PMEM tool [3] in the FISHY control services (Figure 9). As observed in the figure below, the captured flows contain normal traffic which is sent to the PMEM, and different traffic statistics are shown.

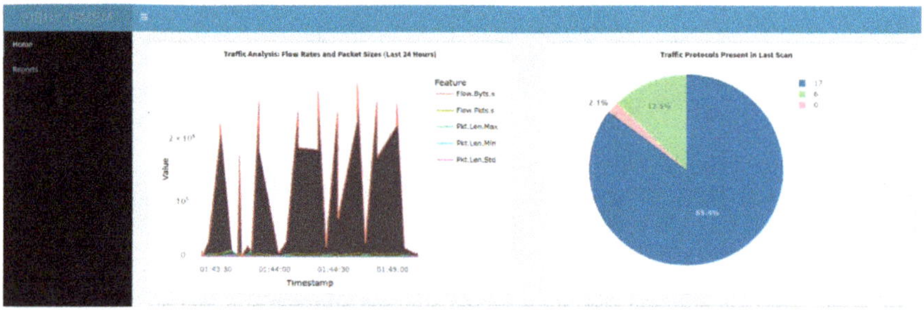

Figure 9. The PMEM dashboard showing the traffic of the system under examination.

PMEM gives information about the different flows in the network as well as different useful statistics about traffic share and severity of the attacks. To test the capability of PMEM to detect a DDoS attack, we intentionally simulate the scenario on the F2F platform. This malicious traffic along with the normal traffic is captured and sent to the PMEM tool. The traffic analysis shows that something abnormal is happening in the network (Figure 10).

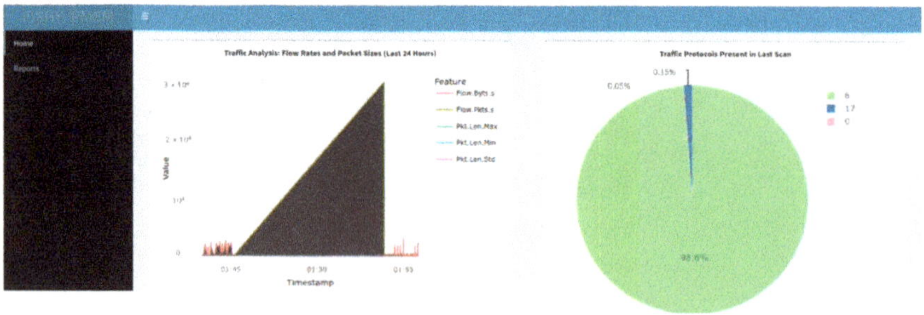

Figure 10. PMEM dashboard showing the statistics which show the results of the machine learning model (which classifies the traffic in benign and suspicious).

The prediction result of the PMEM for the network flows are presented in Figure 11:

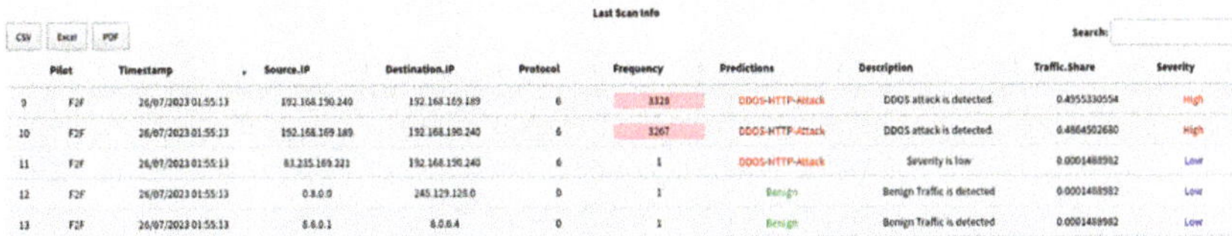

Figure 11. Details of the PMEM prediction results as shown in the PMEM dashboard.

The PMEM has detected the IP addresses which are trying to perform a DDOS attack, also including the frequency of the specific combination of source IP and destination IP. The severity of the attack is related to its computed frequency. The system shows, for instance, that the 3rd row in the table is considered a DDoS attack with low severity because the frequency is only 1, whereas the two first rows are considered real DDoS attacks because the frequencies are higher than a specific threshold.

4.4. Evaluation of FISHY for Detection of Brute Force Attacks

The aim in this section is to demonstrate that FISHY platform detects brute force attacks or in other words prevent an "unauthorized user" that possibly uses a machine to check all potential passwords to access the farm-to-fork platform. We assume that the attacker uses the wrong password or issues a brute force attack to gain access and potentially alter information relevant to specific products either to create a mesh or to diminish the value of specific brands. The attacker tries different combinations of username and password to enter the F2F platform. The F2F platform operator sets a rule in FISHY to check unsuccessful logic attempts from the same IP address and when these exceed a predefined threshold that he/she sets, FISHY notifies the F2F platform operator and suggests the banning of this IP.

In more detail, the FISHY platform detects this event (attack) as shown in Figure 12 through the Wazuh tool.

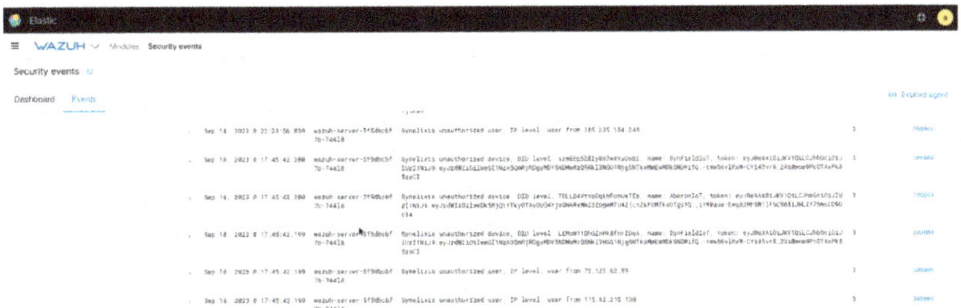

Figure 12. Screenshot from the dashboard of Wazuh that detects the brute force attack.

Next, to the detection, the FISHY platform proposes a ban-IP policy to be enforced. This policy is generated in IRO and turned to low-level policy by EDC which then enforces it in the F2F use case. Finally, the F2F platform displays a message to the attacker that the information he/she tries to register is not accepted.

4.5. Benchmarking FISHY to MITRE ATT&CK Framework

To assess the expandability of the FISHY platform, we use the MITRE ATT&CK framework and we proceed to logical implications to reach useful conclusions. To elaborate, during our research in the area of security attacks, we realized that the MITRE ATT&CK framework [9] plays a vital role. First, we must clarify that ATT&CK stands for adversarial tactics, techniques and common knowledge, and these are what the framework and accompanying ATT&CK knowledge base consist of. This framework aims at addressing the gap left by traditional models which are very focused on the study of attacks rather than their role in risk analysis, where the concern is not how the attack is executed but more on the effects and exploitation opportunities that can impact the system. This is of particular interest in the supply chain environments where the attacks to one of the interconnected IoT islands directly affect other actors in the chain. An additional reason to study this framework is that MITRE table is enriched by the open community that supports it. Thus, regularly inspecting this table can help us (a) continuously upgrade FISHY so that it protects against an ever-increasing set of attack types it handles and (b) verify that the techniques addressed are those reported in this open "literature".

In the farm-to-fork use case, the attacks we identified have been proposed to be detected using logs. To verify our decision, we select as the "control element" the log in the MITRE navigator, and we see the set of attacks that can be detected using logs, shown in green colour in Figure 13. All the attacks shown in green in this figure can be detected based on logs. This implies that should a platform owner be interested in detecting all these attacks, he/she should take care of providing the FISHY platform with the relevant logs in real time (i.e., ensure the provisioning of the relevant information).

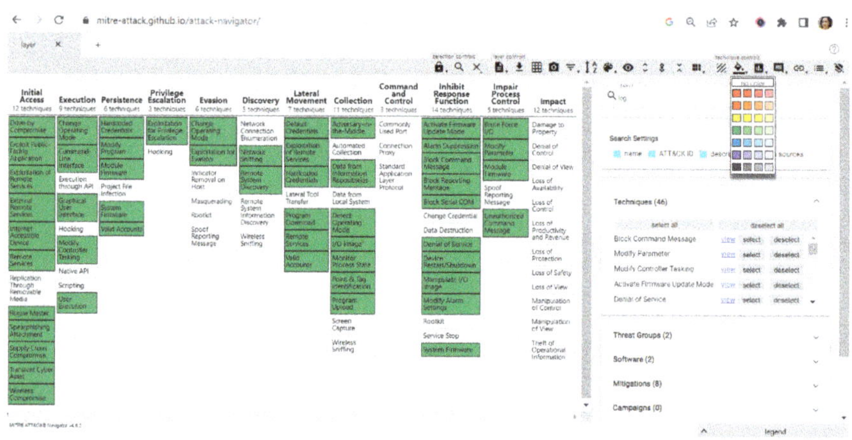

Figure 13. The attacks that can be detected based on logs shown/highlighted in green.

From the green boxes highlighted in the figure, we then select one-by-one the threat most relevant to our system. For example, the "default credentials" attack and the "denial of service" attack. Then, selecting the attack, the MITRE ATT&CK navigator displays all the procedures that an adversary may follow to issue such an attack that have been registered in the framework, the mitigation measures identified so far and the detection alternatives. Then, we check for the cells of interest whether FISHY platform implements a detection technique and whether the mitigation identified (and recommended and/or enforced) in FISHY is aligned with the one suggested by MITRE table. This way we have confirmed that FISHY platform adopts mitigation strategies well recognized in the market.

Another way to use the MITRE ATT&CK framework is the following: to check what can be detected based on specific controls. The rationale behind this choice is the following: in the farm-to-fork system, FISHY is capable of detecting threats based on logs and based on traffic analysis. So, in the MITRE ATT@CK navigator, we first selected "log" and then "traffic analysis", and the result is shown in Figure 14. The attacks that can be detected based on traffic analysis are marked in orange colour while those that can be detected using logs and not on traffic analysis are marked in green colour. (A subset of the orange-coloured threats are also detected using logs as was shown in the previous figure.) Again, as mentioned for the attacks detected based on logs, similarly, for the attacks detected based on traffic analysis information, the security officers of any platform interested in protecting their platform using FISHY, they should only ensure that the appropriate traffic analysis data are passed to the FISHY platform. Then, FISHY integrates all the necessary tools for detecting, recommending and potentially enforcing the mitigation policies.

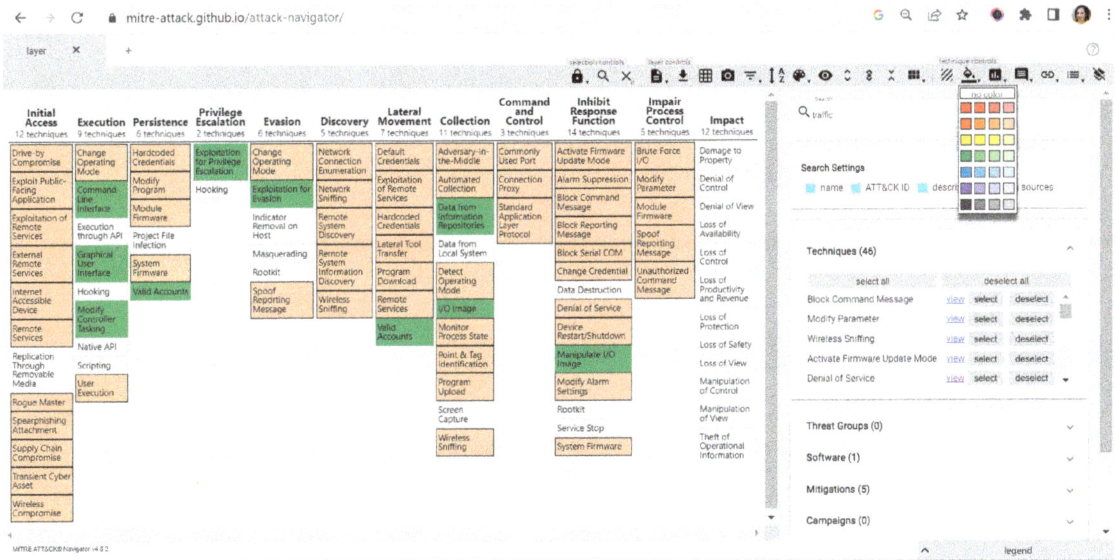

Figure 14. The threats that can be detected based on logs and traffic analysis information are coloured.

This has an important logical implication for FISHY: FISHY components can detect the majority of the identified threats which shows that FISHY is a flexible platform that can be exploited to detect the proliferating attacks that supply chain systems suffer today. With regard to mitigation, the flexible FISHY user interface allows for easy registration of multiple mitigation rules which could be drawn from a MITRE ATT&CK table.

4.6. FISHY-Enabled Security Enhancement in F2F Supply Chain

As has been shown in the previous sections, with the integration of the F2F IT system with FISHY, a set of interesting (to the actors) and important attacks are detected and mitigated. Additionally, we have realised that the different components of the FISHY platform can detect more attacks than those presented above: generating additional security probes, FISHY platform can detect attacks to additional points in the supply chain IT platform based on SACM and also, analysing traffic at different network levels or network islands, based on PMEM additional parts of the supply chain system can be protected. Analysing log information and performing machine learning-based traffic analysis enables the detection of a variety of attacks.

To assess the FISHY platform as objectively as possible, we presented the platform and asked colleagues outside the project teams to experiment with the features of the platform during a workshop that we held with seven people. The alpha version of the FISHY platform was released to the select group of testers from the consortium partners for evaluation and feedback. This process focused on identifying fundamental issues such as bugs, glitches and major functionality gaps, ensuring that the core features of the software were operational, and collecting feedback on performance and stability. The feedback collected during the workshop served as a valuable resource for refining the software before progressing to more extensive testing, where a larger and more diverse user base will be involved. Although this is not a large and statistically representative sample, due to the high expertise of the participants we consider their opinion valuable, carrying their extensive experience in the farm-to-fork sector and more specifically from the

IT system vendors. During this workshop, the user group answered/commented on the following topics:

- Easiness to use and user friendliness: the Average rating was 4.1 (using a five-point Likert scale), which was considered very good for a platform resulting from a research project.
- Security improvement: The question we asked was this: "what would you say if you were to quantify how much more secure is now the platform?". From the discussion that was raised, the answers converged towards the following key points:
    - The platform seems to efficiently detect the main attacks of interest.
    - The flexibility provided by the dashboard makes the operators feel they control what happens in the platform they operate.
    - The flexibility in detection offered by the different tools make the operators feel they can defend a wide range of attacks.
    - The FISHY dashboard with its clear presentation of events leaves time to the operator to focus on configuring the platform to detect additional attacks.
    - The immutability of the events guaranteed by the introduction of the blockchain technology and the registration of events in the blockchain network open the door to IoT vendors to persuade IT platform vendors to consider integrating IoT devices by less popular vendors, thus fostering competition.
    - To assess whether the multiple deployment options are of interest to the buyers, we asked the group: "deployment options: are they important?". They all found that they are very important as the deployment in each supply chain is different and tailored to the actors of the chain. One of the main business lines of Entersoft S.A. is software customisation company providing services to big supply chain actors. So, having the option to deploy on premise or on hybrid approach the platform and decide the split of components offers huge and valuable flexibility.

Other comments we received include the following: "At the beginning, it was not easy for us to understand how the platform is connected to the IT platform of the supply chain. The user manual helped but needs to be accompanied by a video". And it is "not easy to understand the flexibility of the platform. Somebody needs to delve into the details to find out".

4.7. FISHY Scalability and Potential Enhancements

The FISHY platform has been shown to efficiently detect a set of attacks. Additionally, it has been proven (based on the MITRE ATT&CK navigation tool) that it can potentially expand to detect other attacks. This would require the implementation of security probes on the side of the supply chain platform and on the configuration of appropriate rules in the FISHY platform. Furthermore, the architecture of the FISHY platform can flexibly integrate additional (open source or not) tools which can use the information captured by FISHY platform, they can also use the central repository and finally exploit the user-friendly FISHY user interface. With respect to the number of IT platforms that FISHY can protect, there is no limitation on this as it has been designed with scalability in mind. To sum up, the FISHY platform is both scalable and expandable with respect to the number of attack it is capable of detecting, with respect to the IT platform it can protect and with respect to the threat detection tools it can integrate. Its designers have pointed out that potential enhancement would follow two directions: the design of a very easy-to-use front end (so that it can be used not only by security officers) and the integration of tools that may be optimized for other IoT threats.

5. Conclusions

To sum up, we have shown that it is possible to have a platform that can detect and recommend the mitigation of multiple attack of different types (from network configuration to blockchain specific threats) and, at the same time, be expandable to be able to detect attacks that may be defined in the future. The FISHY platform efficiently protects the considered supply chain IT systems against multiple type of attacks, while with almost straightforward configurations, it can protect against a really large set (almost 85%) of the supply chain attacks reported in the MITRE ATT@CK framework. Apart from configuration of the components, in certain cases, some development of the appropriate mechanism to provide FISHY with the required supply chain platform details and data may be needed, but this is considered minor once the components and their user interface to the administrators are ready. Additionally, the flexible deployment of the FISHY platform is well appreciated from external end users. The authors anticipate that security platforms like FISHY have a strong potential not only in the supply chain but also in interconnected IT systems as, for example, the connected health care systems and applications [17], which are of very high importance to the quality and reliability of the health services provided.

Author Contributions: Conceptualization and methodology, H.C.L.; software, A.L. and P.A.K.; MITRE ATT&CK analysis, H.M.D.S. and H.C.L.; resources, E.M.T. and A.A.R.; writing—original draft preparation, P.A.K.; writing—review and editing, A.A.R. and J.P.C.; supervision, E.M.T.; project administration, A.A.R. and E.M.T. All authors have read and agreed to the published version of the manuscript.

Funding: This article has partially been supported by the EU funded H2020 FISHY Project (Grant agreement ID: 952644).

Data Availability Statement: Data are available upon request.

Acknowledgments: The authors would like to acknowledge all FISHY project partners for their technical contributions to the definition and development of the FISHY platform. And would like to acknowledge Ayaz Hussain for his contribution in improving the quality of the article.

Conflicts of Interest: The authors declare no conflicts of interest.

References

1. Karamitsios, K.; Orphanoudakis, T. Efficient IoT data aggregation for connected health applications. In Proceedings of the 2017 IEEE Symposium on Computers and Communications (ISCC), Heraklion, Greece, 3–6 July 2017; pp. 1182–1185. [CrossRef]
2. European Commission. TERAFLOW [Online]. 2023. Available online: https://www.teraflow-h2020.eu/ (accessed on 7 November 2023).
3. European Commission. ENSURESEC [Online]. 2023. Available online: https://www.ensuresec.eu/ (accessed on 7 November 2023).
4. Yin, H.-L.; Fu, Y.; Li, C.-L.; Weng, C.-X.; Li, B.-H.; Gu, J.; Lu, Y.-S.; Huang, S.; Chen, Z.-B. Experimental quantum secure network with digital signatures and encryption. *Natl. Sci. Rev.* **2023**, *10*, nwac228. [CrossRef] [PubMed]
5. Zhou, L.; Lin, J.; Xie, Y.-M.; Lu, Y.-S.; Jing, Y.; Yin, H.-L.; Yuan, Z. Experimental Quantum Communication Overcomes the Rate-Loss Limit without Global Phase Tracking. *Phys. Rev. Lett.* **2023**, *130*, 250801. [CrossRef] [PubMed]
6. Bulla, L.; Pivoluska, M.; Hjorth, K.; Kohout, O.; Lang, J.; Ecker, S.; Neumann, S.P.; Bittermann, J.; Kindler, R.; Huber, M. Nonlocal Temporal Interferometry for Highly Resilient Free-Space Quantum Communication. *Phys. Rev. X* **2023**, *13*, 021001. [CrossRef]
7. Lella, I.; Theocharidou, M.; Tsekmezoglou, E.; Malatras, A.; Garcia, S.; Valeros, V. *Enisa Threat Landscape for Supply Chain Attacks*; ENISA: Athens, Greece, 2021; ISBN 978-92-9204-509-8. [CrossRef]
8. Trakadas, P.; Karkazis, P.; Leligou, H.C.; Gonos, A.; Zahariadis, T. Farm to fork: Securing a supply chain with direct impact on food security. In Proceedings of the IEEE International Conference on High Performance Switching and Routing, Paris, France, 7–10 June 2021. [CrossRef]
9. Available online: https://attack.mitre.org/ (accessed on 3 October 2023).
10. Jukan, A.; Dizdarević, J.; Carpio, F. D2.4 Final Architectural Design and Technology Radar. Available online: https://fishy-project.eu/sites/fishy/files/public/content-files/deliverables/D2.4%20Final%20Architectural%20design%20and%20technology%20radar_v1.0.pdf (accessed on 10 September 2023).
11. Bensalem, M.; Dizdarević, J.; Carpio, F.; Jukan, A. The role of intent-based networking in ict supply chains. In Proceedings of the 2021 IEEE 22nd International Conference on High Performance Switching and Routing (HPSR), Paris, France, 7–10 June 2021; pp. 1–6.
12. Santos, H.; Oliveira, A.; Soares, L.; Satis, A.; Santos, A. Information Security Assessment and Certification within Supply Chains. In Proceedings of the 16th International Conference on Availability, Reliability and Security, Vienna, Austria, 17–20 August 2021; pp. 1–6.
13. Bensalem, M.; Dizdarević, J.; Jukan, A. Benchmarking various ML solutions in complex intent-based network management systems. In Proceedings of the 2022 45th Jubilee International Convention on Information, Communication and Electronic Technology (MIPRO), Opatija, Croatia, 23–27 May 2022.

14. Settanni, F.; Regano, L.; Basile, C.; Lioy, A. A Model for Automated Cybersecurity Threat Remediation and Sharing. In Proceedings of the 2023 IEEE 9th International Conference on Network Softwarization (NetSoft), Madrid, Spain, 19–23 June 2023; pp. 492–497.
15. Gonzalez, L.F.; Vidal, I.; Valera, F.; Lopez, D.R. Link Layer Connectivity as a Service for Ad-Hoc Microservice Platforms. *IEEE Netw.* **2022**, *36*, 10–17. [CrossRef]
16. Available online: https://www.sofie-iot.eu/ (accessed on 3 October 2023).
17. Hussain, A.; Aguiló-Ghost, F.; Simó-Mezquita, E.; Marín-Tordera, E.; Masip-Bruin, X. An NIDS for Known and Zero-Day Anomalies. In Proceedings of the 2023 19th International Conference on the Design of Reliable Communication Networks (DRCN), Vilanova i la Geltru, Spain, 17–20 April 2023. [CrossRef]

electronics

MDPI

Article

Resilient Electricity Load Forecasting Network with Collective Intelligence Predictor for Smart Cities†

Mohd Hafizuddin Bin Kamilin * and Shingo Yamaguchi *

Graduate School of Sciences and Technology for Innovation, Yamaguchi University, Yamaguchi 753-8511, Japan
* Correspondence: d010wcu@yamaguchi-u.ac.jp (M.H.B.K.); shingo@yamaguchi-u.ac.jp (S.Y.)

† This paper is an extended version of our paper published in Bin Kamilin, M.H.; Yamaguchi, S.; Bin Ahmadon, M.A. Fault-Tolerance and Zero-Downtime Electricity Forecasting in Smart City. In Proceedings of the 2023 IEEE 12th Global Conference on Consumer Electronics (GCCE), Nara, Japan, 10–13 October 2023; pp. 298–301.

Abstract: Accurate electricity forecasting is essential for smart cities to maintain grid stability by allocating resources in advance, ensuring better integration with renewable energies, and lowering operation costs. However, most forecasting models that use machine learning cannot handle the missing values and possess a single point of failure. With rapid technological advancement, smart cities are becoming lucrative targets for cyberattacks to induce packet loss or take down servers offline via distributed denial-of-service attacks, disrupting the forecasting system and inducing missing values in the electricity load data. This paper proposes a collective intelligence predictor, which uses modular three-level forecasting networks to decentralize and strengthen against missing values. Compared to the existing forecasting models, it achieves a coefficient of determination score of 0.98831 with no missing values using the base model in the Level 0 network. As the missing values in the forecasted zone rise to 90% and a single-model forecasting method is no longer effective, it achieves a score of 0.89345 with a meta-model in the Level 1 network to aggregate the results from the base models in Level 0. Finally, as missing values reach 100%, it achieves a score of 0.81445 by reconstructing the forecast from other zones using the meta-model in the Level 2 network.

Keywords: electricity load forecasting; internet of things; machine learning; security

Citation: Bin Kamilin, M.H.; Yamaguchi, S. Resilient Electricity Load Forecasting Network with Collective Intelligence Predictor for Smart Cities. *Electronics* **2024**, *13*, 718. https://doi.org/10.3390/electronics13040718

Academic Editors: Tomasz Rak and Dariusz Rzońca

Received: 31 December 2023
Revised: 1 February 2024
Accepted: 8 February 2024
Published: 9 February 2024

1. Introduction

With digital technologies becoming more incorporated into smart city management systems, machine learning (ML) is widely proposed as a forecasting model to predict the electricity load with high accuracy in smart cities [1]. Having the capability to accurately forecast the electricity load is necessary to allow the smart grids to distribute the electric power in advance to avoid overloading the electricity delivery network [2], better renewable energy integration with traditional energy to generate electricity [3], and minimize the operation loss during the peak hours [4]. As the scale of electricity infrastructure and reliability directly correlates to economic growth, it is crucial to maintain reliable service to avoid financial losses and interruption of other essential services [5].

However, digitalizing essential infrastructures in smart cities opens up new problems, such as cyberattacks against the infrastructures in smart cities. IBM Security observes this trend where 10.7% of cyberattacks in 2022 happened in the energy sector alone [6]. Looking deeper into distributed denial-of-service (DDoS) attacks that could cause packet loss and bring the server offline [7], the Azure Network Security Team reported that 89% of DDoS attacks span up to one hour [8], which may add missing values (MV) in the electricity load data and disrupt a centralized forecasting system. Due to the importance of energy services, the attack on electricity infrastructure in Ukraine during the Russo–Ukrainian War in 2016 shows the potential weakness of the current system that enemies could exploit [9]. Hence, it is necessary to create a decentralized and resilient forecasting method to solve these issues.

Still, recent studies in forecasting the electricity load showed most ML implementations overlooked the issue posed by MV [10,11], which could occur due to the packet loss and potentially impacting forecasting accuracy in real-world applications. Several methods exist to tackle this problem. Jung et al. [12] proposed a novel imputation technique to fill the MV accurately. In addition, there are also lightweight alternatives that sacrifice accuracy to train and evaluate MLs that use artificial neural networks (ANN), such as padding, which replaces MV with a placeholder value, and masking, which excludes MV from the computation, as noted by Rodenburg et al. [13]. As well as inadequate MV handling, recent studies also disregarded the single point of failure (SPoF) vulnerability, which could bring the entire forecasting system down when the server hosting the centralized ML architecture is offline [14]. Although existing distributed ML architectures could solve this [15], they are inefficient, and the data heterogeneity could negatively impact the accuracy [16].

In this study, we tackle the issues with MV in the electricity load data due to packet loss and SPoF due to the server hosting the forecasting system being taken offline from the DDoS attacks by proposing the Collective Intelligence Predictor (CIP) implementation, forming modular three-level forecasting networks of distributed MLs shown in Figure 1 to forecast the next one hour of electricity load data, matching the DDoS duration. Although weather and calendar data are proven to improve the electricity load forecasting accuracy in existing studies [17,18], this paper focuses solely on the electricity load data to investigate how well the design in CIP could perform against existing methods without relying on external data to negate the accuracy penalty when forecasting with MV.

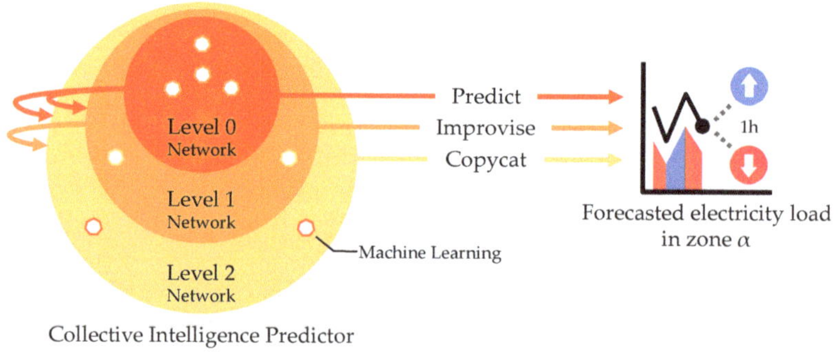

Figure 1. Generalized overview of a Collective Intelligence Predictor forming modular three-level forecasting networks to forecast the electricity load.

Each level in CIP represents three forecasting methods that it can use to forecast the electricity load. The modularity comes from CIP behaviors in activating the networks based on the MV percentages in the electricity load data used as independent variables, reducing unnecessary computation to forecast the electricity load. In addition, it increases the effective range the CIP can handle the MV to forecast the electricity load.

During regular operations where the independent variables have no MV, CIP relies exclusively on the base model trained with 0% MV in Level 0 to "predict" the electricity load in the zone CIP was assigned. As there is no MV, the forecasting accuracy from a single base model trained with 0% MV is sufficient to forecast the electricity load accurately. When the MV percentages in the independent variables range from 1% to 90%, CIP uses the meta-model in Level 1 to "improvise" the forecast by combining and refining the predictions from the base models in Level 0. Each base model is trained with different percentages of MV to contribute diversity in handling different MV percentages, allowing a broader effective range of CIP to forecast the electricity load as the MV percentages rise. Finally, when the MV percentages in the independent variables range from 91% to 100%, and it is no longer potent to use Level 1 to forecast, CIP uses the meta-model in Level 2 to create a "copycat" by reconstructing the forecasts taken from other CIPs meta-models in Level 1. Figure 2 summarized the CIP behaviors in activating the networks.

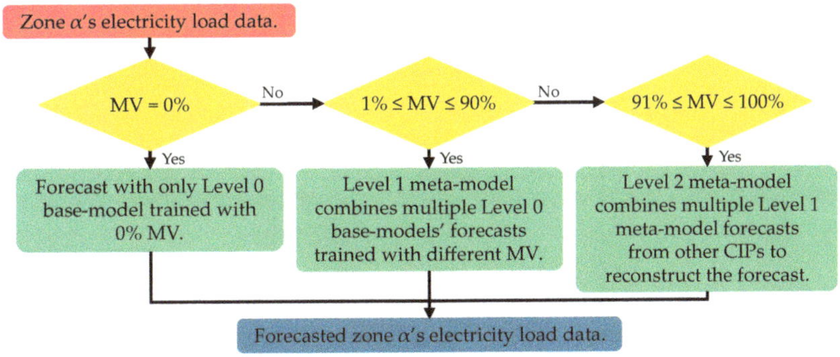

Figure 2. Collective Intelligence Predictor behaviors in activating the networks to handle different missing values percentages.

The primary contribution of this paper lies in developing a decentralized multi-level network of MLs that has modularity in its structure, the capability to handle a broader range of MV percentages, and a failsafe mechanism in Level 2 to reconstruct the forecast when handling MV the MV percentages is too high, which is unattainable with existing electricity load forecasting methods. In addition, with the implementation of multiple levels of networks, CIP could reduce unnecessary computation to forecast the electricity load by activating only the necessary MLs to forecast and increase the effective range of MV percentages CIP can handle when needed. Furthermore, CIP uses two feature selections to choose the best electric load data to improve forecasting accuracy and reconstruction. The contributions are significant in pioneering research predicting the electricity load to address security and reliability issues.

After the introduction in Section 1, Section 2 provides the preliminary for the dataset, feature selection algorithms, hyperparameter optimization, network construction, and comparison with the previous studies in this field. Section 3 provides the concept, application, and model training to implement CIP. Section 4 presented the evaluation of CIP with different MV percentages and compared forecasting accuracy with the existing centralized model architectures. Finally, the work is summarized, and we conclude the future planning for this research in Section 5.

2. Related Works

2.1. Overview

This section presents the preliminaries for the dataset, two feature selections to choose the other electricity load zones that may improve the CIP forecasting accuracy or reconstruct the forecast, the hyperparameter optimization algorithm to tune the base model in the Level 0 network of CIP, the multi-layer stacking ensemble learning that the CIP takes the inspiration to construct the networks, and the comparison against the previous studies to forecast the electricity load. Figure 3 shows the high-level summary for the related works implemented to create CIP and compares it against existing methods.

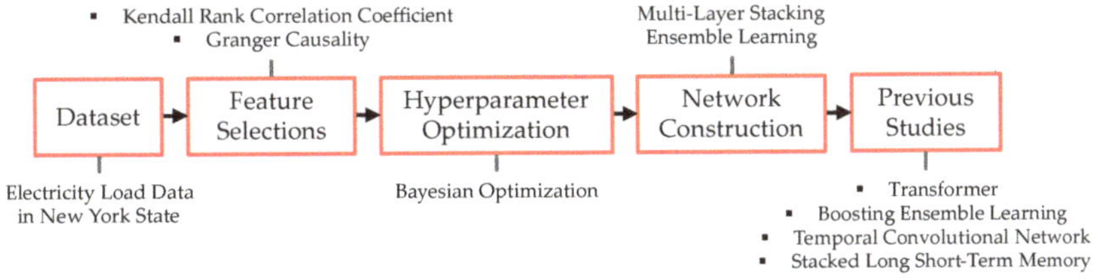

Figure 3. High-level summary of the related works implemented in this study to build CIP and compare it against existing forecasting methods.

2.2. Dataset

The dataset used to evaluate CIP in this study is publicly available electricity load data sourced from the New York Independent System Operator (NYISO) repository data [19]. It consists of an actual load sampled in real time at 5-minute intervals from 11 zones shown in Figure 4.

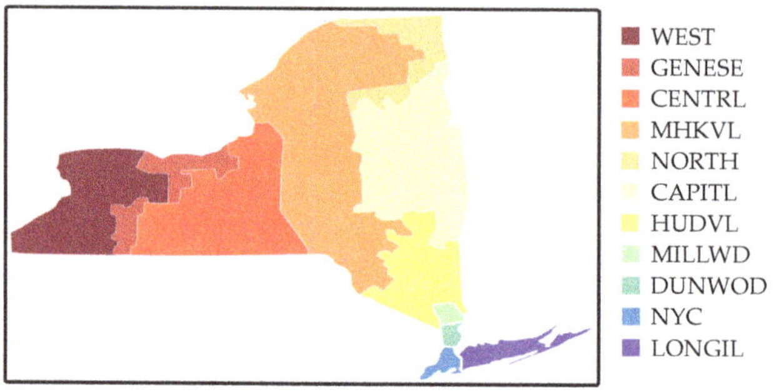

Figure 4. The electricity load zones and their corresponding zone codes managed by the New York Independent System Operator in New York state.

The electricity load data taken from the repository to train and evaluate the CIP against existing methods span from 1 January 2018 until 31 December 2020, which is exactly three years, as shown in Figure 5. Once the MV imputed with a polynomial interpolation by order of 2, the training and evaluation datasets split to the ratio 2:1, with the training dataset covering the period from 1 January 2018 to 31 December 2019 and the evaluation dataset from 1 January 2020 to 31 December 2020.

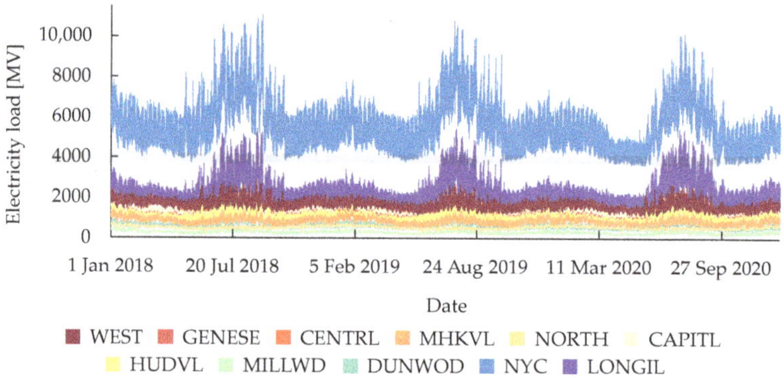

Figure 5. The electricity load data sampled in real time at 5-minute intervals in 11 zones from 1 January 2018 to 31 December 2020.

The training dataset normalized from −1 to 1 using the min-max normalization method. Then, using the same minimum and maximum values found in each electricity load zone from the training dataset, the same scale was applied to the evaluation dataset to be normalized for simulating the real-world application, where the latest data are not always used to update the models. After that, sequencing was applied to the datasets to forecast the next hour using the current one hour of electricity load data, which translates to 12 steps in independent and dependent variables. Although the training dataset used a sliding window moving 1 step per sequence to capture intricate patterns, the evaluation data used a sliding window moving 12 steps instead to ease the evaluation. Finally, MV simulated in the independent variables where $MV = \{0\%, 10\%, 20\%, \ldots, 90\%\}$ to create additional ten independent variables sequences for each zone.

2.3. Feature Selections

2.3.1. Kendall Rank Correlation Coefficient

CIP utilizes the Kendall rank correlation coefficient in Level 0 and Level 1 networks to find additional electricity load zones that may improve the forecasting accuracy for the electricity load zone CIP will be assigned to forecast.

Kendall rank correlation coefficient measures the strength and direction of the association between two sets of ranked data [20]. Since it is non-parametric, the data distribution does not affect the result [21]. To calculate, find the number of concordant pairs C and discordant pairs D in the ranked data of the training dataset before using Equation (1), where n represents the number of observations.

$$\tau = \frac{C - D}{\frac{1}{2}n(n-1)}, \quad -1 \leq \tau \leq 1 \tag{1}$$

For interpretation, as τ approaches 1, it indicates a strong positive correlation between two sets of ranked data. Similarly, as τ approaches -1, it indicates a strong negative correlation. However, if $\tau \approx 0$, it indicates weak or no correlation. Using pandas library [22], we compute the correlation for each zone, convert the values into absolute values, and sort it in descending order to choose the zones with high correlation.

2.3.2. Granger Causality

CIP utilizes Granger causality in the Level 2 network to find other electricity load zones that may improve the reconstruction of the forecast in the zone where the MV percentage is 91% and above. The motives for using a different feature selection in Level 2 are to avoid selecting the matching zones in the Kendall rank correlation coefficient to ensure redundancy and, as Granger causality is better in reconstructing the forecast.

Granger causality is a statistical hypothesis test to evaluate if a time series y_t possesses causality for another time series x_t [23]. With "var" represents the variance of a random variable, $\mathcal{H}_{<t}$ as the history of all relevant information up to $t-1$ and $\mathcal{P}(x_t|\mathcal{H}_{<t})$ as the optimal prediction for x_t given $\mathcal{H}_{<t}$, y is causal to x if it met the condition shown in Equation (2). As we want to analyze the causality with the time lag from 1 to 6, Equation (2) was rewritten as Equation (3) to reflect this change.

$$\text{var}[x_t - \mathcal{P}(x_t|\mathcal{H}_{<t})] < \text{var}[x_t - \mathcal{P}(x_t|\mathcal{H}_{<t} \setminus y_{<t})] \tag{2}$$

$$\text{var}[x_t - \mathcal{P}(x_t|\mathcal{H}_{<t})] < \text{var}[x_t - \mathcal{P}(x_t|\mathcal{H}_{<t} \setminus \{y_{t-1}, y_{t-2}, y_{t-3}, \ldots y_{t-6}\})] \tag{3}$$

After calculating the differences between consecutive observations in the dataset two times, we perform the Granger causality tests using statsmodels library [24] to obtain the p-values, which we sort in ascending order to find zones with high causality.

2.4. Hyperparameter Optimization

One of the challenges in designing the CIP is to tune the hyperparameters for the base model in Level 0, as there is no single best set of hyperparameters due to the complex relationships between them, requiring trial and error to find the best combination [25]. CIP utilizes Bayesian optimization as it utilizes new parameter combinations and exploits known promising regions to navigate the optimization space efficiently [26], which could shorten the computation time and guarantee an optimized outcome. With A as the search space of z, Equation (4) describes the optimization goal to find the maximum value at the sampling point of an unknown function f.

$$z^{+} = \arg\max_{z \in A} f(z) \tag{4}$$

We use Keras Tuner library [27] to implement the Bayesian optimization to optimize the hyperparameters in the base model, as it has good integration and ease of application with the TensorFlow library [28] used to build the ML models in CIP.

2.5. Network Construction

CIP takes inspiration to construct the networks from the multi-layer stacking ensemble learning. Stacking ensemble learning is a methodology to combine heterogenous base models to create a superior model compared to its components [29]. Compared to other ensemble techniques that use a deterministic way to combine the base models, stacking ensemble learning relies on a non-deterministic algorithm to combine the base models with a meta-model. Figure 6 shows the implementation example of a multi-layer stacking ensemble learning to forecast the electricity load in zone α. A set of i base models that use different algorithms are implemented in Level 0 to add diversity in capturing different patterns in zone α. In Level 1, a set of j meta-models combines the forecasting outcomes from the base models in Level 0 to improve the accuracy. In Level 2, a final meta-model refines the forecasts further by combining the output from the meta-models in Level 1. Most multi-layer stacking ensemble learning is limited to three layers, as the accuracy improvement greatly diminishes when adding a new layer.

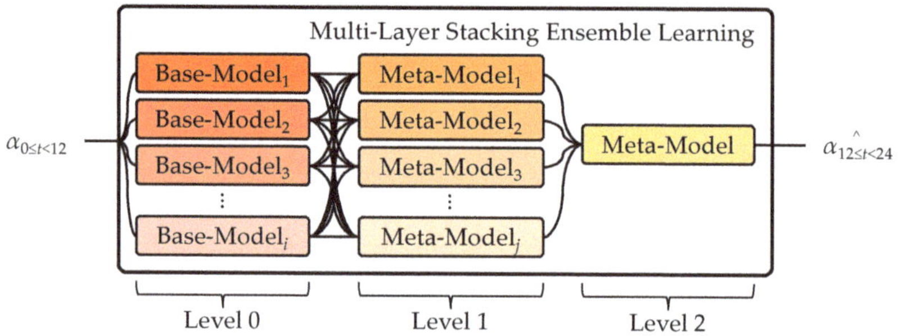

Figure 6. Implementation example using ensemble learning with multi-layer stacking to forecast the electricity load in zone α.

However, CIP does not fully adhere to the conventional method of deploying multi-layer stacking ensemble learning. Instead of using heterogeneous base models that use different algorithms in each base model, the base models in CIP are homogeneous, where the diversity in handling different percentages of MV is from training each base model with various percentages of MV. In addition, the independent variables are exposed to the meta-models in Level 1 and Level 2 networks to help the meta-models grasp the amount of MV they need to consider when combining the forecasts. Finally, the meta-model in the Level 2 network uses the forecasting outcomes taken from other CIPs Level 1 networks to reconstruct the prediction. Section 3.2 will discuss more in detail on CIP network implementation.

2.6. Previous Studies

The most commonly used models to forecast the electricity load rely on the Recurrent Neural Networks (RNN)-based implementation and its derivatives, such as long short-term memory (LSTM) and Gated Recurrent Unit (GRU), due to their capability to capture long-term dependencies in sequential data with high accuracy [30]. In addition, the advancement of technologies and techniques in ANN brings new model architectures that show promising results in forecasting the electricity load. For example, forecasting models that rely on temporal convolutional network architecture (TCN) can capture the local dependencies in the data via convolutional layers, which contributes to shorter interfacing time when compared to the RNN-based derivatives [31,32]. As well as TCN, forecasting models that rely on Transformer architecture show parallelization capabilities when compared to the RNN-based derivatives with the implementation of attention mechanism to capture the relationships between the elements in the sequential data [33,34]. However, they share issues mentioned in Section 1, where the forecasting models that rely on RNN-based derivatives, TCN, and Transformer cannot directly handle the MV due to packet loss from the DDoS attack without relying on some form of imputation, masking, and padding.

As well as relying on masking, padding, or imputation to make existing models capable of handling MV, several forecasting model designs could directly handle the MV. Stratigakos et al. [35] proposed handling the MV with Linear Programming (LP) to formulate a robust regression model that minimizes the worst-case loss when a subset of the independent variables has MV. The authors noted that their method can handle up to 50% of MV. In addition, Mienye et al. [36] found that an ensemble learning that utilizes boosting is effective when handing MV. Grotmol et al. [37] expand further using stacking ensemble learning that implemented boosting and other ML models as the base models to harden against MV. The authors noted that using heterogeneous base models in stacking could improve the Mean Absolute Error (MAE) score by 10.7%. Although these methods could effectively reduce the accuracy penalty when forecasting with MV, the centralized ML architectures make them suffer from SPoF vulnerability due to the server hosting the forecasting system being taken offline from the DDoS attacks.

Although several distributed computing methods could solve the SPoF vulnerability when the server hosting the forecasting system is taken offline by the DDoS attacks, recent studies show that federated learning is the favorable method due to the capability to train the model in independent sessions without sharing the datasets that may contain sensitive information [38–40]. However, in addition to the inefficiency and data heterogeneity negatively impacting the accuracy mentioned in Section 1, the studies in federated learning did not consider the countermeasures against MV.

We previously developed multivariate models to create a distributed forecasting network. Using the electricity load data from the zones with high Kendall rank correlation coefficient values negates the accuracy penalty when using padding to replace MV. In addition, it can substitute the forecast from an offline model by averaging the predictions from other models. However, as the models trained with only 25% of MV in the dataset to avoid overfitting where the models exhibit behaviors where the accuracy will only improve as the MV percentages rise, the effective range it could perform well before the coefficient of determination (r^2) score dropped below 0.95 is limited to 40% of MV. Furthermore, the network does not fully solve the SPoF vulnerability, as it does not have the countermeasure when one of the nodes supplying the electricity loads is offline. Table 1 summarized the previous studies and their capabilities in handling MV and SPoF.

Table 1. Comparison between previous studies to forecast the electricity load and their capabilities to handle missing values and the single point of failure.

Methodology	Missing Values	Single Point of Failure
Robust Model [35]	✓	×
Transformer [33,34]	△	×
Forecasting Network	✓	△
Federated Learning [38–40]	×	✓
Boosting Ensemble Learning [36,37]	✓	×
Temporal Convolutional Network [31,32]	△	×
Recurrent Neural Network Derivatives [30]	△	×

✓ = good, △ = moderate, × = bad.

In this study, we compare CIP against four existing electricity load forecasting models shown in the list below:

- TCN + padding
- Transformer + padding
- Stacked LSTM + padding
- Boosting ensemble learning

We use padding to help some existing models forecast with MV, as masking may change the sequence length during computation, negatively affecting the forecasting model to capture the dependencies in sequential data. In addition, without imputation to augment the sequence, we can analyze the strength of each model in handling MV.

3. Implementation

3.1. Overview

This section presents the CIP concept to implement modular three-level forecasting networks, its application on feature selections, hyperparameter optimization, and network construction to forecast the electricity load in zone $WEST$ of New York State, and the training methods used to train the ML models in Level 0, Level 1, and Level 2 networks in CIP. Figure 7 shows the high-level summary for the CIP concept, implementation, and training to implement CIP.

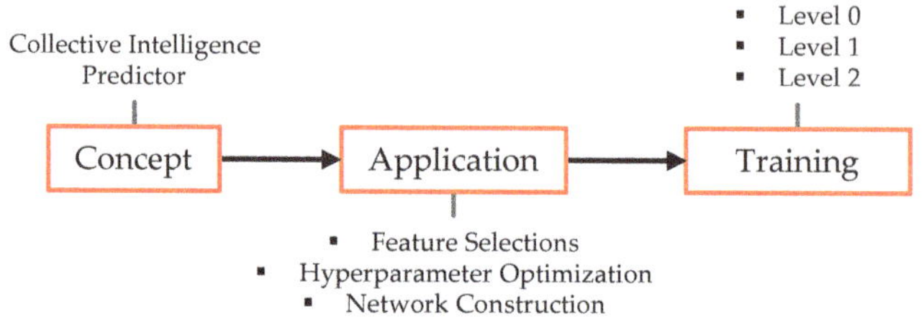

Figure 7. High-level summary of Collective Intelligence Predictor concept, application to forecast the electricity load in zone $WEST$ of New York State, and the methods used to train the models.

3.2. Concept

Referring to the generalized overview of CIP in Figure 1, CIP utilizes multi-level networks to distribute the ML models as a countermeasure against SPoF vulnerability. The models are connected to form forecasting networks similar to the multi-layer stacking ensemble learning, shown in Figure 6 to reduce the accuracy penalty when forecasting with MV. Figure 8 represented the CIP networks architecture to forecast the electricity load in zone α, where we define the CIP_α's forecast using the "predict" method as $\alpha_{\hat{Predict}_{12\leq t<24}}$, "improvise" method as $\alpha_{\hat{Improvise}_{12\leq t<24}}$, and "copycat" method as $\alpha_{\hat{Copycat}_{12\leq t<24}}$.

Referring to the summarized CIP behavior in Figure 2, CIP has a hierarchical network structure of Level 0, Level 1, and Level 2 to handle different MV percentages accordingly. With Predict(), Improvise(), and Copycat() functions representing "predict," "improvise," and "copycat" forecasting methods in CIP_α, Algorithm 1 shows the pseudocode to choose either "predict", "improvise", or "copycat" forecasting method based on the total MV percentage in the independent variables ($\alpha_{0\leq t<12}, \beta_{0\leq t<12}, \gamma_{0\leq t<12}$).

Algorithm 1 Networks activation in CIP_α.

Input: $\alpha_{0\leq t<12}, \beta_{0\leq t<12}, \gamma_{0\leq t<12}$
Output: $\alpha_{12\hat{\leq} t<24} \in \left\{\alpha_{\hat{Predict}_{12\leq t<24}}, \alpha_{\hat{Improvise}_{12\leq t<24}}, \alpha_{\hat{Copycat}_{12\leq t<24}}\right\}$

$concatenate \leftarrow \alpha_{0\leq t<12} + \beta_{0\leq t<12} + \gamma_{0\leq t<12}$
$mv\_count \leftarrow |\{c_i \in concatenate : c_i = \text{null}\}|$
$mv\_percentage \leftarrow mv\_count / |concatenate| \times 100\%$
if $mv\_percentage = 0$ **then**
    $\alpha_{\hat{Predict}_{12\leq t<24}} \leftarrow$ Predict()
    return $\alpha_{\hat{Predict}_{12\leq t<24}}$
else if $1 \leq mv\_percentage \leq 90$ **then**
    $\alpha_{\hat{Improvise}_{12\leq t<24}} \leftarrow$ Improvise()
    return $\alpha_{\hat{Improvise}_{12\leq t<24}}$
else if $91 \leq mv\_percentage \leq 100$ **then**
    $\alpha_{\hat{Copycat}_{12\leq t<24}} \leftarrow$ Copycat()
    return $\alpha_{\hat{Copycat}_{12\leq t<24}}$
end if

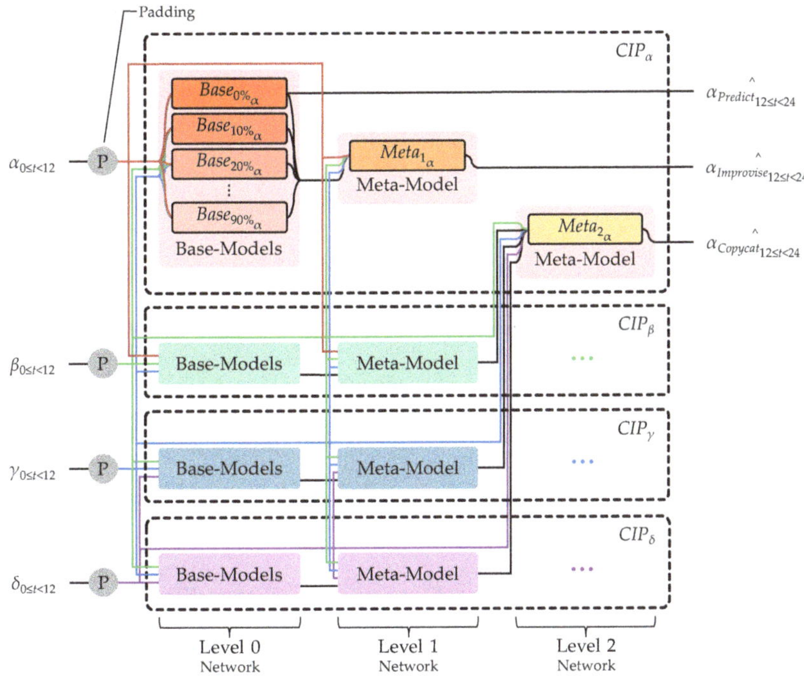

Figure 8. Collective Intelligence Predictor implementation CIP_α to forecast the electricity load in zone α using "predict", "improvise", and "copycat" methods.

3.2.1. Level 0

When the sum of MV percentages in the independent variables is 0%, CIP relies on the "predict" forecasting method in Level 0, where CIP uses only the $Base_{0\%_\alpha}$ base model in Level 0 to obtain $\alpha_{\hat{Predict}_{12\leq t<24}}$, reducing unnecessary computation and operation cost in regular operation to forecast the electricity load during normal operation. The only time CIP will activate all the base models in Level 0 is when the MV percentages in the independent variables are more than 1%, as the meta-model in Level 1 needs to combine the forecasts from the base models to obtain $\alpha_{\hat{Improvise}_{12\leq t<24}}$.

CIP uses ten base models $Base_{MV_\alpha}$ in CIP_α trained using the dataset simulated with MV in Section 2.2 to introduce diversity in handling a wide range of MV percentages during deployment. The base-model architecture shown in Figure 9 is a multivariable stacked LSTM that utilizes hyperbolic tangent (TanH) as the activation function in each layer, where k and l represent the number of LSTM units in the first and second layers of the base-model.

Assuming the electricity load zones from β and γ could improve the electricity load forecast in zone α, we choose multivariable stacked LSTM as the architecture in the base model due to its capability to grasp the dependencies from the independent variable CIP wants to forecast ($\alpha_{0\leq t<12}$) and independent variables that have strong correlation to improve the forecasting accuracy ($\beta_{0\leq t<12}$, $\gamma_{0\leq t<12}$) in zone α, allowing each of the base model in Level 0 to have a broader range of MV percentages it can handle before the forecasting accuracy degrade.

With $Base_{0\%_\alpha}()$ representing the base model trained with dataset that has 0% of MV, Algorithm 2 shows the pseudocode for Predict() function to obtain $\alpha_{\hat{Predict}_{12\leq t<24}}$.

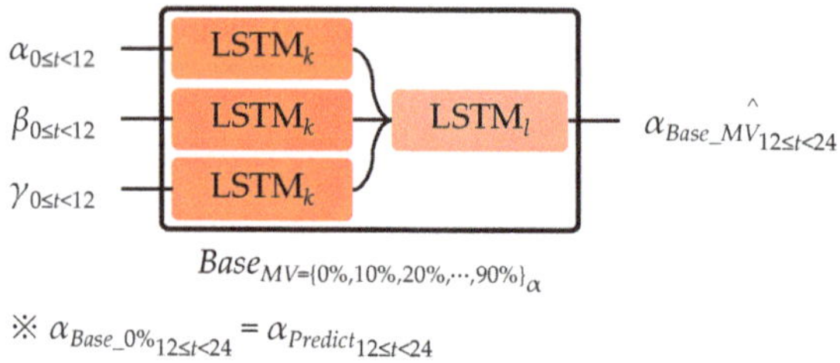

Figure 9. Multivariable stacked Long Short-Term Memory architecture implementation for the base model in Level 0 network.

Algorithm 2 Predict() function in CIP_α.

Input: $\alpha_{0\le t<12}, \beta_{0\le t<12}, \gamma_{0\le t<12}$
Output: $\hat{\alpha}_{Predict_{12\le t<24}}$
1: $\hat{\alpha}_{Predict_{12\le t<24}} \leftarrow Base_{0\%_\alpha}(\alpha_{0\le t<12}, \beta_{0\le t<12}, \gamma_{0\le t<12})$
2: **return** $\hat{\alpha}_{Predict_{12\le t<24}}$

3.2.2. Level 1

When the MV percentages in the independent variables ranged from 1% to 90%, CIP relies on the "improvise" forecasting method in Level 1, where all $Base_{MV_\alpha}$ in CIP_α's Level 0 are activated for the $Meta_{\alpha_1}$ meta-model in Level 1 to combine their forecasts. As each $Base_{MV_\alpha}$ has its effective MV percentage range to forecast the electricity load, combining the result with $Meta_{\alpha_1}$ ensures minimal forecasting accuracy degradation as the MV percentage rises, which is impossible with the bagging ensemble learning that averages the forecasts from the base models.

Following the same assumption in Level 0, CIP combine the forecasts from all $Base_{MV_\alpha}$ in Level 0 ($\hat{\alpha}_{Base\_0\%_{12\le t<24}}, \hat{\alpha}_{Base\_10\%_{12\le t<24}}, \hat{\alpha}_{Base\_20\%_{12\le t<24}}, \ldots, \hat{\alpha}_{Base\_90\%_{12\le t<24}}$) and the same electricity load data ($\alpha_{0\le t<12}, \beta_{0\le t<12}, \gamma_{0\le t<12}$) used by $Base_{MV_\alpha}$ using $Meta_{\alpha_1}$. The meta-model architecture shown in Figure 10 uses a multivariable deep neural network (DNN) model that utilizes TanH as the activation function in each dense layer. The numbers 156, 75, and 75 represent the dense unit numbers in the first, second, and third layers of $Meta_{\alpha_1}$. We choose multivariable DNN as the architecture in the meta-model due to its capability to fine-tune the combined forecasts from the $Base_{MV_\alpha}$ by the amount of MV that exists in the electricity load data used to forecast in zone α, which is impossible with other algorithms that do not consider the amount of MV in the independent variables.

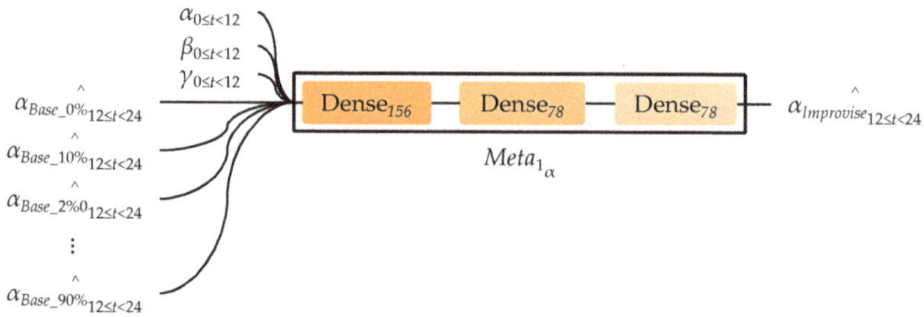

Figure 10. Multivariable Deep Neural Network architecture implementation for the meta-model in Level 1 network.

With $Meta_{\alpha_1}()$ function representing the multivariable DNN meta-model in Level 1 network, Algorithm 3 shows the pseudocode to concatenate the forecasts from all $Base_{MV_\alpha}$ and the electricity load data used by $Base_{MV_\alpha}$ to obtain $\hat{\alpha}_{Improvise_{12\le t<24}}$.

Algorithm 3 Improvise() function in CIP_α.

Input: $\alpha_{0\leq t<12}, \beta_{0\leq t<12}, \gamma_{0\leq t<12}$
and $\hat{\alpha}_{Base\_0\%_{12\leq t<24}}, \hat{\alpha}_{Base\_10\%_{12\leq t<24}}, \hat{\alpha}_{Base\_20\%_{12\leq t<24}}, \ldots, \hat{\alpha}_{Base\_90\%_{12\leq t<24}}$
Output: $\hat{\alpha}_{Improvise_{12\leq t<24}}$
$concat\_a \leftarrow \alpha_{0\leq t<12} + \beta_{0\leq t<12} + \gamma_{0\leq t<12}$
$concat\_b \leftarrow \hat{\alpha}_{Base\_0\%_{12\leq t<24}} + \hat{\alpha}_{Base\_10\%_{12\leq t<24}} + \hat{\alpha}_{Base\_20\%_{12\leq t<24}} + \ldots + \hat{\alpha}_{Base\_90\%_{12\leq t<24}}$
$concat\_c \leftarrow concat\_a + concat\_b$
$\hat{\alpha}_{Improvise_{12\leq t<24}} \leftarrow Meta_{\alpha_1}(concat\_c)$
return $\hat{\alpha}_{Improvise_{12\leq t<24}}$

3.2.3. Level 2

When the MV percentages in the independent variables exceed 90%, CIP relies on the "copycat" forecasting method in Level 2, where $Meta_{\alpha_2}$ in Level 2 reconstruct the forecasts in zone α by combining the forecast from the $Meta_1$ meta-models in Level 1 taken from CIP_β, CIP_γ, and CIP_δ. Although it is inferior in accuracy, it performs well in high MV environments where "predict" and "improvise" failed.

Similar to the $Meta_{\alpha_1}$ in Level 1, the meta-model $Meta_{\alpha_2}$ shown in Figure 11 uses a multivariable DNN model that utilizes TanH as the activation function in each dense layer. The only differences are the number of dense units in the first, second, third, and fourth layers, which are 144, 144, 72, and 36. With the assumption that electricity load zones β, γ, and δ could reconstruct the electricity load forecast in zone α, $Meta2_\alpha$ combines the forecasts with strong causality taken from the $Meta_1$ in Level 1 of CIP_β, CIP_γ, and CIP_δ ($\hat{\beta}_{Improvise_{12\leq t<24}}, \hat{\gamma}_{Improvise_{12\leq t<24}}, \hat{\delta}_{Improvise_{12\leq t<24}}$) together with the electricity load data corresponding to the zones other CIPs are assigned ($\beta_{0\leq t<12}, \gamma_{0\leq t<12}, \delta_{0\leq t<12}$) to reconstruct the electricity load forecast in zone α as $\hat{\alpha}_{Copycat_{12\leq t<24}}$. As redundancy is necessary for reconstruction, $Meta2_\alpha$ uses Granger causality to avoid selecting the same data chosen by the Kendall rank correlation coefficient used in Level 0 and Level 1 networks.

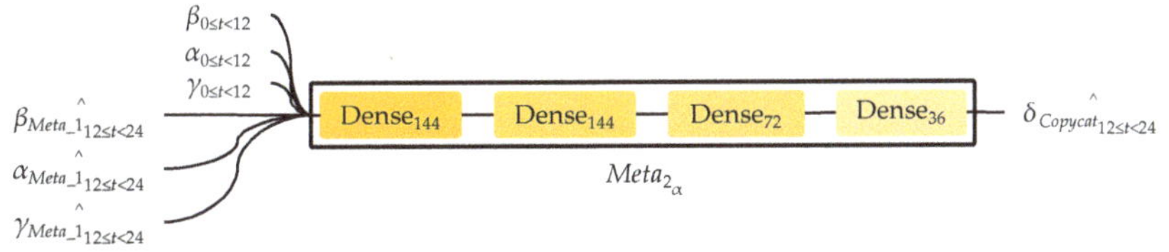

Figure 11. Multivariable Deep Neural Network architecture implementation for the meta-model in Level 2 network.

With $Meta2_\alpha()$ function representing the multivariable DNN meta-model in Level 2 network, Algorithm 4 shows the pseudocode to concatenate the electricity load data and the forecasts from the $Meta_1$ in Level 1 in CIP_β, CIP_γ, and CIP_δ to obtain $\hat{\alpha}_{Copycat_{12\leq t<24}}$.

Algorithm 4 Copycat() function in CIP_α.

Input: $\beta_{0\leq t<12}, \gamma_{0\leq t<12}, \delta_{0\leq t<12}$
and $\hat{\beta}_{Meta\_1_{12\leq t<24}}, \hat{\gamma}_{Meta\_1_{12\leq t<24}}, \hat{\delta}_{Meta\_1_{12\leq t<24}}$
Output: $\hat{\alpha}_{Improvise_{12\leq t<24}}$
$concat\_a \leftarrow \beta_{0\leq t<12} + \gamma_{0\leq t<12} + \delta_{0\leq t<12}$
$concat\_b \leftarrow \hat{\beta}_{Meta\_1_{12\leq t<24}} + \hat{\gamma}_{Meta\_1_{12\leq t<24}} + \hat{\delta}_{Meta\_1_{12\leq t<24}}$
$concat\_c \leftarrow concat\_a + concat\_b$
$\hat{\alpha}_{Copycat_{12\leq t<24}} \leftarrow Meta2_\alpha(concat\_c)$
return $\hat{\alpha}_{Copycat_{12\leq t<24}}$

3.3. Application

3.3.1. Feature Selections

In this study, CIP_{WEST} was implemented to forecast the electricity load in zone $WEST$ of the New York State. To construct the Level 0, Level 1, and Level 2 networks in CIP_{WEST}, Kendall rank correlation coefficient and Granger causality introduced in Section 2.3 used on the training dataset prepared in Section 2.2 with 0% of MV. Figures A1 and A2 shown in Appendix A are the generated feature selection heatmap on the training dataset. Using the Kendall rank correlation coefficient, zones $GENESE$ and $CENTRL$ are used to construct the Level 0 and Level 1 networks in CIP_{WEST}. Using Granger causality, zones $GENESE$, $NORTH$, and $MHKVL$ are suggested to construct Level 2 network in CIP_{WEST}. However, as the Kendall rank correlation coefficient has selected $GENESE$, we replace it with $NORTH$ as the next zone with high causality to ensure redundancy.

Figure 12 shows the CIP_{WEST} network implementation based on the zones selected by Kendall rank correlation coefficient and Granger causality to construct the Level 0, Level 1, and Level 2 networks. As the $Meta2_{WEST}$ in CIP_{WEST} requires the $Meta_1$ forecasts taken from CIP_{NORTH}, CIP_{MHKVL}, and CIP_{CAPITL}, we implemented the CIPs up to Level 1 network, where the selected zones for each CIP network shown in Table 2.

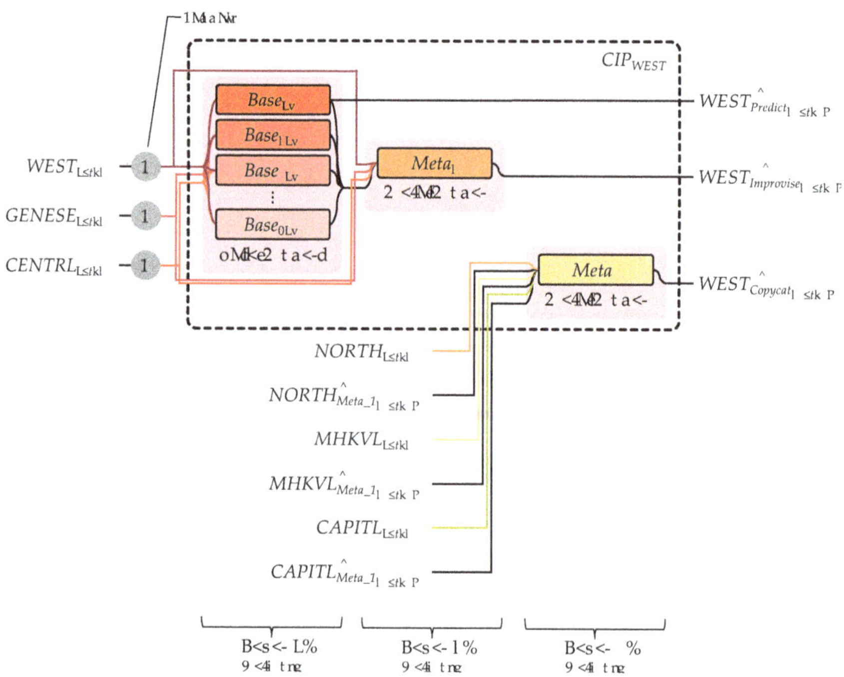

Figure 12. CIP_{WEST} networks implementation based on the recommendation zones that may improve the forecasting accuracy and reconstruction in zone $WEST$.

Table 2. The zones selected by Kendall rank correlation coefficient to create the Level 0 and Level 1 networks in CIP_{NORTH}, CIP_{MHKVL}, and CIP_{CAPITL}.

Forecaster	Required Independent Variable	Selected Independent Variables
CIP_{NORTH}	$NORTH$	$MHKVL$, $CENTRL$
CIP_{MHKVL}	$MHKVL$	$CENTRL$, $CAPITL$
CIP_{CAPITL}	$CAPITL$	$HUDVL$, $GENESE$

3.3.2. Hyperparameter Optimization

Using the Keras Tuner introduced in Section 2.4, we optimized the hyperparameters for the base models implemented in CIP_{WEST}, CIP_{NORTH}, CIP_{MHKVL}, and CIP_{CAPITL} with Bayesian optimization. Using fixed randomization, we tuned the base models using the training dataset with 0% of MV prepared in Section 2.2, optimization objective set to minimize the root-mean-square error (RMSE) score, five initial random points to start, and a maximum number of trials set to 5. Furthermore, we set the search range for the first and second LSTM layers units from 32 to 256 with 32 steps and the learning rate for Adam to choose from 0.001, 0.0001, and 0.00001.

Table 3 shows the hyperparameter optimization outcome. Most base models share the same hyperparameters, except the base model in CIP_{NORTH}. Most likely, it is due to most zones having a weak correlation zone *North*, leading to a different optimization outcome.

Table 3. The hyperparameters obtained for the base models in CIP_{WEST}, CIP_{NORTH}, CIP_{MHKVL}, and CIP_{CAPITL} with Bayesian optimization.

Base Model	First LSTM Layer Units	Second LSTM Layer Units	Adam's Learning Rate
CIP_{WEST}	192	96	0.001
CIP_{NORTH}	128	128	0.001
CIP_{MHKVL}	192	96	0.001
CIP_{CAPITL}	192	96	0.001

3.4. *Training*

3.4.1. Level 0

To train the $Base_{MV_{WEST}}$ in the Level 0 network of CIP_{WEST}, ten untrained base models are prepared based on the hyperparameters defined in Table 3, where the first and second LSTM layers use TanH with 192 units in the first layer, and 96 units in the second layer, and 0.001 as the learning rate for Adam optimizer. Using the random seed to replicate the weight initialization, each base model trained with a dataset with different MV percentages prepared in Section 2.2, where $MV = \{0\%, 10\%, 20\%, \ldots, 90\%\}$, 1000 batch size, 100 training epoch, and the early stop set to 3 with 0.0001 as the minimum observable improvement on mean squared error (MSE).

The same method are used to train the $Base_{MV}$ in the Level 0 network of CIP_{NORTH}, CIP_{MHKBL}, and CIP_{CAPITL} for the $Meta2_{WEST}$ to use in reconstructing the forecast in zone $WEST$.

3.4.2. Level 1

To train the $Meta1_{WEST}$ in the Level 1 network of CIP_{WEST}, the forecasts from the base models $Base_{MV_{WEST}} = \{Base_{0\%_{WEST}}, Base_{10\%_{WEST}}, Base_{20\%_{WEST}}, \ldots, Base_{90\%_{WEST}}\}$ done with different MV percentages in the training dataset are aggregated. Using the same hyperparameters described in Section 3.2 for $Meta1_{\alpha}$, $Meta1_{WEST}$ is prepared and trained with the training dataset, and the aggregated forecasts from the $Base_{MV_{WEST}}$, where the batch size is 1000, 100 training epoch, 0.0001 learning rate for Adam, and the early stop set to 3 with 0.0001 as the minimum observable improvement on MSE.

The same method are used to train the $Meta_1$ in the Level 1 network of CIP_{NORTH}, CIP_{MHKBL}, and CIP_{CAPITL} for the $Meta2_{WEST}$ to use in reconstructing the forecast in zone $WEST$.

3.4.3. Level 2

To train the $Meta2_{WEST}$ in the Level 2 network of CIP_{WEST}, the forecasts from the $Meta1_{NORTH}$, $Meta1_{MHKVL}$, and $Meta1_{CAPITL}$ taken from the Level 1 networks of CIP_{NORTH}, CIP_{MHKBL}, and CIP_{CAPITL} done with different MV percentages in the training dataset are aggregated. Using the same hyperparameters described in Section 3.2 for $Meta2_{\alpha}$,

$Meta2_{WEST}$ is prepared and trained with the training dataset and the aggregated $Meta_1$ forecasts from the CIP_{NORTH}, CIP_{MHKBL}, and CIP_{CAPITL}, where the batch size is 1000, 100 training epoch, 0.0001 learning rate for Adam, and the early stop set to 3 with 0.0001 as the minimum observable improvement on MSE.

4. Evaluation

4.1. Overview

This section presents the Transformer, boosting ensemble learning, TCN, and stacked LSTM as the previous methods to compare against CIP in forecasting the electricity load in zone $WEST$, forecasting outcome on different percentages of simulated MV, and the forecasting outcome when part of the CIP networks was offline due to the DDoS attack. Figure 13 shows the high-level summary for the previous methods, various MV percentages simulation and compromised network simulation.

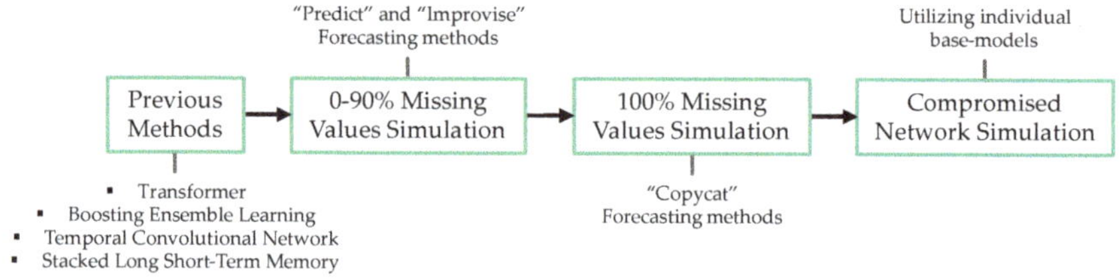

Figure 13. High-level summary of the previous forecasting methods, forecasting outcome on various simulated missing values percentages, and forecasting outcome with compromised network.

4.2. Previous Methods

4.2.1. Transformer

Figure 14 shows the Transformer model implementation to forecast the electricity load in zone $WEST$, where the head_size represents the size of the attention heads, the num_head represents the number of attention heads in the multi-head attention layer, the ff_dim represents the size of the feed-forward layer inside the Transformer block, and the num_transformer_blocks as the number of Transformer blocks stacked in the model. In addition, the mlp_units represents the number of units in each fully connected layer of the multi-layer perceptron (MLP) following the Transformer blocks, mlp_dropout represents the dropout rate in the output of each fully connected layer in the MLP, and ovl_dropout represents the dropout rate in the output of the multi-head attention layer in each Transformer block.

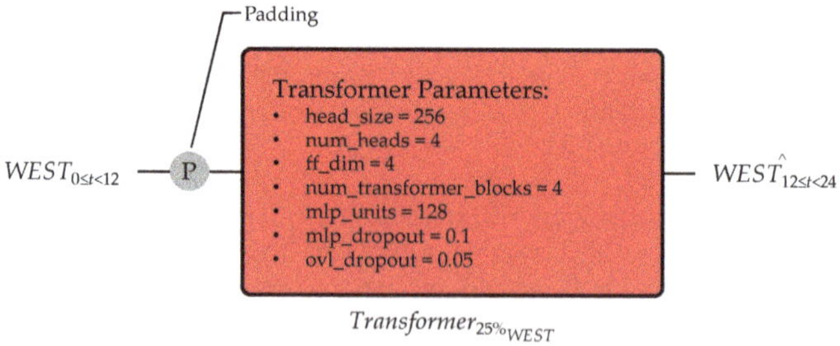

Figure 14. Transformer-based electricity load forecasting model to forecast the electricity load in zone $WEST$.

As the Transformer model tends to overfit when trained with ten training datasets that have the same electricity load data with varying MV percentages prepared in Section 2.2, we took the training dataset with 0% of MV and simulated 25% of MV in it instead, which is the technique we used in our previous study to prevent overfitting. Models that exhibit overfitting will show unexpected behavior, where the forecasting accuracy will only increase as the MV percentages increase, making it unpractical for normal operations.

We trained the Transformer model with a batch size of 1000, 100 training epoch, 0.0001 learning rate for Adam, and the early stop set to 3 with 0.0001 as the minimum observable improvement on MSE.

4.2.2. Boosting Ensemble Learning

Figure 15 shows the boosting ensemble learning model implementation to forecast the electricity load in zone $WEST$. We implemented the boosting ensemble learning based on eXtreme Gradient Boosting (XGBoost) [41], where the max_depth represents the maximum depth of each tree in the boosting process, the learning_rate represents the step size at each iteration while moving toward a minimum of the loss function, and the objective as reg:squarederror represents the specified learning task and objective function, which show the model trained for regression problem to minimize the MSE.

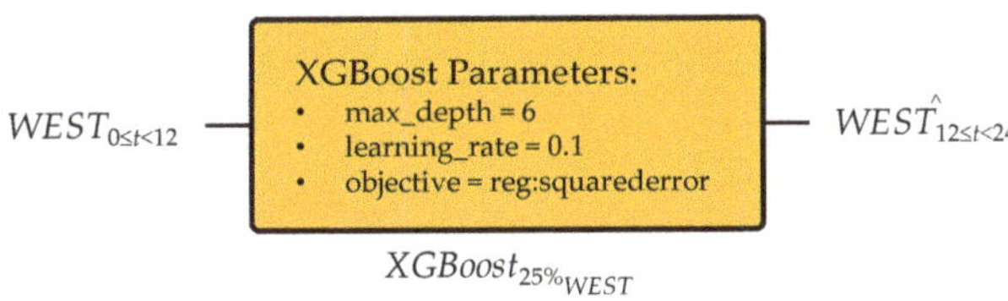

Figure 15. Boosting ensemble learning-based electricity load forecasting model to forecast the electricity load in zone $WEST$.

Similar to the Transformer model, even with the early stop function set to stop the training when the MSE score no longer improves after three times, the XGBoost model exhibits overfitting tendencies when trained with ten training datasets with varying MV percentages prepared in Section 2.2. We solved this issue using the training dataset with 25% of MV used on the Transformer model to train the XGBoost model in 500 epochs.

4.2.3. Temporal Convolutional Network

Figure 16 shows the TCN-based model implementation to forecast the electricity load in zone $WEST$, where the first and second convolutional layers use 64 filters, kernel size set to 3, and padding set to causal to ensure the current output depends only on the current and past input, while the third dense layer has 50 units. The convolutional and the dense layers use rectified linear units (ReLU) as the activation function.

As the TCN model does not exhibit the overfitting behavior shown in the Transformer and XGBoost model, we used ten training datasets that have the same electricity load data with varying MV percentages prepared in Section 2.2, concatenated into one long sequence to train the TCN model with a batch size of 1000, 100 training epoch, 0.0001 learning rate for Adam, and the early stop set to 3 with 0.0001 as the minimum observable improvement on MSE.

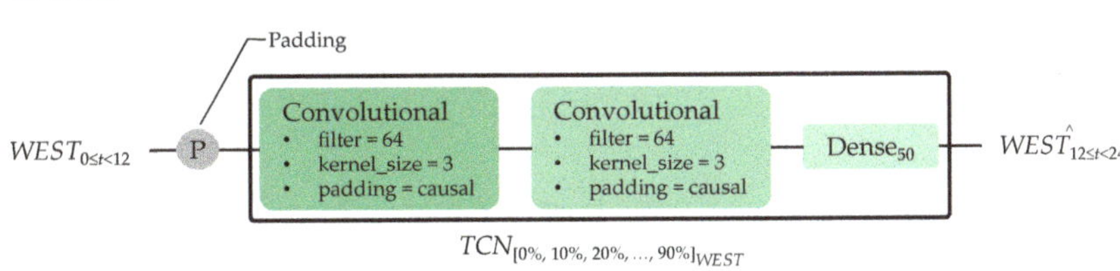

Figure 16. Temporal Convolutional Network-based electricity load forecasting model to forecast the electricity load in zone $WEST$.

4.2.4. Stacked Long Short-Term Memory

Figure 17 shows the stacked LSTM-based model implementation to forecast the electricity load in zone *WEST*, where the first and second LSTM layers use 32 units and TanH as the activation function.

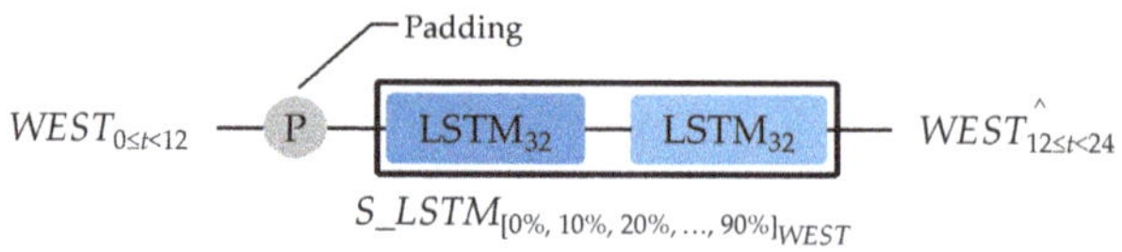

Figure 17. Stacked Long Short-Term Memory-based electricity load forecasting model to forecast the electricity load in zone *WEST*.

Similar to the TCN model, we use ten training datasets that have the same electricity load data with varying MV percentages prepared in Section 2.2, concatenated into one long sequence to train the LSTM model with a batch size of 1000, 100 training epoch, 0.0001 learning rate for Adam, and the early stop set to 3 with 0.0001 as the minimum observable improvement on MSE.

4.3. 0–90% Missing Values Simulation

In this test, we used evaluation datasets with different MV percentages prepared in Section 2.2 to evaluate the forecasting accuracy of CIP, TCN, boosting ensemble learning, Transformer, and stacked LSTM models on the electricity load in zone *WEST*. Table 4 and Figure 18 show the r^2 forecasting scores on zone *WEST*. We used r^2 to calculate the forecasting accuracy, as the ease of interpretability gives us a generalized idea of how similar the forecast would match with the plotted real values [42].

With 0% of MV in zones *WEST*, *GENESE*, and *CENTRL*, CIP utilizes the "predict" forecasting method in the Level 0 network to forecast the electricity load, which achieves the highest r^2 score of 0.98831 when compared to the previous forecasting methods. As the MV percentages in *WEST*, *GENESE*, and *CENTRL* rise from 1% to 90%, CIP utilizes the "improvise" forecasting method to combine the forecast from base models in Level 0 network and fine-tune them into one forecast using the meta-model in Level 1 network, which yields r^2 score of 0.96225 with 80% of MV in the independent variables. In contrast, none of the previous forecasting methods achieve r^2 score of 0.9 and above with 80% of MV. Even with 90% of MV, the r^2 score on CIP only falls to 0.89345, showing the resilience of our proposed method against MV, as the r^2 scores for the previous methods already fall below 0.7.

Table 4. Coefficient of determination (r^2) scores comparison between multiple forecasting methods on different missing values percentages in zone *WEST*.

MV [%]	CIP	TCN	Boosting	Transformer	Stacked LSTM
0	0.98831	0.98567	0.98523	0.93445	0.98626
10	0.98501	0.98465	0.98478	0.91115	0.98579
20	0.98498	0.98381	0.98392	0.89104	0.98518
30	0.98492	0.98273	0.98241	0.87363	0.98421
40	0.98478	0.98159	0.97655	0.85770	0.98311
50	0.98410	0.97856	0.95719	0.36864	0.98029
60	0.98214	0.97292	0.91396	−1.53278	0.97468
70	0.97727	0.95687	0.75946	−11.5464	0.95892
80	0.96225	0.88543	0.33560	−74.1568	0.88781
90	0.89345	0.65085	−0.66817	−296.831	0.65736
Average	0.97272	0.93631	0.72109	−37.92314	0.93836

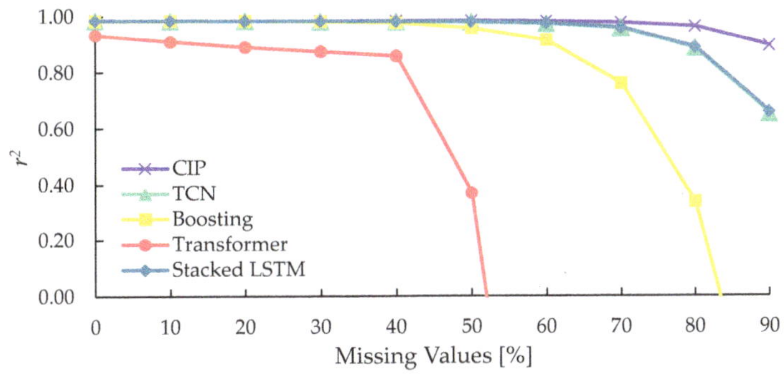

Figure 18. Plotted coefficient of determination (r^2) scores comparison between multiple forecasting methods on different missing values percentages in zone *WEST*.

Examining the forecasting accuracies for the previous forecasting methods, TCN and stacked LSTM models are the only previous methods that equally perform well and could maintain a r^2 score of 0.95 with 70% of MV in the independent variable. These results show the TCN and stacked LSTM capability in capturing the dependencies in the independent variables with either convolutional layers or gating mechanisms without MV negatively affecting the forecast.

Table 5 and Figure 19 show the RMSE forecasting scores on zone *WEST*, which support the results shown in Table 4 and Figure 18 where CIP surpass previous forecasting methods in resiliency against MV.

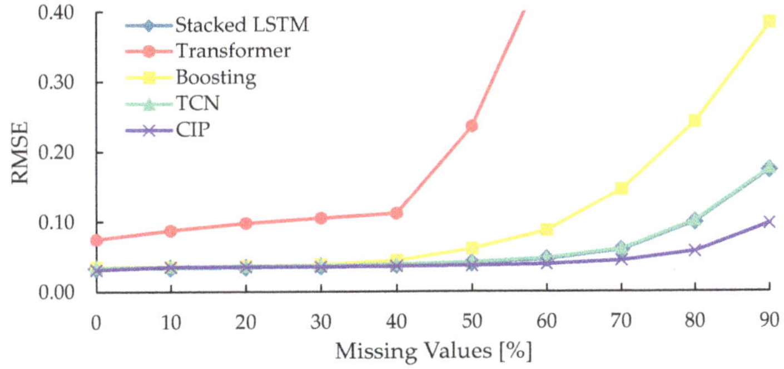

Figure 19. Plotted root-mean-square error scores comparison between multiple forecasting methods on different missing values percentages in zone *WEST*.

Table 5. Root-mean-square error scores comparison between multiple forecasting methods on different missing values percentages.

MV [%]	CIP	TCN	Boosting	Transformer	Stacked LSTM
0	0.03137	0.03433	0.03480	0.07452	0.03355
10	0.03555	0.03557	0.03535	0.08756	0.03413
20	0.03565	0.03656	0.03637	0.09733	0.03488
30	0.03573	0.03779	0.03814	0.10501	0.03605
40	0.03588	0.03903	0.04443	0.11154	0.03733
50	0.03670	0.04223	0.06089	0.23587	0.04044
60	0.03899	0.04779	0.08686	0.47234	0.04616
70	0.04409	0.06083	0.14554	1.05115	0.05935
80	0.05712	0.10024	0.24196	2.57265	0.09918
90	0.09663	0.17537	0.38340	5.12129	0.17371
Average	0.04477	0.06097	0.11077	0.99293	0.05948

4.4. 100% Missing Values Simulation

In this test, we set the MV percentages to 100% for zones $WEST$, $GENESE$, and $CENTRL$. As it is impossible to forecast with 100% of MV, CIP relies on the "copycat" forecasting method to reconstruct the forecast for zone $WEST$, where the MV percentages in each zone rise from 0% to 90% with a 10% increment. Table 6 and Figure 20 show the results where CIP obtained an r^2 score of 0.81445 with 0% of MV. In addition, the score only drops to 0.74013 with 90% of MV, which is a 9.56142% degradation.

Table 6. Coefficient of determination (r^2) and root-mean-square error scores obtained from the reconstructed electricity load forecast for zone $WEST$.

MV [%]	r^2	RMSE
0	0.81445	0.12785
10	0.76918	0.14261
20	0.76517	0.14385
30	0.76375	0.14428
40	0.75983	0.14548
50	0.75710	0.14630
60	0.75839	0.14591
70	0.76295	0.14453
80	0.76616	0.14354
90	0.74013	0.15129

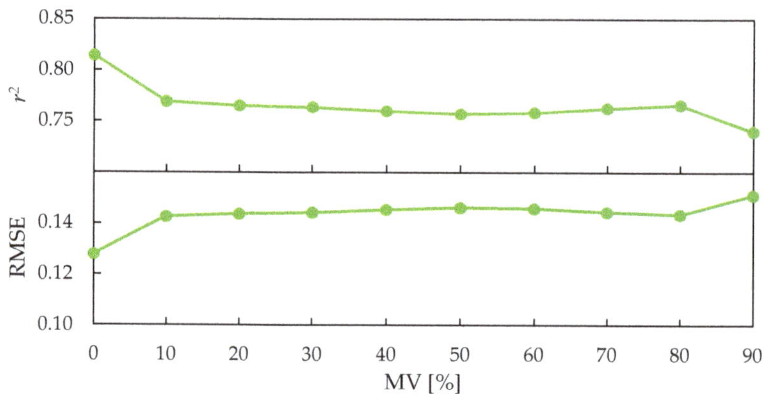

Figure 20. Plotted coefficient of determination (r^2) and root-mean-square error scores obtained from the reconstructed electricity load forecast for zone $WEST$.

Although the forecast accuracy using the "copycat" method is inferior, it constructed the forecast in zone $WEST$ even with 100% of MV in $WEST$, $GENESE$, and $CENTRL$, which is unattainable with previous methods.

4.5. Compromised Network Simulation

In the final test, we simulated a scenario where Level 1 and Level 2 networks in CIP_{WEST} are offline. Using the base model trained with an MV percentage close to the MV percentage in the input data, we could obtain an accurate prediction similar to the meta-model in Level 1. Table 7 and Figure 21 show the forecasting outcome using the individual base models.

Table 7. Coefficient of determination (r^2) and root-mean-square error scores obtained from the individual base-model load forecast for zone *WEST*.

MV [%]	r^2	RMSE
0	0.98826	0.03141
10	0.98739	0.03245
20	0.98704	0.03294
30	0.98440	0.03614
40	0.98409	0.03644
50	0.98235	0.03846
60	0.97132	0.04917
70	0.97086	0.04997
80	0.95744	0.06045
90	0.88504	0.10029

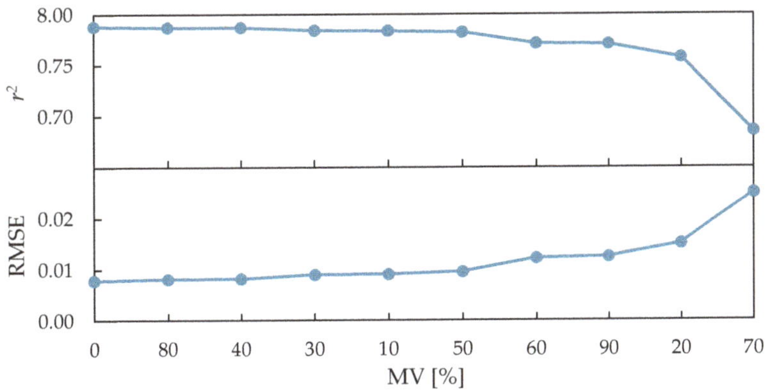

Figure 21. Plotted coefficient of determination (r^2) and root-mean-square error scores obtained from the individual base-model load forecast for zone *WEST*.

5. Conclusions

The digitalization of essential infrastructures in smart cities introduces new challenges. With the increasing threat of cyberattacks targeting the electricity infrastructure, we must design countermeasures to ensure the service will not be interrupted, which could negatively impact the economy and other essential services. We proposed CIP, a distributed forecasting network that could handle a high percentage of MV and solve the SPoF vulnerability to prevent interruption. CIP works by utilizing multi-level networks to forecast the electricity load based on the MV percentage in the input sequence. When there is no MV, we rely solely on the base model in Level 0 to "predict" the electricity load to reduce unnecessary computation, with an r^2 score of 0.98831. As the MV rises from 1% to 90%, CIP utilizes the meta-model in the Level 1 network to "improvise" the "prediction" from the base models from Level 0, which allows our proposed method to handle up to 80% of MV while maintaining r^2 score of 0.96225. Even when one of the data sources providing the electricity load data is offline, we reconstruct the forecast using a meta-model in Level 2 to create a "copycat" forecast, which CIP reconstructs from electricity load data from other zones with r^2 score of 0.81445. Finally, as our proposed forecast method is modular, the predictions from the individual base models trained with the MV percentage close to the input data are accessible with comparable accuracy with the meta-model in Level 1.

For future works, we aim to expand the capability of our CIP to handle concept drift by integrating our previous research using radian scaling [43], detecting data falsification, and improving the forecasting accuracy in Level 2 using different types of data, as our current research is limited only to the electricity load data from other zones.

Author Contributions: Data curation, M.H.B.K.; Formal analysis, M.H.B.K.; Funding acquisition, S.Y.; Investigation, M.H.B.K.; Methodology, M.H.B.K. and S.Y.; Project administration, S.Y. Resources, M.H.B.K.; Software, M.H.B.K.; Supervision, S.Y.; Validation, M.H.B.K.; Writing—original draft, M.H.B.K.; Writing—review and editing, S.Y. All authors have read and agreed to the published version of the manuscript.

Funding: This work was supported by JST SPRING, Grant Number JPMJSP2111, and Interface Corporation, Japan.

Data Availability Statement: Data presented in this study are openly available from New York Independent System Operator at https://www.nyiso.com/load-data (accessed on 6 December 2023).

Conflicts of Interest: The authors declare no conflict of interest.

Abbreviations

The following abbreviations are used in this manuscript:

ML	Machine Learning
DDoS	Distributed Denial-of-Service
MV	Missing Values
ANN	Artificial Neural Networks
SPoF	Single Point of Failure
CIP	Collective Intelligence Predictor
NYISO	New York Independent System Operator
RNN	Recurrent Neural Networks
LSTM	Long Short-Term Memory
GRU	Gated Recurrent Unit
TCN	Temporal Convolutional Network
LP	Linear Programming
MAE	Mean Absolute Error
TanH	Hyperbolic Tangent
DNN	Deep Neural Networks
RMSE	Root-Mean-Square Error
MSE	Mean squared Error
MLP	Multi-Layer Perceptron
XGBoost	eXtreme Gradient Boosting
ReLU	Rectified Linear Units

Appendix A

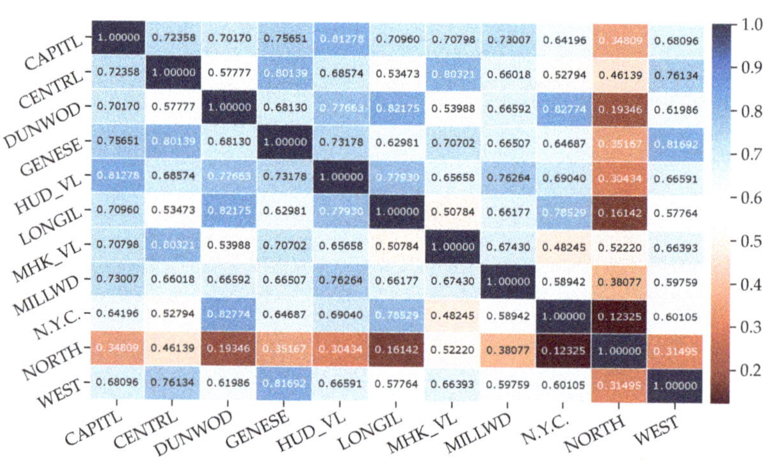

Figure A1. Kendall rank correlation coefficient heatmap on the New York Independent System Operator's dataset.

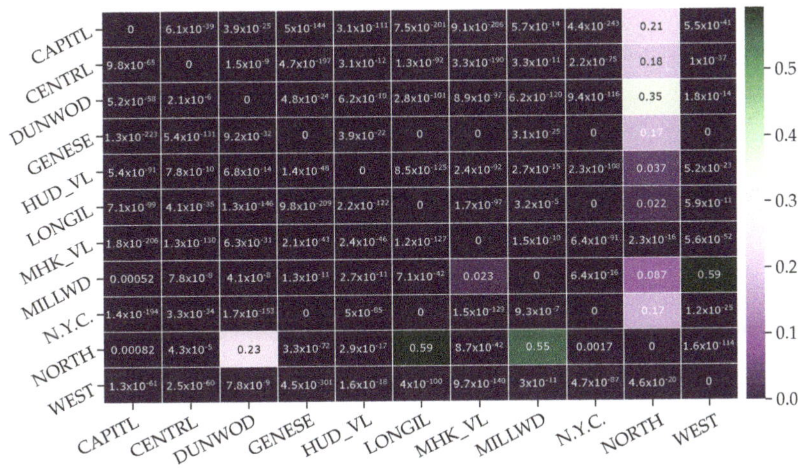

Figure A2. Granger causality heatmap on the New York Independent System Operator's dataset.

References

1. Nti, I.K.; Teimeh, M.; Nyarko-Boateng, O.; Adekoya, A.F. Electricity load forecasting: A systematic review. *J. Electr. Syst. Inf. Technol.* **2020**, *7*, 13. [CrossRef]
2. Kruse, J.; Schäfer, B.; Witthaut, D. Predictability of Power Grid Frequency. *IEEE Access* **2020**, *8*, 149435–149446. [CrossRef]
3. Sweeney, C.; Bessa, R.J.; Browell, J.; Pinson, P. The future of forecasting for renewable energy. *WIREs Energy Environ.* **2020**, *9*, e365. [CrossRef]
4. Klyuev, R.V.; Morgoev, I.D.; Morgoeva, A.D.; Gavrina, O.A.; Martyushev, N.V.; Efremenkov, E.A.; Mengxu, Q. Methods of Forecasting Electric Energy Consumption: A Literature Review. *Energies* **2022**, *15*, 8919. [CrossRef]
5. Sue Wing, I.; Rose, A.Z. Economic consequence analysis of electric power infrastructure disruptions: General equilibrium approaches. *Energy Econ.* **2020**, *89*, 104756. [CrossRef]
6. IBM Security. X-Force Threat Intelligence Index 2023. Available online: https://www.ibm.com/reports/threat-intelligence/ (accessed on 21 November 2023).
7. Li, Y.; Liu, Q. A comprehensive review study of cyber-attacks and cyber security; Emerging trends and recent developments. *Energy Rep.* **2021**, *7*, 8176–8186. [CrossRef]
8. Azure Network Security Team. 2022 in Review: DDoS Attack Trends and Insights. Microsoft. Available online: https://www.microsoft.com/en-us/security/blog/2023/02/21/2022-in-review-ddos-attack-trends-and-insights/ (accessed on 10 August 2023).
9. Gjesvik, L.; Szulecki, K. Interpreting cyber-energy-security events: Experts, social imaginaries, and policy discourses around the 2016 Ukraine blackout. *Eur. Secur.* **2023**, *32*, 104–124. [CrossRef]
10. Rodrigues, F.; Cardeira, C.; Calado, J.M.F.; Melicio, R. Short-Term Load Forecasting of Electricity Demand for the Residential Sector Based on Modelling Techniques: A Systematic Review. *Energies* **2023**, *16*, 4098. [CrossRef]
11. Wazirali, R.; Yaghoubi, E.; Abujazar, M.S.S.; Ahmad, R.; Vakili, A.H. State-of-the-art review on energy and load forecasting in microgrids using artificial neural networks, machine learning, and deep learning techniques. *Electr. Power Syst. Res.* **2023**, *225*, 109792. [CrossRef]
12. Jung, S.; Moon, J.; Park, S.; Rho, S.; Baik, S. W.; Hwang, E. Bagging Ensemble of Multilayer Perceptrons for Missing Electricity Consumption Data Imputation. *Sensors* **2020**, *20*, 1772. [CrossRef] [PubMed]
13. Rodenburg, F.J.; Sawada, Y.; Hayashi, N. Improving RNN Performance by Modelling Informative Missingness with Combined Indicators. *Appl. Sci.* **2019**, *9*, 1623. [CrossRef]
14. Myllyaho, L.; Raatikainen, M.; Männistö, T.; Nurminen, J.K.; Mikkonen, T. On misbehaviour and fault tolerance in machine learning systems. *J. Syst. Softw.* **2022**, *183*, 111096. [CrossRef]
15. Dehghani, M.; Yazdanparast, Z. From distributed machine to distributed deep learning: A comprehensive survey. *J. Big Data* **2023**, *10*, 158. [CrossRef]
16. Drainakis, G.; Pantazopoulos, P.; Katsaros, K.V.; Sourlas, V.; Amditis, A.; Kaklamani, D.I. From centralized to Federated Learning: Exploring performance and end-to-end resource consumption. *Comput. Netw.* **2023**, *225*, 109657. [CrossRef]
17. Aguilar Madrid, E.; Antonio, N. Short-Term Electricity Load Forecasting with Machine Learning. *Information* **2021**, *12*, 50. [CrossRef]
18. Jiang, W. Deep learning based short-term load forecasting incorporating calendar and weather information. *Internet Technol. Lett.* **2022**, *5*, e383. [CrossRef]
19. New York Independent System Operator. Load Data. Available online: https://www.nyiso.com/load-data/ (accessed on 18 July 2023).

20. Puth, M.-T.; Neuhäuser, M.; Ruxton, G.D. Effective use of Spearman's and Kendall's correlation coefficients for association between two measured traits. *Anim. Behav.* **2015**, *102*, 77–84. [CrossRef]
21. Makowski, D.; Ben-Shachar, M.S.; Patil, I.; Lüdecke, D. Methods and algorithms for correlation analysis in R. *J. Open Source Softw.* **2020**, *5*, 2306. [CrossRef]
22. Pandas 2.1.3. 2023. Available online: https://pandas.pydata.org (accessed on 18 November 2023).
23. Shojaie, A.; Fox, E.B. Granger Causality: A Review and Recent Advances. *Annu. Rev. Stat. Its Appl.* **2022**, *9*, 289–319. [CrossRef] [PubMed]
24. Statsmodels 0.14.0. 2023. Available online: https://www.statsmodels.org (accessed on 17 June 2023).
25. Kadhim, Z.S.; Abdullah, H.S.; Ghathwan, K.I. Artificial Neural Network Hyperparameters Optimization: A Survey. *Int. J. Online Biomed. Eng.* **2022**, *18*, 59–87. [CrossRef]
26. Wu, J.; Chen, X.-Y.; Zhang, H.; Xiong, L.-D.; Lei, H.; Deng, S.-H. Hyperparameter Optimization for Machine Learning Models Based on Bayesian Optimization. *J. Electron. Sci. Technol.* **2019**, *17*, 26–40.
27. Keras Tuner 1.4.6. 2023. Available online: https://github.com/keras-team/keras-tuner (accessed on 3 December 2023).
28. TensorFlow 2.13.1. 2023. Available online: https://www.tensorflow.org (accessed on 4 September 2023).
29. Shafieian, S.; Zulkernine, M. Multi-layer stacking ensemble learners for low footprint network intrusion detection. *Complex Intell. Syst.* **2023**, *9*, 3787–3799. [CrossRef]
30. Abumohsen, M.; Owda, A.Y.; Owda, M. Electrical Load Forecasting Using LSTM, GRU, and RNN Algorithms. *Energies* **2023**, *16*, 2283. [CrossRef]
31. Wan, R.; Mei, S.; Wang, J.; Liu, M.; Yang, F. Multivariate Temporal Convolutional Network: A Deep Neural Networks Approach for Multivariate Time Series Forecasting. *Electronics* **2019**, *8*, 876. [CrossRef]
32. Lara-Benítez, P.; Carranza-García, M.; Luna-Romera, J.M.; Riquelme, J.C. Temporal Convolutional Networks Applied to Energy-Related Time Series Forecasting. *Appl. Sci.* **2020**, *10*, 2322. [CrossRef]
33. Zhao, Z.; Xia, C.; Chi, L.; Chang, X.; Li, W.; Yang, T.; Zomaya, A.Y. Short-Term Load Forecasting Based on the Transformer Model. *Information* **2021**, *12*, 516. [CrossRef]
34. L'Heureux, A.; Grolinger, K.; Capretz, M.A.M. Transformer-Based Model for Electrical Load Forecasting. *Energies* **2022**, *15*, 4993. [CrossRef]
35. Stratigakos, A.; Andrianesis, P.; Michiorri, A.; Kariniotakis, G. Towards Resilient Energy Forecasting: A Robust Optimization Approach. *IEEE Trans. Smart Grid* **2024**, *15*, 874–885. [CrossRef]
36. Mienye, I.D.; Sun, Y. A Survey of Ensemble Learning: Concepts, Algorithms, Applications, and Prospects. *IEEE Access* **2022**, *10*, 99129–99149. [CrossRef]
37. Grotmol, G.; Furdal, E.H.; Dalal, N.; Ottesen, A.L.; Rørvik, E.-L.H.; Mølnå, M.; Sizov, G.; Gundersen, O.E. A robust and scalable stacked ensemble for day-ahead forecasting of distribution network losses. In Proceedings of the AAAI Conference on Artificial Intelligence, Washington, DC, USA, 7–14 February 2023; Volume 37, pp. 15503–15511.
38. Gupta, H.; Agarwal, P.; Gupta, K.; Baliarsingh, S.; Vyas, O.P.; Puliafito, A. FedGrid: A Secure Framework with Federated Learning for Energy Optimization in the Smart Grid. *Energies* **2023**, *16*, 8097. [CrossRef]
39. Shi, B.; Zhou, X.; Li, P.; Ma, W.; Pan, N. An IHPO-WNN-Based Federated Learning System for Area-Wide Power Load Forecasting Considering Data Security Protection. *Energies* **2023**, *16*, 6921. [CrossRef]
40. Shi, Y.; Xu, X. Deep Federated Adaptation: An Adaptative Residential Load Forecasting Approach with Federated Learning. *Sensors* **2022**, *22*, 3. [CrossRef] [PubMed]
41. eXtreme Gradient Boosting 2.0.2. 2023. Available online: https://github.com/dmlc/xgboost (accessed on 19 November 2023).
42. Chicco, D.; Warrens, M.J.; Jurman, G. The coefficient of determination R-squared is more informative than SMAPE, MAE, MAPE, MSE and RMSE in regression analysis evaluation. *PeerJ Comput. Sci.* **2021**, *7*, e623. [CrossRef] [PubMed]
43. Bin Kamilin, M.H.; Yamaguchi, S.; Bin Ahmadon, M.A. Radian Scaling: A Novel Approach to Preventing Concept Drift in Electricity Load Prediction. In Proceedings of the 2023 IEEE International Conference on Consumer Electronics-Asia (ICCE-Asia), Busan, Republic of Korea, 23–25 October 2023; pp. 1–4.

electronics

MDPI

Article

Reconfigurable CAN Intrusion Detection and Response System

Rachit Saini and Riadul Islam *

Department of Computer Science and Electrical Engineering, University of Maryland, Baltimore County, MD 21250, USA; r98@umbc.edu
* Correspondence: riaduli@umbc.edu

Abstract: The controller area network (CAN) remains the de facto standard for intra-vehicular communication. CAN enables reliable communication between various microcontrollers and vehicle devices without a central computer, which is essential for sustainable transportation systems. However, it poses some serious security threats due to the nature of communication. According to caranddriver.com, there were at least 150 automotive cybersecurity incidents in 2019, a 94% year-over-year increase since 2016, according to a report from Upstream Security. To safeguard vehicles from such attacks, securing CAN communication, which is the most relied-on in-vehicle network (IVN), should be configured with modifications. In this paper, we developed a configurable CAN communication protocol to secure CAN with a hardware prototype for rapidly prototyping attacks, intrusion detection systems, and response systems. We used a field programmable gate array (FPGA) to prototype CAN to improve reconfigurability. This project focuses on attack detection and response in the case of bus-off attacks. This paper introduces two main modules: the multiple generic errors module with the introduction of the error state machine (MGEESM) module and the bus-off attack detection (BOAD) module for a frame size of 111 bits (BOAD111), based on the CAN protocol presenting the introduction of form error, CRC error, and bit error. Our results show that, in the scenario with the transmit error counter (TEC) value 127 for switching between the error-passive state and bus-off state, the detection times for form error, CRC error, and bit error introduced in the MGEESM module are 3.610 ms, 3.550 ms, and 3.280 ms, respectively, with the introduction of error in consecutive frames. The detection time for BOAD111 module in the same scenario is 3.247 ms.

Keywords: controller area network (CAN); bus-off attack; CAN attack detection; CAN attack response

Citation: Saini, R.; Islam, R. Reconfigurable CAN Intrusion Detection and Response System. *Electronics* **2024**, *13*, 2672. https://doi.org/10.3390/electronics13132672

Academic Editors: Dariusz Rzońca and Tomasz Rak

Received: 15 June 2024
Revised: 3 July 2024
Accepted: 4 July 2024
Published: 7 July 2024

1. Introduction

Intelligent connected vehicles (ICVs) are currently in a phase of rapid advancement, with intelligence and connectivity being the prevailing trends. A recent study indicates that over 86% of vehicles by the year 2023 will be outfitted with network control systems [1–4], offering a broader selection of advanced features [5], including vehicle management and adaptive cruise control, as depicted in Figure 1. This figure represents the CAN layout in cars with the CAN bus for linear and star topology connecting various electronic control units (ECUs) through CAN nodes to the CAN bus. The transmission control, adaptive cruise control, and comfort control CAN modules are connected to the CAN bus with linear topology, and rear control and safety control CAN modules are connected to the CAN bus with star topology, where various ECUs are connected to CAN modules as control units.

CAN enables reliable communication between microcontrollers and vehicle devices without a central computer. This efficiency is crucial for electric vehicles (EVs) and hybrid vehicles, where precise control over battery management systems, motor controllers, and other subsystems is essential for optimal performance and energy efficiency and is key to sustainable transportation systems. By allowing multiple microcontrollers to communicate over a single or dual-wire network, CAN reduces the need for complex wiring harnesses. This not only reduces the weight of the vehicle, leading to improved fuel efficiency and reduced emissions, but also lowers production costs and the environmental impact of manufacturing. Moreover, in electric and hybrid vehicles, CAN networks integrate renewable energy sources, such as solar panels, with the vehicle's energy system. This integration is a crucial aspect of making transportation more sustainable.

Moreover, automobiles establish links with diverse external networks, such as vehicle-to-vehicle (V2V) and vehicle-to-infrastructure (V2I) communication networks, as depicted in Figure 2. This figure exemplifies the communication network consisting of vehicles, cellular base stations, an internet unit, and a roadside unit. This shift turns present-day vehicles into interconnected systems rather than operating in isolation. The more sophisticated the system is and the more connected the vehicle is, the more exposed it is to attacks as mentioned in the Detroit Free Press [6]. To meet the requirements for interfacing with the external networks, the number of ECUs within cars is steadily increasing. Consequently, the complexity of IVNs is also on the rise [5,7,8].

Considering factors such as data volume, response time, reliability, application needs, and other system criteria, there are five frequently employed IVNs: the local interconnect network (LIN), CAN, FlexRay, media-oriented system transport (MOST), and Ethernet. Among these, the CAN protocol is the most widely used, primarily due to its cost-efficiency, reliable performance, and fault tolerance [9].

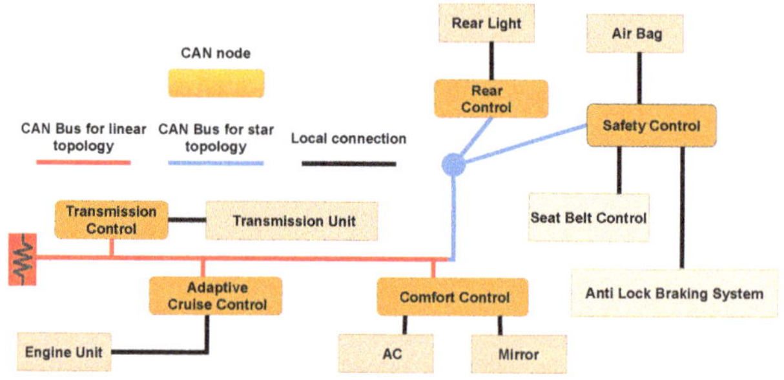

Figure 1. The layout of the CAN network used for ECU communication in cars connects various units within the vehicle. The linear and star topologies for the CAN network are widely used, connecting regular and safety-critical nodes together.

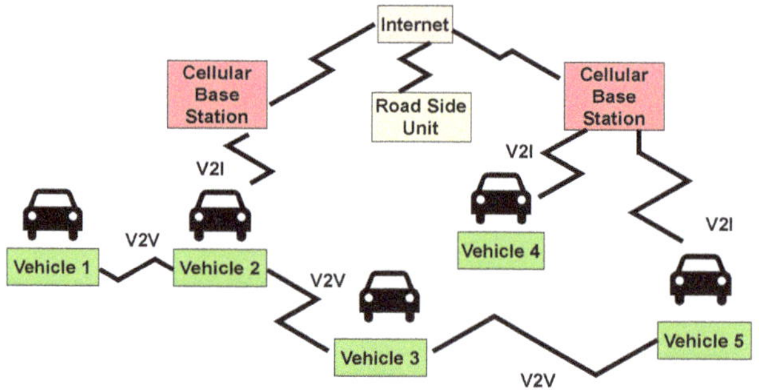

Figure 2. The communication between vehicles and external infrastructure denoted by V2V and V2I links connecting cars to each other and roadside units for sharing information.

The CAN communication mentioned above utilizes a bus topology known as the CAN bus to facilitate communication among ECUs, which was originally developed by Bosch for vehicle communication networks. This system allows ECUs to connect without relying on a central host computer. The CAN system enables real-time control by enabling direct message exchange between any pair of nodes and is known for its robust error tolerance [10,11].

Nevertheless, the advantages resulting from improved connectivity and added functionalities do expose evident security weaknesses, including potential threats such as suspension attacks, flooding attacks, spoofing attacks, replay attacks, fuzzing attacks, and masquerade attacks, as outlined in references [5,11–14].

One of the strategies discussed to counter CAN attacks is the employment of an intrusion detection system (IDS) [13,15]. IVN IDSs are introduced with multiple goals in mind concerning the security of automotive systems. These include the ability to swiftly identify abnormal intrusions (from the adversary or malicious user), furnish accurate reference data for intrusion prevention systems (IPSs), and the capability to prevent further damage resulting from IVN attacks. Early alerts provided by IVN IDS can help mitigate risks posed by malicious adversaries, making it especially suitable for IVN environments with constrained computing and bandwidth resources, as referenced in [16–18].

This paper employs a hierarchical approach to building, emulating, and implementing modules for prototyping IDS for CAN structure. For this purpose, the Xilinx Vivado tool is used along with the Nexys A7 board while using Verilog hardware description language (HDL). Here, we calculated the time it takes for the compromised module to enter a 'bus-off' state and recover from it, and we presented it in a graphical format.

Under conditions where errors are introduced in every consecutive frame and every alternate frame, these cases are generated considering the transmit error counter (TEC) value for error state transition between the error-passive state and bus-off state switching between 255 and 127.

The main contributions of this paper are as follows :

- Create a real scenario environment for an embedded system showcasing a bus-off assault on the CAN accompanied by a method for detecting such an attack.
- Devise a safeguarding mechanism for CAN communication with a response system designed to counteract potential intrusions.
- Explore different configurations of CAN communication protocol error states on reconfigurable platforms forming part of intrusion detection and intrusion response systems.
- Introduce a reconfigurable CAN protocol based on a field programmable gate array (FPGA).

The rest of the paper is organized as follows: Section 2 provides background information, Section 3 presents the proposed methodology, and Section 4 provides the experimental setup and results. Finally, Section 5 summarizes the contributions of this work.

2. Background

In this chapter, we first provide an overview of the concept of CAN. Then, we discuss the characteristics and vulnerabilities of IVNs. Additionally, we review the associated attacks. Then, we discuss the constraints of IVN IDSs. Next, we present countermeasures such as IDSs to detect the vulnerabilities. Finally, we discuss the advantages of implementing CAN using the FPGA.

2.1. CAN Preliminaries

The CAN operates as a broadcast-message communication protocol, utilizing bitwise arbitration for contention resolution on the CAN bus. In cases of simultaneous frame transmission by different nodes, the node with the highest priority continues, while the other nodes retry later [19].

The CAN frame includes data, remote, error, and overload frames. A data frame provides data transmission (can be a standard data CAN frame or extended data CAN frame), a remote frame requests data, an error frame signals an error, and an overload frame delays the following message until the current one is processed [20].

A standard data CAN frame composition consists of the following components: start-of-frame (SOF-1 bit), identifier (11 bits), remote transmission request (RTR-1 bit), control field (6 bits), data field (ranging from 0 to 8 bytes), cyclic redundancy check (CRC) field along with CRC delimiter (16 bits), acknowledge (ACK) field along with ACK delimiter (2 bits), end-of-frame (EOF-7 bits), and inter-frame space (3 bits) [21], as shown in Figure 3a. The extended data CAN frame employs 29 bits for identifier arbitration, which includes an identifier field (11 bits) and an extended identifier field (18 bits). Furthermore, the extended data CAN frame also has substitute remote request (SRR-1 bit) and identifier extension (IDE-1 bit), which differentiates standard data CAN frames from extended data CAN frames, and RTR (1 bit) after the extended identifier field [22], as shown in Figure 3b. The remote frame closely resembles the extended data CAN frame but lacks the data field, as shown in Figure 3c. Figure 3 illustrates these three frame types, in addition to the error and overload frames. The error frame consists of the following fields: error flag (6 bits), error echo flag (6 bits), and error delimiter (8 bits), as shown in Figure 3d. Five types of errors can be generated within the CAN frame. These include acknowledge (ACK) error, bit error, CRC error, form error, and stuff error. This paper focuses on the generation and detection of bit error, CRC error, and form error to formulate an attack on the CAN frames. Moreover, bit stuffing is also taken into account in certain cases. The overload frame encompasses the following fields: overload flag (6 bits) and overload delimiter (8 bits), as shown in Figure 3e.

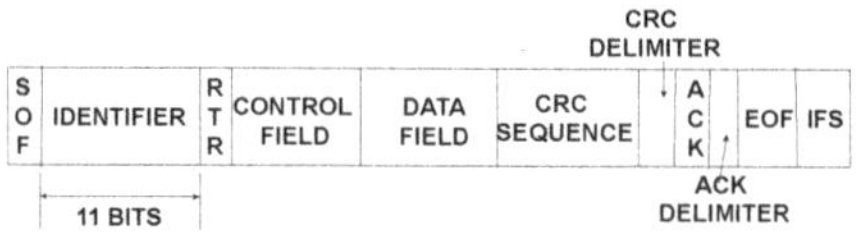

(**a**) Standard data frame.

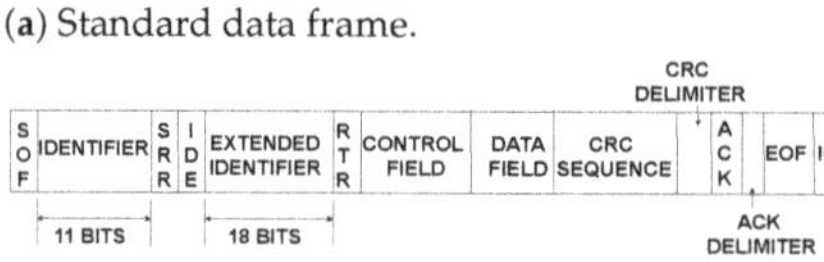

(**b**) Extended data frame.

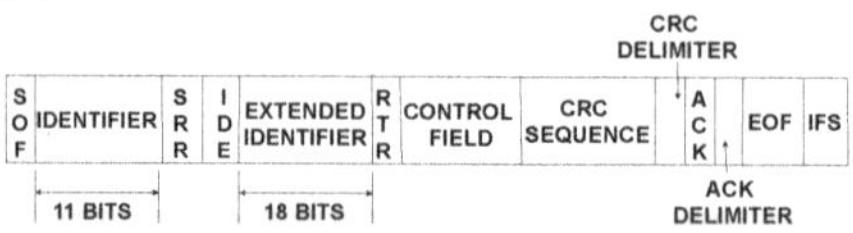

(**c**) Remote frame.

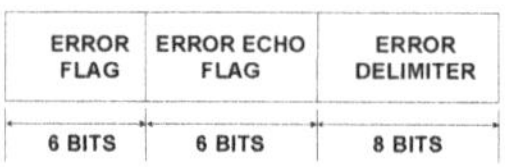

(**d**) Error frame.

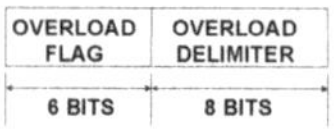

(**e**) Overload frame.

Figure 3. Different frames integral to the CAN protocol, facilitating communication among multiple CAN nodes. (**a**) The standard data frame with size varying (i.e., 0 to 8 bytes) from 47 bits to 111 bits, (**b**) the extended data frame with size varying from 67 bits to 131 bits, (**c**) the remote frame with frame size of 67 bits, (**d**) the error frame with frame size 20 bits, and (**e**) the overload frame with frame size of 14 bits.

The CAN frame handles up to 8 bytes of data [23], featuring collision detection, error detection, signaling, and fault confinement. The CAN protocol employs static, fixed priority non-preemptive scheduling and accommodates periodic, sporadic, or aperiodic messages [24].

2.2. Characteristics and Vulnerabilities of CAN IVNs

2.2.1. IVN Characteristics

The automotive electronic system functions as a diverse distributed real-time system, with multiple ECUs connected through an IVN that communicates via a central gateway. The IVN is characterized by a heterogeneous distributed real-time system environment, numerous external interfaces, a multi-function safety-critical level system, and a lack of cybersecurity design [16].

2.2.2. Vulnerabilities in CAN-Based IVNs

The CAN bus lacks fundamental security mechanisms in its protocol, leaving vehicles susceptible to malicious adversaries. Six vulnerabilities exist according to the confidentiality, integrity, availability (CIA) security model. These vulnerabilities involve the lack of encryption, authentication, and integrity-checking in CAN bus traffic. Additionally, protocol characteristics such as broadcast transmission, priority-based arbitration, and limited bandwidth introduce vulnerabilities [25]. These vulnerabilities expose IVNs to various attacks, as elaborated in the following section.

2.3. Types of CAN Attacks

The six categories of CAN attack scenarios can be described as follows:

Suspension Attack: To mount a suspension attack, the adversary needs only one weakly compromised ECU. As one type of denial-of-service (DoS) attack, the objective of this attack is to suspend the weakly compromised ECU's message transmissions, thus preventing the delivery of information it acquired to other ECUs [12].

Flooding Attack: In this attack scenario, an adversary seeks to initiate a DoS attack by inundating the network with a high volume of CAN packets, often with high priority (e.g., CAN ID of 0×000) [26].

Spoofing Attack: To disrupt specific vehicle functions (such as gear control or RPM), an adversary injects control packets based on prior knowledge of the target vehicle [27].

Replay Attack: An adversary records regular CAN bus traffic and subsequently replays it onto the CAN bus [28].

Fuzzing Attack: In a fuzzing attack, the adversary generates CAN packets randomly. This attack can lead to unexpected and erratic behavior in the targeted vehicle [5].

Masquerade Attack: In this scenario, a normal ECU's transmission is halted, allowing a compromised ECU to assume the role of the original ECU by mimicking its CAN IDs and transmission patterns [29].

Out of the six categories of CAN attack scenarios described above, this paper focuses on the detection of a suspension attack to emulate a bus-off condition.

2.4. Constraints of CAN IVN IDS

Constraints in the context of IDSs for IVNs encompass limitations related to hardware, cost, detection accuracy, response time, and standardized construction [13].

2.5. Categories of IVN IDSs

The IVN IDS for CAN can be categorized into three techniques: statistical-based, machine learning-based, and neural network-based.

2.5.1. Statistical-Based IDS for IVN

The IDS, which relies on statistical analysis, assesses message sequences statistically. This approach involves comparing two sets of messages using statistical metrics like cosine similarity, Pearson correlation, and the chi-squared test [30,31]. Suppose there is a notable alteration in message frequencies or sequences indicated by metric values surpassing specified thresholds. In that case, the system predicts the occurrence of intrusions in the subsequent message interval [32]. Another aspect of the statistical analysis for intrusion detection involves assessing message entropy [33,34].

2.5.2. Machine Learning-Based IDS for IVN

In machine learning, three main models are generally employed for prediction: the regression model, the classification model, and the clustering model. The classification-based or clustering-based models find application in real-time intrusion detection scenario prediction [14,35]. Specifically, the classification-based model is suitable for supervised problems, while the clustering-based model is more relevant for unsupervised problems [36].

Supervised machine learning models can be further divided into single classifiers and ensemble learning models. Decision trees (DT) and the k-nearest neighbor (KNN) algorithm serve as examples of single classifiers, while random forest (RF) and extreme gradient boosting (XGBoost) are chosen for ensemble learning models. In the context of semi-supervised learning methods, robust covariance (RC), local outlier factor (LOF), and isolation forest (IF) are selected as baselines [37].

Another study outlined in [38] employed unsupervised learning, a method that operates without the need for labeled data. This unsupervised approach adopted a two-stage process involving deep learning and a probabilistic model.

2.5.3. Neural Network-Based IDS for IVN

Deep and machine learning algorithms have made significant progress and been proven highly effective in anomaly detection [39], demonstrating excellent performance [40]. The neural networks employed for this purpose encompass a range of architectures, including convolutional neural networks (CNNs), long short-term memory (LSTM) neural networks, and advanced models such as the residual neural network (ResNet) and leCun network (LeNet) based on deep transfer learning, as proposed by Mehedi et al. [40]. These models are considered baseline models in the context of anomaly detection [41].

Deep transfer learning (DTL) addresses issues such as limited data availability and the prevalence of application-specific intrusion detection system (IDS) models. The concept revolves around integrating knowledge from a pre-trained source model into a target model. Through this process, DTL facilitates more efficient information amalgamation, potentially yielding superior outcomes compared to training models anew [42]. However, due to a lack of computational power in FGPA, these efforts are limited to GPU-based implementations.

2.6. Advantages of Implementing CAN Protocol on Reconfigurable Computing Platform

FPGAs are highly prized for their ample resources and adaptability as specialized integrated circuits. They play a crucial role in digital electronic design and offer three main benefits [43]. Firstly, FPGA vendors provide robust and user-friendly electronic design tools (EDA), extensive documentation, and personalized support to assist with design and verification. Secondly, unlike application-specific integrated circuits (ASICs) [44], the manufacturing costs for demonstration examples are low [45]. Thirdly, modifications can be implemented at any stage of the design process, thanks to advanced systems that enable dynamic hardware reconfiguration [46,47].

In aerospace and military/aviation critical systems, where programming errors are intolerable, FPGAs' early-stage design verification feature becomes indispensable. FPGA verification encompasses various processes, such as coding rule checks, manual walkthroughs, functional and timing simulations, static timing analysis, cross-clock domain checks, and logical equivalence checks. Functional simulation, in particular, holds significant importance in ensuring design reliability, a critical consideration given the exponential growth of verification cases with increasing design scale [48]. Implementation of CAN protocol on FPGA allows researchers to prototype different IDS quickly and allows adaptability with varying CAN speeds.

3. Proposed Methodology

3.1. CAN Architecture

Figure 4 shows the basic architecture of the CAN module interacting with the ECU on one end and the CAN transceiver connected to the CAN bus on the other end. The CAN module comprises a transmission buffer unit (TX buffer unit) and a reception buffer unit (RX buffer unit). The data are fed into the transmission buffer unit from the ECU with a frame size of 111 bits (for standard frame size) and are received from the reception buffer unit with a frame size of 111 bits (for standard frame size) into the ECU. In addition, there is a transmitting unit (TX Unit), a reception unit (RX Unit), and an error detection unit. The clock unit maintains synchronization by connecting to transmission, reception, and error detection units along with the TX buffer unit and RX buffer unit. The data are transmitted between various units within the CAN module one bit at a time with respect to the clock signal. The flow of data is from the RX unit to the RX buffer unit. For data flow on the transmission side, there is a contention between data from the TX buffer unit and error frame based on the error generation signal from the error detection unit. The data are passed onto the TX unit. The data flows between the TX unit and the CAN transceiver and also between the CAN transceiver and RX Unit. On the other side of the CAN transceiver is the CAN bus.

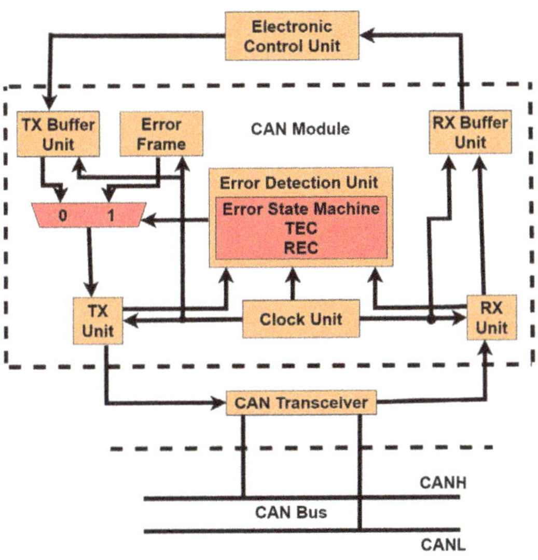

Figure 4. The essential components of the CAN architecture show the interaction of the CAN module with the ECU on one end and with the CAN bus through the CAN transceiver on the other end.

3.2. Communication among CAN Nodes over CAN Bus

The communication network of the CAN modules over the CAN bus is shown in Figure 5. Here, we present N nodes with one compromised node (the adversary has access to CAN bus through this node) and $N-1$ normal nodes. The identifier values highlighted in different colors indicate which identifiers among different nodes will be considered at a respective time stamp for arbitration, as can be seen in Figure 5. Node 1, which the adversary compromises, has the lowest arbitration ID values at all the time stamps. So, communication is dominated by the data from this node.

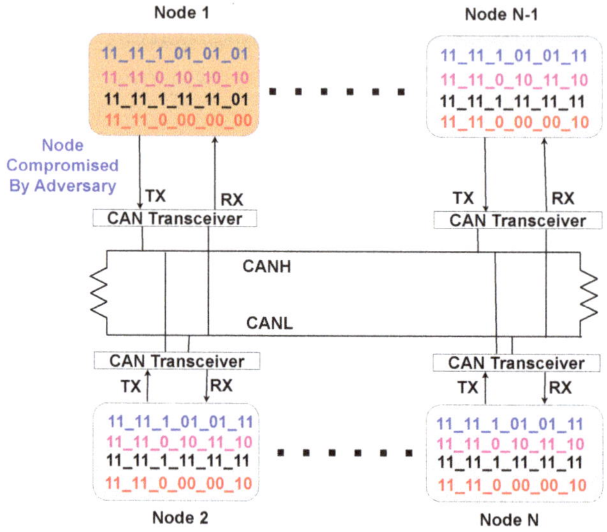

Figure 5. The CAN communication network comprises N CAN modules interacting over the CAN bus. Here, node one is compromised by the malicious adversary for communication with other nodes. The arbitration IDs to be considered at each time stamp are color-coded. The IDs used by the compromised node have a lower value at all time stamps, indicating that this node will win the arbitration every time and put its content on the CAN bus, which can lead to a bus-off attack through this compromised node.

3.3. Proposed Intrusion Detection and Intrusion Response Systems

We utilized the concept of a bus-off state, which is associated with a scenario when a node fails to transmit data frames and the associated error counter reaches a specified value. In order to detect a bus-off attack, the CAN module needs to enter the bus-off state. Furthermore, the CAN module also comes out of the bus-off state after the transmission of a specific number of recessive bits. The detection time is the time for the CAN module to enter the bus-off state. The response time is the time for the CAN module to come out of the bus-off state.

The transition of the CAN node from the error-passive state into the bus-off state and back into the error-active state is represented in two error state diagrams based on the values of the transmit error counter (TEC) and the receive error counter (REC) [21]. Figure 6 illustrates respective error state diagrams. In Figure 6a, a TEC value of 127 facilitates the transition from the error-active state to the error-passive state. A TEC value of 255 is required to shift from the error-passive state to the bus-off state. The transition from bus-off to error-active states involves the transmission of 128 $\times$ 11 recessive bits. Similarly, in Figure 6b, the TEC value for moving from error-active to error-passive states is 63, while transitioning from error-passive to bus-off states requires a TEC value of 127. The shift from bus-off to error-active states involves the transmission of 64 $\times$ 11 recessive bits. Hence, using two error state diagrams for the threat models signifies the reconfigurability of the CAN prototype on the FPGA.

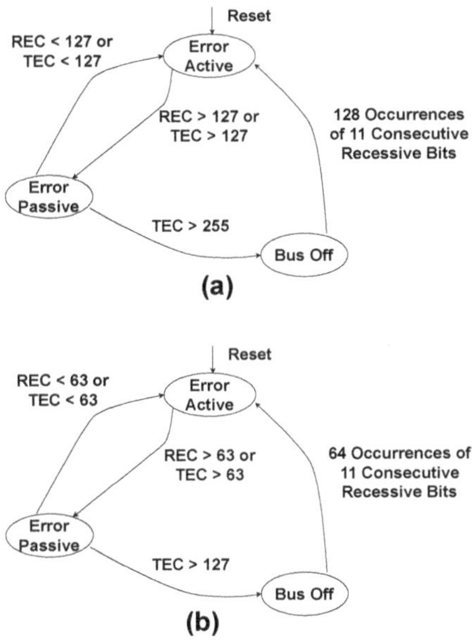

Figure 6. The error state diagrams for a CAN depict the various states that the network can enter due to communication errors. These state diagrams illustrate how the CAN protocol responds to errors by entering specific error states and implementing error recovery mechanisms. (**a**) Error state diagram with error state transitions based on TEC values of 127 and 255. (**b**) Error state diagram with error state transitions based on TEC values of 63 and 127.

Setting the TEC value at 255 as the threshold for transitioning from error-passive to bus-off in the CAN protocol aims to establish a distinct separation between these error states. This choice signifies a severe and persistent communication issue triggered after detecting a significant number of errors. The 8-bit TEC counter ranges from 0 to 255, and the transition to bus-off occurs when TEC reaches the maximum value, providing a clear signal of persistent communication problems.

While a TEC value of 127 allows configurability, values lower than 127 are avoided to prevent frequent entries into the bus-off state. This precaution guards against heightened sensitivity to transient errors, maintaining a balance between error sensitivity and system robustness. Lowering the threshold too much could prompt quicker error responses but might also increase the likelihood of nodes being excluded due to false positives or transient issues.

In summary, the entry of the CAN module into the bus-off state is represented as intrusion detection, for which detection time is computed. Furthermore, exiting the CAN module from the bus-off state is represented as an intrusion response for which response time is calculated.

3.4. Threat Model for Individual CAN Nodes Interacting over CAN Bus

The threat model is shown in Figure 7. This threat model in the research is consistent with the existing literature, as mentioned in [49]. The assumption here is that the adversary can eavesdrop on the TX signal coming out of the CAN module and going into the CAN transceiver from that CAN module. Due to the adversary's access to the TX signal, the adversary manipulates the logic value placed on the line going into the CAN transceivers.

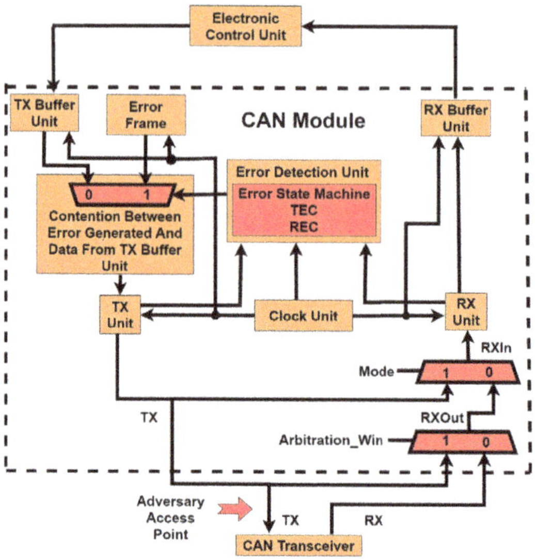

Figure 7. Threat model showing adversary taking charge of the CAN bus in communication of individual CAN nodes over CAN bus.

The threat model is built on the basic architecture shown in Figure 4. In this threat model, data transmission and reception happens one bit at a time with respect to the clock signal. However, transmission from and reception to the ECU from the CAN module is considered for a 111 bit frame size (standard frame size). The TX signal outputted with the adversary access point is sent to the CAN transceiver from the TX Unit. This signal and the RX signal from the CAN transceiver are input to a multiplexer within the CAN module with $Arbitration\_Win$ as the select signal. The output of this multiplexer, $RXOut$, is put as input to the second multiplexer, which has its second input coming from the TX line of the CAN module, and $RXIn$ is its output with $Mode$ as a select signal sent to the RX Unit. The error is generated based on comparing TX and $RXIn$ signals within the error detection unit. There is contention between data from the TX buffer unit and error frame with respect to the error signal generated from the error detection unit. The contented data are put onto the TX unit, from where the data are sent as input to the CAN transceiver and multiplexer with $Arbitration\_Win$ as the select signal.

The types of errors introduced include form, CRC, and bit errors. When the bus-off attack comes into the picture through multiple occurrences of any of the errors, the communication on the CAN bus is stopped. However, an inner transition from TX to $RXIn$ still occurs (based on the value of the $Mode$ signal). The communication happens bit by bit in each clock cycle. The transfer of 11×128 recessive bits in one case and 11×64 recessive bits in the second case puts the node back into the network for communication (transmission from bus-off state to error-active state) on the CAN bus for transmission and reception.

3.5. Threat Model for Interaction of Multiple CAN Nodes

Figure 8 shows communication among N nodes over the CAN bus presented in Figure 5. In this threat model, data transmission and reception occurs one bit at a time with respect to the clock signal in all the respective CAN modules, with the transmission from and reception to ECU from the CAN modules happening for a frame size of 111 bits (standard frame size considered). When multiple CAN nodes interact with the CAN bus, each CAN module outputs the signal from the TX Unit to the CAN transceiver. The adversary has access to the TX line of CAN module 1, on which it injects an inverted signal with respect to the CAN signal from the respective CAN module at a specific time stamp. The compromised output is sent to the arbitration process unit, which also has signals from all other CAN modules ($N - 1$ modules). Based on the arbitration process, after the contention between the signals from all the CAN modules, one signal wins the arbitration, and that signal is broadcast to all CAN transceivers. The RX signals from all the CAN transceivers are sent to CAN modules. The error frame is generated based on the comparison between received and transmitted signals within the CAN modules. Upon generation of enough error frames, the $NV(NodeVictimized)$ signal is set, and it puts that particular node in the bus-off state in which the recessive bit is passed through from the reception line into the RX Unit of the respective CAN module.

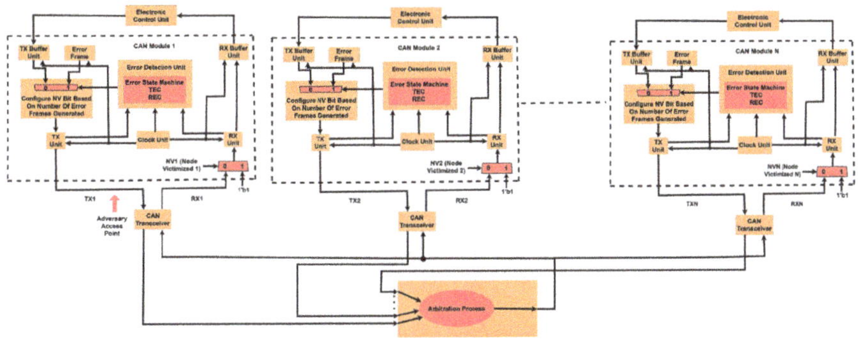

Figure 8. Threat model showing adversary gaining access to a CAN node interacting with multiple nodes in communication over the CAN bus.

Algorithm 1 explains Figure 8. The input in Line 1 of the algorithm consists of N standard CAN messages. The output of the algorithm is the data being fed into the RX units. In Line 3, the *TEC* values and *node victimized (NV)* values are set to zero for all N CAN modules. Iterating over N CAN modules in lines 4–9, the *TX* signal is fed in the signal from the respective standard CAN messages in Line 5. Next, in Line 6, the *TXCompromised* signal is fed with the *TX* value for all modules except the compromised one. For the compromised module, there is a bit flip with respect to the *TX* signal at a specific position in the standard CAN message that is being provided to the *TXCompromised* signal. Finally, the *TXCompromised* signals are placed onto the respective *TXCANTransceivers* for all N modules in Line 7. In Line 8, the $RX_i^{'}$ signal is given the result of the $arbitrationpriority()$ function, where the signals from all CAN transceivers contest for the bus and only one signal with the highest arbitration priority is selected as output. In Lines 10 to 15, the "For loop" iterates over all N modules and checks for if the condition does not match the $RX_i^{'}$ signal to the $TXCompromised_i$ signal in Line 11. Based on this if statement, the NV_i signal is assigned a value of zero in Line 12. The $RXUnit_i$ is fed with either a recessive bit stream (RBS) or an $RX_i^{'}$ based on the NV_i signal using the $fetchdata()$ function, as shown in Line 14. The RBS consists of a stream of logic one values. The next "For loop", in Lines 16 to 33, again iterates over all N modules, conditionally updating TEC values using the $errorgeneration()$ function. The condition on $TEC_i^{'}$ being greater than 255 sets the $NV_i^{'}$ value to one. Based on the $NV_i^{'}$, the RXUnit$_i$ is fed with either a RBS or a $RX_i^{'}$ using the $fetchdata()$ function in Line 30.

Algorithm 1 shows the transfer of a node to a bus-off state for the TEC value exceeding a value of 255. The value for TEC chosen for transition between states is large enough to separate the attack from a malfunction in terms of false positives.

Algorithm 1 The bus-off attack detection and response algorithm.

Input : Standard CAN messages
Output : Data into RXUnits
Initialize : $TEC_i \longleftarrow 0$ and $NV_i \longleftarrow 0$ for i from 1 to N ▷ Transmit error counter$_i$ (TEC$_i$), node victimized$_i$ (NV$_i$)
for $i \longleftarrow 1$ **to** N **do**
  TX$_i$ ⟵ transmitframe(Standard CAN message$_i$) ▷ Transmitting Standard CAN message
  TXCompromised$_i$ ⟵ adversarysccess(TX$_i$) ▷ Transmitting TX signal with compromised value at a specific position within the message frame for a specific module and without a compromised value for rest of the modules.
  TXCANTransceiver$_i$ ⟵ TXCompromised$_i$ ▷ Value assigned to CAN transceiver from TX signal
  $RX_{\mathrm{i}}^{'} \longleftarrow$ arbitrationpriority(TXCANTransceiver$_i$) ▷ Result of arbitration process moved into RX signal
end for
for $i \longleftarrow 1$ **to** N **do**
  if $RX_{\mathrm{i}}^{'}\ != TXCompromised_i$ **then**
    NV$_i$ ⟵ 0
  end if
  RXUnit$_i$ ⟵ fetchdata(NV_i, RBS, $RX_{\mathrm{i}}^{'}$) ▷ Putting data into CAN RXUnit$_i$ based on NV$_i$ from either recessive bit stream (RBS) or $RX_{\mathrm{i}}^{'}$
end for
for $i \longleftarrow 1$ **to** N **do**
  if $RX_{\mathrm{i}}^{'} == TXCompromised_i$ **then**
    while $TEC_{\mathrm{i}}^{'} <= 255$ **do**
      Error$_{\mathrm{i}}^{'}$ ⟵ errorgeneration(TX_i, $RX_{\mathrm{i}}^{'}$)
      if $Error_{\mathrm{i}}^{'} == 1$ **then**
        $TEC_{\mathrm{i}}^{'} \longleftarrow TEC_{\mathrm{i}}^{'} + 8$
      else
        $TEC_{\mathrm{i}}^{'} \longleftarrow TEC_{\mathrm{i}}^{'} - 1$
      end if
      if $TEC_{\mathrm{i}}^{'} > 255$ **then**
        $NV_{\mathrm{i}}^{'} \longleftarrow 1$
      else
        $NV_{\mathrm{i}}^{'} \longleftarrow 0$
      end if
      RXUnit$_i$ ⟵ fetchdata($NV_{\mathrm{i}}^{'}$, RBS, $RX_{\mathrm{i}}^{'}$) ▷ Putting data into CAN RXUnit$_i$ based on $NV_{\mathrm{i}}^{'}$ from either RBS or $RX_{\mathrm{i}}^{'}$
    end while
  end if
end for

4. Experimental Results

4.1. Experimental Setup

The Xilinx Vivado tool is used for coding in Verilog and seeing the simulation results for the modules created to emulate the behavior of CAN. The implementation of CAN functionality is observed on the NEXYS A7 Digilent board, which is coded using the Xilinx Vivado tool and passes through synthesis, implementation, and bitstream generation phases before programming the board through the hardware manager. The hardware setup used in this project is shown in Figure 9. The figure shows the interaction between Arduino and the CAN shield and FPGA, in which CAN logic is prototyped. The clock period used for the simulation of modules is 1 microsecond (to match the 1 Mbps speed of CAN protocol).

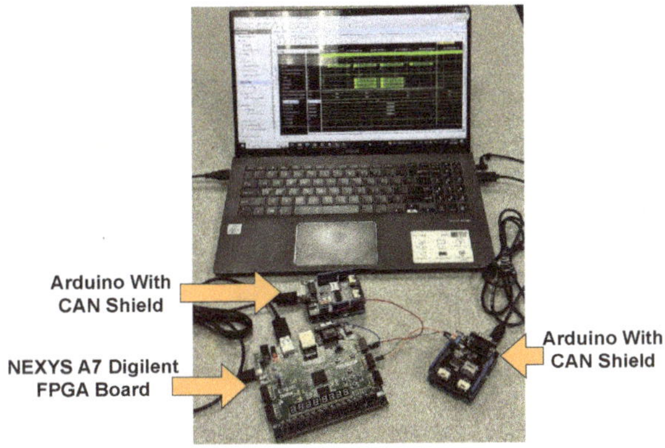

Figure 9. Our hardware setup for emulating CAN controller logic on FPGA and its interaction with other CAN modules over the CAN bus.

Figure 10 illustrates an example of simulation results for the MGEESM module with the transmission of form error in every alternate standard frame with a length of 111 bits by showing the transition between the error-passive state and the bus-off state as the transmit error counter exceeds value 255. Similarly, Figure 11 illustrates an example of simulation results for the MGEESM module with the transmission of form error in every alternate standard frame with a length of 111 bits by showing the transition between the bus-off state and the error-active state after transmission of 128 occurrences of 11 consecutive recessive bits. The form error occurs at a position of 20 bits after the end of the data field within the standard frame.

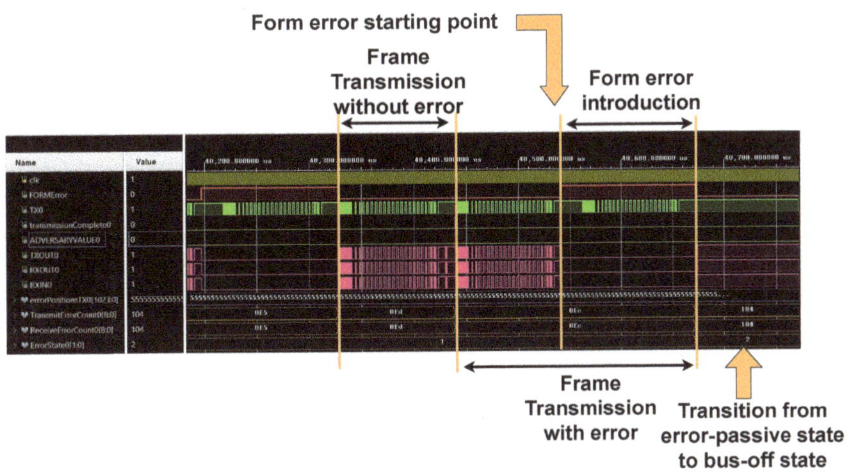

Figure 10. The simulation waveform for the MGEESM module shows the transition of the CAN node from the error-passive state (denoted by value 1) to the bus-off state (denoted by value 2) with the introduction of form error in alternate transmission frames. The form error occurs 20 bits after the end of the data field within the standard frame.

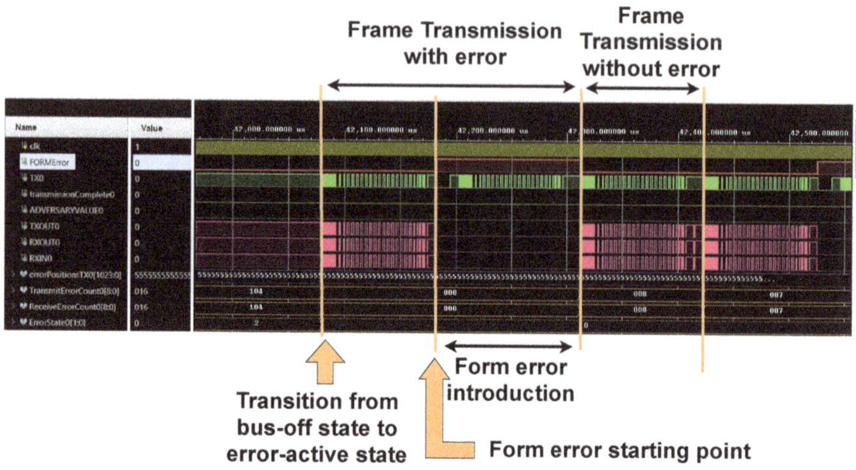

Figure 11. The simulation waveform for the MGEESM module shows the transition of the CAN node from the bus-off state (denoted by value 2) to the error-active state (denoted by value 0) with the introduction of form error in alternate transmission frames. The form error occurs 20 bits after the end of the data field within the standard frame.

We designed several modules to implement a configurable CAN protocol and attack detection and response system, which are listed in Table 1. The TX module is the primary transmission module with options to output a standard frame, extended frame, or remote frame, with the length of the message varying from zero to eight bytes. The RX module is the basic reception module for receiving frames of standard, extended, or remote types as input and storing them in receiver buffers based on specific criteria. The GE module stands for the generic error module, which introduces the form error, CRC error, and bit error within the single frame outputted from the CAN node. The GE module is built on top of the TX and RX modules.

Next, we have the MGE module. The MGE module stands for the multiple generic errors module presenting form errors, CRC errors, and bit errors in various frames within a single CAN node. This module is built hierarchically on top of the GE module. The MGEESM module represents multiple generic errors, including multiple form errors, multiple CRC errors, and multiple bit errors, respectively, with the introduction of the error state machine. This module is put in place based on the MGE module.

The next set of modules indicates the communication between multiple nodes over the CAN bus, as shown in Figure 8. For error-free processing of the CAN protocol in interactions among a network of multiple CAN nodes, MCIWOERROR111 refers to multiple CAN nodes' interaction without error for a frame size of 111 bits.

Conversely, the multiple CAN nodes interaction with error introduction for a frame size of 111 bits (MCIWITHERROR111 module) is employed to introduce errors in communication among multiple nodes over the CAN bus. Lastly, the bus-off attack detection module for a frame size of 111 bits (BOAD111 module) incorporates an error state machine to represent error generation within a CAN node communicating among multiple CAN nodes over the CAN bus.

Table 1. The required building blocks and their descriptions for implementing configurable CAN protocol, attack/error detection, and response systems.

Modules	Description
TX	Basic CAN transmission module.
RX	Basic CAN reception module.
GE	A module that introduces form error, CRC error, and bit error in a single frame within a single CAN node built on the combination of transmission and reception modules.
MGE	A module that presents form error, CRC error, and bit error in multiple frames within a single CAN node built on top of the GE module.
MGEESM	A module that introduces form error, CRC error, and bit error in multiple frames and introduces an error state machine within a single CAN node built based on the MGE module.
MCIWOERROR111	A module that interacts with multiple nodes without error introduction for a frame size of 111 bits.
MCIWITHERROR111	A module that interacts with multiple nodes and considers error introduction for a frame size of 111 bits.
BOAD111	A module that interacts with multiple nodes and considers the introduction of errors and error state machine for a frame size of 111 bits.

4.2. Results

We define the attack/error detection time as the time for the victim node to enter the bus-off state. The response time is the time for the victim node to come out of the bus-off state. Both detection and response times are measured for the victim node in two scenarios. In both scenarios, four sub-cases were examined with TEC value for switching between error-passive state and bus-off state. In the initial sub-case, an error introduction was simulated in every frame with a TEC value of 255. For the second sub-case, an error introduction with a TEC value of 255 was applied in every alternate frame. In the third sub-case, the error was introduced in every frame as modeled with a TEC value of 127. Finally, the fourth sub-case involves error introduction in every alternate frame, utilizing a TEC value of 127.

In the first scenario, only one node (victim node) interacts over the CAN bus. In this case, specific errors are introduced (within the MGEESM module), which are form error, CRC error, and bit error. The purpose of these errors is to induce a bus-off attack within the CAN modules. Here, the data length of the frame considered is eight bytes for the standard frame with the inclusion of bit stuffing violation. The results for this scenario are shown in Figure 12. For form error, an error is introduced 20 positions after the end of the data field within the frame. For CRC error, an error is introduced 42 positions before the end of the data field within the frame. For bit error, an error is introduced two positions before the end of the data field within the frame.

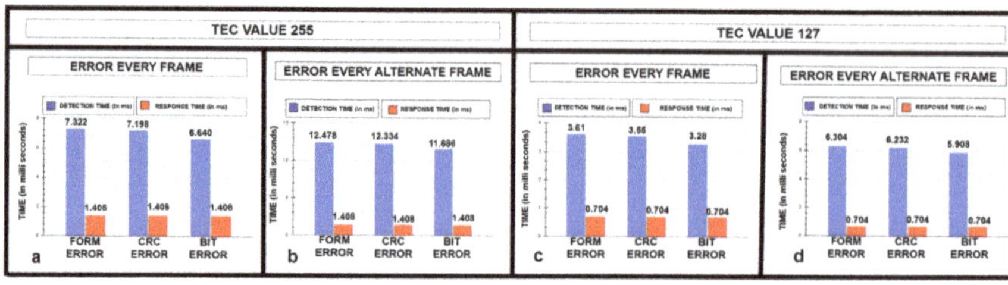

Figure 12. The detection and response times for form error, CRC error, and bit error in the MGEESM module were compared in four cases. (**a**) TEC value 255 with error introduced in every frame indicates a 1.69% lower value for CRC error and 9.31% lower value for bit error in terms of detection time with respect to form error introduction. (**b**) TEC value 255 with error introduced in every alternate frame indicating 1.15% lower value for CRC error and 6.35% lower value for bit error in terms of detection time concerning form error introduction. (**c**) TEC value 127 with error introduced in every frame indicating 1.66% lower value for CRC error and 9.14% lower value for bit error in terms of detection time concerning form error introduction. (**d**) TEC value 127 with error introduced in every alternate frame indicating 1.14% lower value for CRC error and 6.28% lower value for bit error in terms of detection time concerning form error introduction.

In the initial sub-case of the first scenario, the detection times for form error, CRC error, and bit error are 7.322 ms, 7.198 ms, and 6.640 ms, respectively, as shown in Figure 12a. In the second sub-case of the first scenario (TEC = 255), the detection times for form error, CRC error, and bit error are 12.478 ms, 12.334 ms, and 11.686 ms, respectively, as shown in Figure 12b. In the third sub-case of the first scenario, the detection times are 3.610 ms, 3.550 ms, and 3.280 ms for form error, CRC error, and bit error, respectively, as shown in Figure 12c. Moving onto the fourth sub-case in the first scenario, the detection times for form, CRC, and bit errors are 6.304 ms, 6.232 ms, and 5.908 ms, as shown in Figure 12d. Figure 12 shows that form error requires the highest detection time for all four sub-cases in the first scenario, and bit error requires the lowest detection time. However, the response time remains constant across all sub-cases with a value of 1.408 ms for a TEC value of 255 and 0.704 ms for a TEC value of 127. Though the response time is constant with respect to TEC value across all four sub-cases, it is included to give a comprehensive view of the result generated for the four sub-cases for the first scenario.

In the second scenario, the victim node interacts with other nodes over the CAN bus. The focus is on emulating the entire network. Here, a frame size of 111 bits (for the BOAD111 module) is considered. The arbitration IDs considered are 11 bits. The results for this scenario are shown in Figure 13. In this case, the error is introduced at position 60 within the frame with an error field size of 20 bits. The purpose of the error introduced here is to induce a bus-off attack in the CAN module communicating with multiple CAN modules. No bit stuffing violation is considered for this scenario.

In the first sub-case of the second scenario, the detection time for the BOAD111 module is 6.303 ms, as shown in Figure 13a. The detection time is 11.174 ms for the BOAD111 module for the second sub-case, as shown in Figure 13b. In the third sub-case of the second scenario, detection times of 3.247 ms are observed for the BOAD111 module, as shown in Figure 13c. In the fourth sub-case of the second scenario, detection times of 5.738 ms are noted for the BOAD111 module, as shown in Figure 13d.

The response time remains constant in all sub-cases: 1.408 ms for the sub-case with a TEC value of 255 and 0.704 ms for the sub-case with a TEC value of 127. Again, though the response time is constant regarding TEC value across all four sub-cases, it is included to give a comprehensive view of the result generated for the four sub-cases for the second scenario.

For the two modules (MGEESM and BOAD111), this analysis presents utilization parameters (Slice LUTs, Slice Registers, Slice, LUT as Logic, Bonded IOB, BUFGCTRL, F7 Muxes, and F8 Muxes). Moreover, this analysis presents latency values, power metrics, and energy values across four sub-cases for two modules (with the introduction of form error, CRC error, and bit error in the MGEESM module).

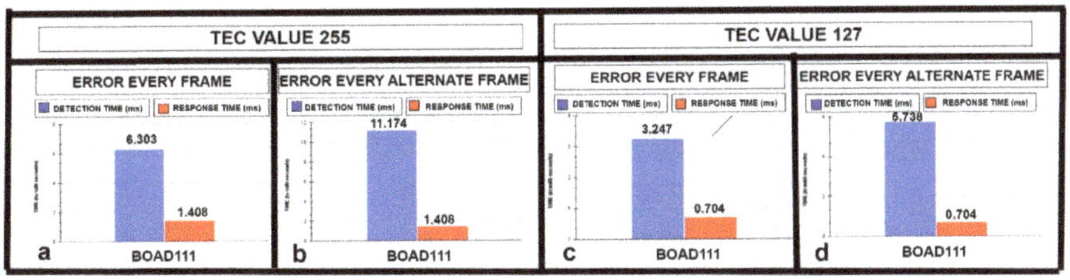

Figure 13. The detection time and response time are presented for the BOAD111 module across all four sub-cases: (**a**) TEC value 255 with error introduced in every frame with response time 77.66% lower than detection time. (**b**) TEC value 255 with error introduced in every alternate frame with response time 87.40% lower than detection time. (**c**) TEC value 127 with error introduced in every frame with response time 78.32% lower than detection time. (**d**) TEC value 127 with error introduced in every alternate frame with response time 87.73% lower than detection time.

Table 2 presents detailed information concerning the utilization parameters linked to the MGEESM and BOAD111 modules. The BUFGCTRL utilization parameter has the same value for both modules. Moreover, the results for LUT as a logic utilization parameter are the same as those for Slice LUTs utilization parameters for both modules. The Slice LUTs utilization parameter has a value of 2299 for the MGEESM module. This parameter has a value of 2663 for the BOAD111 module. The Slice Registers utilization parameter has a value of 595 for the MGEESM module, while this parameter has a value of 662 for the BOAD111 module. Moreover, the Slice utilization parameter has values of 794 and 897 for the MGEESM and BOAD111 modules, respectively. The Bonded IOB utilization parameter has a value of 26 for the MGEESM module and 21 for the BOAD111 module. In addition, module MGEESM has values of 92 and 1 for F7 Muxes and F8 Muxes utilization parameters, respectively.

Table 2. The proposed configurable CAN system design metrics: utilization values for MGEESM and BOAD111 modules are presented.

Modules	Slice LUTs	Slice Registers	Slice	LUT as Logic	Bonded IOB	BUFGCTRL	F7 Muxes	F8 Muxes
MGEESM	2299	595	794	2299	26	1	92	7
BOAD111	2663	662	897	2663	21	1	-	-

Table 3 provides latency, power, and energy data for four sub-cases pertaining to the two modules mentioned before. The table includes details related to form error, CRC error, and bit error introduction in the MGEESM module, along with the results for the BOAD111 module.

In the first set of comparisons, CRC error exhibits a latency of 1.42% lower than form error, while bit error demonstrates a 7.81% decrease in latency compared to form error. Conversely, BOAD111 shows a latency of 7.711 ms. Power consumption for CRC error and bit error is the same as that of form error. The power consumption for BOAD111 is 0.115 W. However, CRC error consumes 1.42% less energy than form error, and bit error consumes 7.81% less energy. The energy consumption value for BOAD111 is 0.887 mJ.

In the second sub-case, CRC error demonstrates a 1.04% decrease in latency compared to form error, with bit error showing a 5.70% reduction. BOAD111 exhibits a latency of 12.582 ms. Power consumption for CRC error and bit error is the same as that of form error. Power consumption for BOAD111 is 0.115 W. However, CRC error consumes 1.04% less energy, and bit error consumes 5.70% less energy than form error. BOAD111 has an energy consumption of 1.447 mJ.

Table 3. The proposed configurable CAN system design metrics: latency, power, and energy values for four sub-cases for MGEESM (form error, CRC error, and bit error), and BOAD111 modules are presented. Across all sub-cases for MGEESM with the introduction of form error, CRC error, and bit error, the latency is highest for form error and lowest for bit error. For BOAD111, across all sub-cases, the latency is lower with respect to errors introduced in MGEESM. The same is valid for energy metrics for both modules across all 4 sub-cases with comparable values for power numbers.

Modules	Sub-Cases	Latency	Power	Energy
Form Error in MGEESM	TEC value 255. Error introduced every frame.	8.730 ms	0.113 W	0.986 mJ
	TEC value 255. Error introduced every alternate frame.	13.886 ms	0.113 W	1.569 mJ
	TEC value 127. Error introduced every frame.	4.314 ms	0.113 W	0.487 mJ
	TEC value 127. Error introduced every alternate frame.	7.008 ms	0.113 W	0.792 mJ
CRC Error in MGEESM	TEC value 255. Error introduced every frame.	8.606 ms	0.113 W	0.972 mJ
	TEC value 255. Error introduced every alternate frame.	13.742 ms	0.113 W	1.553 mJ
	TEC value 127. Error introduced every frame.	4.254 ms	0.113 W	0.481 mJ
	TEC value 127. Error introduced every alternate frame.	6.936 ms	0.113 W	0.784 mJ
Bit Error in MGEESM	TEC value 255. Error introduced every frame.	8.048 ms	0.113 W	0.909 mJ
	TEC value 255. Error introduced every alternate frame	13.094 ms	0.113 W	1.480 mJ
	TEC value 127. Error introduced every frame.	3.984 ms	0.113 W	0.450 mJ
	TEC value 127. Error introduced every alternate frame.	6.612 ms	0.113 W	0.747 mJ
BOAD111	TEC value 255. Error introduced every frame.	7.711 ms	0.115 W	0.887 mJ
	TEC value 255. Error introduced every alternate frame.	12.582 ms	0.115 W	1.447 mJ
	TEC value 127. Error introduced every frame.	3.951 ms	0.115 W	0.454 mJ
	TEC value 127. Error introduced every alternate frame.	6.442 ms	0.115 W	0.741 mJ

In the third sub-case, CRC error and bit error demonstrate latency reductions of 1.39% and 7.65%, respectively, compared to form error. BOAD111 shows a latency of 3.951 ms. Power consumption for CRC error and bit error remains the same as for form error, with CRC error consuming 1.39% less energy and bit error consuming 7.65% less energy. BOAD111's power consumption is 0.115 W, with an energy consumption of 0.454 mJ.

In the fourth set of comparisons, CRC error and bit error demonstrate latency reductions of 1.03% and 5.65%, respectively, compared to form error. BOAD111 shows a latency of 6.442 ms. CRC error and bit error consume the same power as form error. CRC error energy consumption is 1.03% lower, and bit error is 5.65% lower than form error. BOAD111's power consumption is 0.115 W, but its energy consumption is 0.741 mJ.

5. Conclusions

This research project aimed to assess the susceptibility of the CAN to bus-off attacks by emulating them on an FPGA. The configurability and security of the CAN communication protocol were investigated in this project. The MGEESM module with the introduction of form error, CRC error, and bit error was covered in the first threat model. Furthermore, the BOAD111 module was covered in the second threat model.

This paper also experimentally examined the detection and response times for both the modules covered in both threat models.

These times were compared for respective modules within the threat models. Moreover, the latency, utilization parameters, power, and energy were compared for respective modules considering two threat models. The advantage of this implementation of the CAN protocol and attack scenarios using FPGAs is that changes in clock speed can be easily accommodated within the design without changes in the overall structure of the modules. This is useful for further investigation of the CAN protocol based on varying CAN speeds and other threat models and considering different attacks. Furthermore, in electric and hybrid vehicles, CAN networks integrate renewable energy sources, making transportation more sustainable.

Author Contributions: Conceptualization, R.I.; methodology, R.I. and R.S.; software, R.I. and R.S.; validation and analysis, R.I. and R.S.; investigation, R.I. and R.S.; writing—original draft preparation, R.I. and R.S.; writing—review and editing, R.I.; supervision, R.I.; project administration, R.I.; funding acquisition, R.I. All authors have read and agreed to the published version of the manuscript.

Funding: This research was funded in part by a UMBC start up grant and the National Science Foundation (NSF) award, number: 2138253.

Data Availability Statement: The original contributions presented in the study are included in the article material, further inquiries can be directed to the author with correspondence email.

Conflicts of Interest: The authors declare no conflicts of interest.

References

1. Wei, H.; Ai, Q.; Zhai, Y.; Zhang, Y. Automotive Security: Threat Forewarning and ECU Source Mapping Derived From Physical Features of Network Signals. *IEEE Trans. Intell. Transp. Syst.* **2023**, 25, 2479–2491. [CrossRef]
2. Tan, Z.; Dai, N.; Su, Y.; Zhang, R.; Li, Y.; Wu, D.; Li, S. Human—Machine interaction in intelligent and connected vehicles: A review of status quo, issues, and opportunities. *IEEE Trans. Intell. Transp. Syst.* **2021**, *23*, 13954–13975. [CrossRef]
3. Siegel, J.E.; Erb, D.C.; Sarma, S.E. A survey of the connected vehicle landscape—Architectures, enabling technologies, applications, and development areas. *IEEE Trans. Intell. Transp. Syst.* **2017**, *19*, 2391–2406. [CrossRef]
4. Su, Z.; Dai, M.; Xu, Q.; Li, R.; Zhang, H. UAV enabled content distribution for internet of connected vehicles in 5G heterogeneous networks. *IEEE Trans. Intell. Transp. Syst.* **2021**, *22*, 5091–5102. [CrossRef]
5. Sunny, J.; Sankaran, S.; Saraswat, V. A Hybrid Approach for Fast Anomaly Detection in Controller Area Networks. In Proceedings of the 2020 IEEE International Conference on Advanced Networks and Telecommunications Systems (ANTS), New Delhi, India, 14–17 December 2020; pp. 1–6. [CrossRef]
6. Blanco, S. Car Hacking Danger Is Likely Closer than You Thinkt. Available online: https://www.caranddriver.com/news/a37453835/car-hacking-danger-is-likely-closer-than-you-think/ (accessed on 1 April 2024).
7. Shin, C. A framework for fragmenting/reconstituting data frame in Controller Area Network (CAN). In Proceedings of the 16th International Conference on Advanced Communication Technology, Pyeongchang, Republic of Korea, 16–19 February 2014; pp. 1261–1264. [CrossRef]
8. Ullah, K. On the Use of Opportunistic Vehicular Communication for Roadside Services Advertisement and Discovery. Ph.D. Thesis, Universidade de São Paulo, São Paulo, Brazil, 2016.
9. Zhang, X.; Cui, X.; Cheng, K.; Zhang, L. A Convolutional Encoder Network for Intrusion Detection in Controller Area Networks. In Proceedings of the 2020 16th International Conference on Computational Intelligence and Security (CIS), Guangxi, China, 27–30 November 2020; pp. 366–369. [CrossRef]
10. Choi, E.; Han, S.; Choi, J.W. Channel capacity analysis for high speed controller area network (CAN). In Proceedings of the 2015 International Conference on Information and Communication Technology Convergence (ICTC), Jeju, Republic of Korea, 28–30 October 2015; pp. 188–190. [CrossRef]
11. Jeong, Y.; Kim, H.; Lee, S.; Choi, W.; Lee, D.H.; Jo, H.J. In-Vehicle Network Intrusion Detection System Using CAN Frame-Aware Features. *IEEE Trans. Intell. Transp. Syst.* **2023**, *25*, 3843–3853. [CrossRef]
12. Cho, K.T.; Shin, K.G. Fingerprinting electronic control units for vehicle intrusion detection. In Proceedings of the 25th USENIX Security Symposium (USENIX Security 16), Austin, TX, USA , 10–12 August 2016; pp. 911–927.
13. Jo, H.J.; Choi, W. A Survey of Attacks on Controller Area Networks and Corresponding Countermeasures. *IEEE Trans. Intell. Transp. Syst.* **2022**, *23*, 6123–6141. [CrossRef]

14. Islam, R.; Devnath, M.K.; Samad, M.D.; Al Kadry, S.M.J. GGNB: Graph-based Gaussian naive Bayes intrusion detection system for CAN bus. *Veh. Commun.* **2022**, *33*, 100442. [CrossRef]
15. Ansari, M.R.; Yu, S.; Yu, Q. IntelliCAN: Attack-resilient Controller Area Network (CAN) for secure automobiles. In Proceedings of the 2015 IEEE International Symposium on Defect and Fault Tolerance in VLSI and Nanotechnology Systems (DFTS), Amherst, MA, USA, 12–14 October 2015; pp. 233–236. [CrossRef]
16. Wu, W.; Li, R.; Xie, G.; An, J.; Bai, Y.; Zhou, J.; Li, K. A Survey of Intrusion Detection for In-Vehicle Networks. *IEEE Trans. Intell. Transp. Syst.* **2020**, *21*, 919–933. [CrossRef]
17. Khandelwal, S.; Shreejith, S. A Lightweight FPGA-based IDS-ECU Architecture for Automotive CAN. In Proceedings of the 2022 International Conference on Field-Programmable Technology (ICFPT), Hong Kong, China, 5–9 December 2022; pp. 1–9.
18. Islam, R.; Refat, R.U.D. Improving CAN bus security by assigning dynamic arbitration IDs. *J. Transp. Secur.* **2020**, *13*, 19–31. [CrossRef]
19. Pollicino, F.; Stabili, D.; Marchetti, M. Performance comparison of timing-based anomaly detectors for Controller Area Network: a reproducible study. *Acm Trans. -Cyber-Phys. Syst.* **2023**, *8*, 1–24. [CrossRef]
20. Tariq, S.; Lee, S.; Woo, S.S. CANTransfer: Transfer learning based intrusion detection on a controller area network using convolutional LSTM network. In Proceedings of the 35th annual ACM symposium on applied computing, Brno, Czech Republic, 30 March–3 April 2020; pp. 1048–1055.
21. Microchip, C. Controller MCP2515 Datasheet. Available online: https://ww1.microchip.com/downloads/aemDocuments/documents/APID/ProductDocuments/DataSheets/MCP2515-Family-Data-Sheet-DS20001801K.pdf (accessed on 1 April 2023).
22. Zhang, L. Intrusion Detection Systems to Secure In-Vehicle Networks. Ph.D. Thesis, University of Michigan-Dearborn, Dearborn, MI, USA, 2023
23. Han, K.; Mun, H.; Balakrishnan, M.; Yeun, C.Y. Enhancing security and robustness of Cyphal on Controller Area Network in unmanned aerial vehicle environments. *Comput. Secur.* **2023**, *135*, 103481. [CrossRef]
24. Olufowobi, H.; Young, C.; Zambreno, J.; Bloom, G. Saiducant: Specification-based automotive intrusion detection using controller area network (can) timing. *IEEE Trans. Veh. Technol.* **2019**, *69*, 1484–1494. [CrossRef]
25. Zhang, H.; Meng, X.; Zhang, X.; Liu, Z. CANsec: A practical in-vehicle controller area network security evaluation tool. *Sensors* **2020**, *20*, 4900. [CrossRef] [PubMed]
26. Park, S.B.; Jo, H.J.; Lee, D.H. Flooding attack mitigator for in-vehicle CAN using fault confinement in CAN protocol. *Comput. Secur.* **2023**, *126*, 103091. [CrossRef]
27. Humayed, A.; Li, F.; Lin, J.; Luo, B. Cansentry: Securing can-based cyber-physical systems against denial and spoofing attacks. In Proceedings of the Computer Security—ESORICS 2020: 25th European Symposium on Research in Computer Security, ESORICS 2020, Guildford, UK, 14–18 September 2020; Proceedings, Part I 25; Springer: Berlin/Heidelberg, Germany, 2020; pp. 153–173.
28. Han, M.L.; Kwak, B.I.; Kim, H.K. Event-triggered interval-based anomaly detection and attack identification methods for an in-vehicle network. *IEEE Trans. Inf. Forensics Secur.* **2021**, *16*, 2941–2956. [CrossRef]
29. Ansari, M.R. Low-Cost Approaches to Detect Masquerade and Replay Attacks on Automotive Controller Area Network. Ph.D. Thesis, University of New Hampshire, Durham, New Hampshire, 2016.
30. Jedh, M.; Othmane, L.B.; Ahmed, N.; Bhargava, B. Detection of message injection attacks onto the can bus using similarities of successive messages-sequence graphs. *IEEE Trans. Inf. Forensics Secur.* **2021**, *16*, 4133–4146. [CrossRef]
31. Islam, R.; Refat, R.U.D.; Yerram, S.M.; Malik, H. Graph-based intrusion detection system for controller area networks. *IEEE Trans. Intell. Transp. Syst.* **2020**, *23*, 1727–1736. [CrossRef]
32. Zhang, H.; Zeng, K.; Lin, S. Federated graph neural network for fast anomaly detection in controller area networks. *IEEE Trans. Inf. Forensics Secur.* **2023**, *18*, 1566–1579. [CrossRef]
33. Müter, M.; Asaj, N. Entropy-based anomaly detection for in-vehicle networks. In Proceedings of the 2011 IEEE Intelligent Vehicles Symposium (IV), Baden-Baden, Germany, 5–9 June 2011; pp. 1110–1115.
34. Marchetti, M.; Stabili, D.; Guido, A.; Colajanni, M. Evaluation of anomaly detection for in-vehicle networks through information-theoretic algorithms. In Proceedings of the 2016 IEEE 2nd International Forum on Research and Technologies for Society and Industry Leveraging a better tomorrow (RTSI), Bologna, Italy, 7–9 September 2016; pp. 1–6.
35. Mithu, M.R.A.; Kholodilo, V.; Manicavasagam, R.; Ulybyshev, D.; Rogers, M. Secure industrial control system with intrusion detection. In Proceedings of the Thirty-Third International Flairs Conference, North Miami Beach, FL, USA, 17–20 May 2020.
36. Moulahi, T.; Zidi, S.; Alabdulatif, A.; Atiquzzaman, M. Comparative performance evaluation of intrusion detection based on machine learning in in-vehicle controller area network bus. *IEEE Access* **2021**, *9*, 99595–99605. [CrossRef]
37. Dong, Y.; Chen, K.; Peng, Y.; Ma, Z. Comparative study on supervised versus semi-supervised machine learning for anomaly detection of in-vehicle CAN network. In Proceedings of the 2022 IEEE 25th International Conference on Intelligent Transportation Systems (ITSC), Macau, China, 8–12 October 2022; pp. 2914–2919.
38. Narasimhan, H.; Vinayakumar, R.; Mohammad, N. Unsupervised deep learning approach for in-vehicle intrusion detection system. *IEEE Consum. Electron. Mag.* **2021**, *12*, 103–108. [CrossRef]
39. Islam, R. Early Stage DRC Prediction Using Ensemble Machine Learning Algorithms. *IEEE Can. J. Electr. Comput. Eng.* **2022**, *45*, 354–364. [CrossRef]
40. Seo, E.; Song, H.M.; Kim, H.K. GIDS: GAN based intrusion detection system for in-vehicle network. In Proceedings of the 2018 16th Annual Conference on Privacy, Security and Trust (PST), Belfast, Ireland, 28–30 August 2018; pp. 1–6.

41. Desta, A.K.; Ohira, S.; Arai, I.; Fujikawa, K. U-CAN: A Convolutional Neural Network Based Intrusion Detection for Controller Area Networks. In Proceedings of the 2022 IEEE 46th Annual Computers, Software, and Applications Conference (COMPSAC), Los Alamitos, CA, USA, 27 June–1 July 2022; pp. 1481–1488.
42. Kheddar, H.; Himeur, Y.; Awad, A.I. Deep transfer learning for intrusion detection in industrial control networks: A comprehensive review. *J. Netw. Comput. Appl.* **2023**, *220*, 103760. [CrossRef]
43. Kulisz, J.; Jokiel, F. A Hardware Implementation of the PID Algorithm Using Floating-Point Arithmetic. *Electronics* **2024**, *13*, 1598. [CrossRef]
44. Islam, R.; Saha, B.; Bezzam, I. Resonant Energy Recycling SRAM Architecture. *IEEE Trans. Circuits Syst. II Express Briefs* **2021**, *68*, 1383–1387. [CrossRef]
45. Islam, R. Feasibility Prediction for Rapid IC Design Space Exploration. *Electronics* **2022**, *11*, 1161. [CrossRef]
46. Joost, R.; Salomon, R. Advantages of FPGA-based multiprocessor systems in industrial applications. In Proceedings of the 31st Annual Conference of IEEE Industrial Electronics Society, 2005. IECON 2005, Raleigh, NC, USA, 6–10 November 2005.
47. Croteau, B.; Kiriakidis, K.; Severson, T.A.; Robucci, R.; Rahman, S.; Islam, R. State Estimation Adaptable to Cyberattack Using a Hardware Programmable Bank of Kalman Filters. *IEEE Trans. Control Syst. Technol.* **2024**, 1–13. [CrossRef]
48. Tang, L.; Li, Y.; Wang, H.; Sun, Y. Verification of CAN bus controller based on VIP. In Proceedings of the 2023 IEEE International Conference on Sensors, Electronics and Computer Engineering (ICSECE), Jinzhou, China, 18–20 August 2023; pp. 1383–1387.
49. Lee, H.; Jeong, S.; Kim, H. *CAN Dataset for Intrusion Detection*; Hacking and Countermeasure Research Lab: Seoul, Republic of Korea. 2018. Available online: https://goo.gl/WiVeFj (accessed on 1 April 2024).

MDPI AG
Grosspeteranlage 5
4052 Basel
Switzerland
Tel.: +41 61 683 77 34
www.mdpi.com

Electronics Editorial Office
E-mail: electronics@mdpi.com
www.mdpi.com/journal/electronics

www.ingramcontent.com/pod-product-compliance
Lightning Source LLC
Chambersburg PA
CBHW040808250926
51850CB00007BA/1113
* 9 7 8 3 7 2 5 8 1 8 6 1 7 *